ENGINEER 1C03

ENGINEERING DESIGN AND GRAPHICS

CUSTOM COURSEWARE

FALL 2019

FACULTY OF ENGINEERING

HAMILTON, ONTARIO, CANADA

DR. T.E. DOYLE AND DR. C.P. MCDONALD

Contents

Overview of Courseware Contents

This custom courseware serves as supplementary material for ENGINEER 1C03. You should take the appropriate time to browse all sections and familiarize yourseif with the overall layout.

Section 1 includes guides for both your weekly Labs and weekly Tutorials. The Guide to Weekly Labs includes an overview of the lab and what you can expect to cover each week. The Guide to Weekly Tutorials includes a week-by-week breakdown of the material covered in Tutorial, the required reading material, and a list of assigned questions for you to complete. These assigned questions will not be collected, nor will they be graded, but they serve as an excellent preparation for your biweekly Tutorial Tests.

Section 2 includes information related to the design project. Complete details on the design project will be provided in the term. However, this section provides a broad overview of what is expected of you, including details on submission deadlines and submission requirements. A student-centric resource available to you is the Experiential Playground and Innovation Classroom (EPIC). This EPIC Lab is equipped with modern tools that you will need to complete your design project, including a number of **3D Printers**. Information on the EPIC Lab and documentation covering how to use the appropriate equipment and software is found in the Guide to the EPIC Lab of this section. This is an especially valuable resource as **you are fully expected to familiarize yourself with these documents before coming to the lab**.

Section 3 provides a guide for designing a mixed gear train (i.e., designing a gear train with different types of gears), which will be a key requirement for your design project. This guide gives consideration for incorporation of functional requirements when faced with typical constraints (e.g., spatial, geometric, practical design, etc.), and is intended to complement our discussions on gear train design that will take place in class and in tutorial, as well as the supplementary reading material provided in Appendix B (Dudley, 1st Edition; Uicker et al., 4th Edition).

Section 4 includes additional reading material from sources other than the textbook, but is highly relevant to the course curriculum. This additional reading material will be important for the lectures, the design project and your tutorials. A course schedule that includes all recommended reading materials has been provided for you and can be found in Appendix A of this courseware.

SECTION 1: Labs and Tutorials

1.1 Guide to Weekly Labs

Labs are 3 hours once a week and are run by an Instructional Assistant Intern (IAI) with Teaching Assistant (TA) support. Weekly labs extend the discussion of topics covered in lecture and allow for the opportunity to apply what you have learned through practice assignment and graded tests.

Topics covered in lab include:

1. Part modelling
2. Assembly modelling
3. Constraints and degrees of freedom
4. Design Accelerator
5. Working drawings and dimensioning
6. Mechanical dissection
7. Motion, animation, and simulation

Labs are mandatory. Labs run a span of 2 weeks and are comprised of a lesson week followed a test week. In total, there are 5 Labs throughout the term.

During lesson weeks the IAI will give a lesson and the students can practice what they have learned through sample assignments (which can be found on Avenue to Learn). This section of your courseware complements the course material. **It is strongly recommended that you dedicate time each week to attempting the recommended practice assignments ahead of your lab test.**

1.2 Guide to Weekly Tutorials

Tutorials are 2 hours once a week and are run by an Instructional Assistant Intern (IAI). Weekly tutorials extend the discussion of topics covered in lecture and allow for the opportunity to apply what you have learned through practice questions and graded tests.

Topics covered in tutorial include:

8. Sketching basics (2D and 3D sketching)
9. Isometric pictorials
10. Multiview drawings
11. Visualization (from 2D to 3D and from 3D to 2D)
12. Design of simple mechanisms
13. Reverse engineering

Tutorials are mandatory. Tutorials run a span of 2 weeks (unless otherwise indicated, please refer to avenue.mcmaster.ca) and are comprised of a lesson week followed a test week. In total, there are 5 Tutorials throughout the term.

During lesson weeks the IAI will give a lesson and the students can practice what they have learned through a series of assigned questions (which can be found later in this section). This section of your courseware complements the course material. **It is strongly recommended that you dedicate time each week to attempting the recommended practice questions ahead of your tutorial test.**

Tutorial 1: Lesson

Tutorial 1 provides an introduction to visualization, beginning with basic sketching practices. In this lesson, you will learn tips and tricks for sketching straight edges and curves without the use of drawing aids. You will also learn how to sketch 3D views known as isometric pictorials. Finally, you will develop your visualization skills by learning to recognize (and re-sketch) 3D objects in different orientations and proportions. **You are not expected, nor required, to have any previous sketching and drawing experience**. Technical sketching is a skill that is learned through practice.

Material covered in will be assessed during your **Tutorial 1 Test**. However, subsequent tutorials will build on the material covered here, so you are strongly encouraged to attempt the assigned questions below.

TUTORIAL TOPIC: SKETCHING BASICS AND ISOMETRIC PICTORIALS

- 2D sketches and sketching practices
- 3D sketches and isometric pictorials
- Perspective and proportions

REQUIRED READING MATERIAL:

- Lieu & Sorby, 2nd Edition: Chapter 2.02 – 2.07
- Lieu & Sorby, 2nd Edition: Chapter 3.06 – 3.08
- Lieu & Sorby, 2nd Edition: Chapter 9.02 – 9.03

SUPPLEMENTARY MATERIAL:

- AvenueToLearn Video 1 – Drawing Straight Lines
- AvenueToLearn Video 2 – Perpendicular Axes and Orthogonal Shapes
- AvenueToLearn Video 3 – Drawing an Ellipse
- AvenueToLearn Video 4 – Sketching the Profile of an Object
- AvenueToLearn Video 5 – Introduction to Proportions
- AvenueToLearn Video 6 – Creating an Isometric Box
- AvenueToLearn Video 7 – Changing Proportions
- AvenueToLearn Video 10 – Creating and using the Isometric Box
- AvenueToLearn Video 11 – Changing the Perspective Rotation
- AvenueToLearn Video 12 – Isometric Pictorial Holes

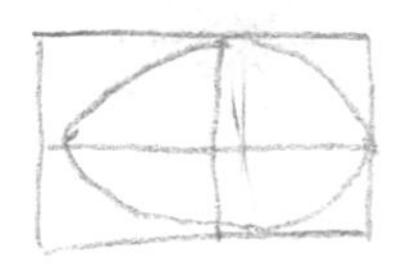

VISUALIZATION QUIZ:

Complete the visualization exercise posted on Avenue.

ASSIGNMENT QUESTIONS:

Attempt as many of the following questions as you can. You should have these questions completed ahead of your Tutorial 1 Test. These questions are for practice and will not be graded. It is strongly recommended, however, that you attempt these questions on your own. Avoid the use of a straight edge or other drawing aids (an isometric underlay is acceptable). Such drawing aids are not permitted for tests.

For your Tutorial 1 Test, you will be tested on a subset of these question and/or question types. For additional practice problems, please refer to the appropriate textbook sections.

1. Re-sketch the profile in Figure 2-B.1
2. Re-sketch the profile in Figure 2-B.2

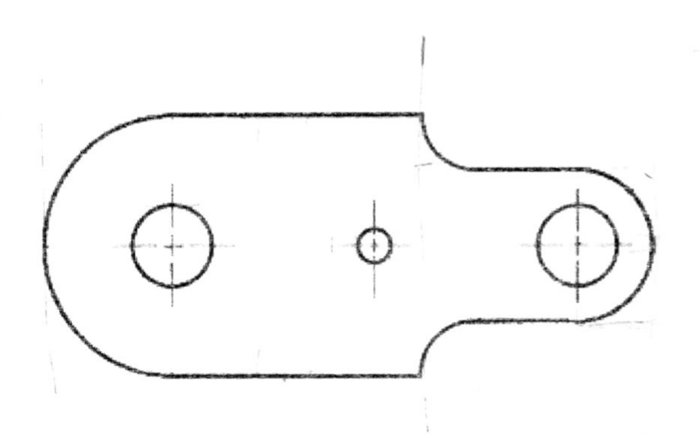

Figure 2-B.1

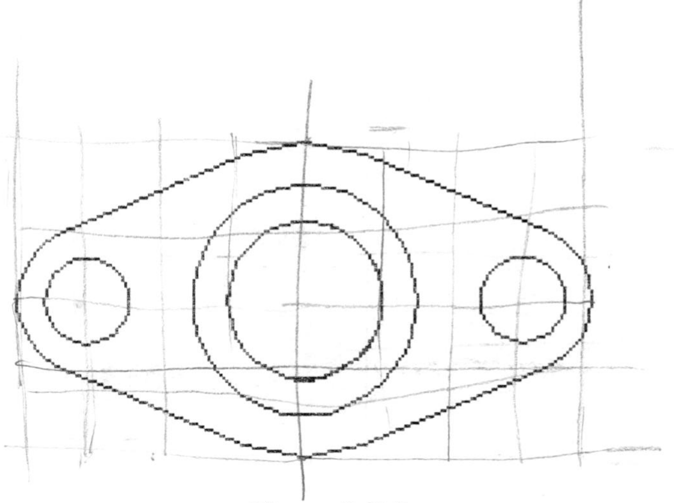

Figure 2-B.2

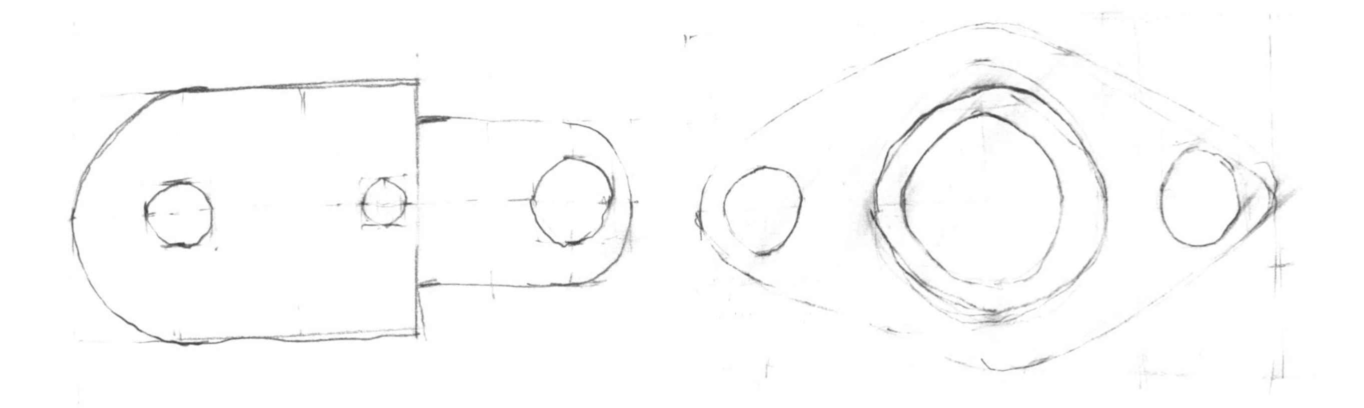

3. Re-sketch the profile in Figure 2-B.3 at a scale of 2:1 (i.e., twice the presented size)
4. Re-sketch the profile in Figure 2-B.4 at a scale of 3:1 (i.e., three times the presented size)

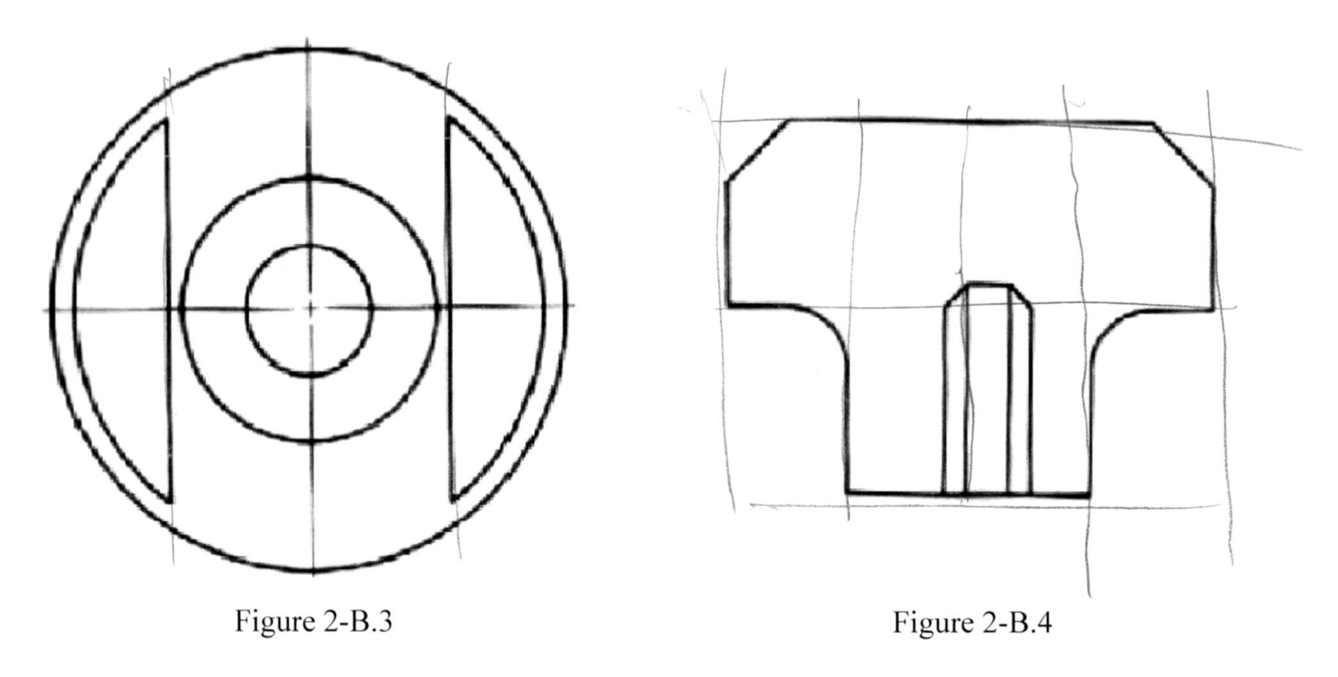

Figure 2-B.3 Figure 2-B.4

5. Re-sketch the profile in Figure 2-B.5 at a scale of 2:1 (i.e., twice the presented size)
6. Re-sketch the profile in Figure 2-B.6 at a scale of 3:1 (i.e., three times the presented size)

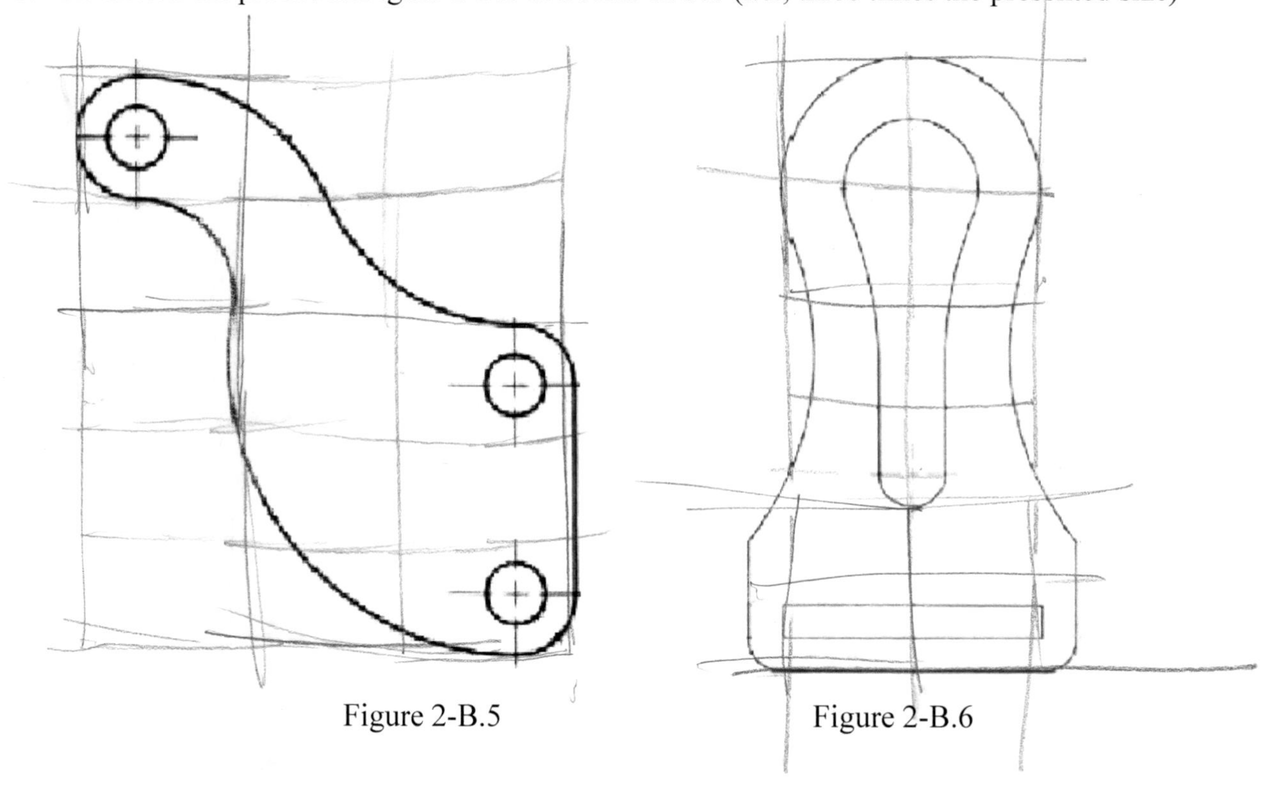

Figure 2-B.5 Figure 2-B.6

7. Sketch the isometric pictorial in Figure 2-B.7. Use the isometric underlay at the end of this assignment as a guide (your sketch should be larger than the provided figure).

8. Sketch the isometric pictorial in Figure 2-B.8. Use the isometric underlay at the end of this assignment as a guide (your sketch should be larger than the provided figure).

Figure 2-B.7

Figure 2-B.8

9. Sketch the isometric pictorial in Figure 2-B.9. Use the isometric underlay at the end of this assignment as a guide (your sketch should be larger than the provided figure).

10. Sketch the isometric pictorial in Figure 2-B.10. Use the isometric underlay at the end of this assignment as a guide (your sketch should be larger than the provided figure).

Figure 2-B.9

Figure 2-B.10

11. Sketch the isometric pictorial in Figure 2-B.11. Use the isometric underlay at the end of this assignment as a guide (your sketch should be larger than the provided figure).

12. Sketch the isometric pictorial in Figure 2-B.12. Use the isometric underlay at the end of this assignment as a guide (your sketch should be larger than the provided figure).

Figure 2-B.11 Figure 2-B.12

13. Sketch the textbook isometric pictorial in **Figure P2.4a** (Lieu and Sorby, Chapter 2.14, Problem 4) using the isometric underlay at the end of this assignment as your guide (your sketch should be larger than the provided figure). Now re-sketch Figure P2.4a from the x:y:z ratio 2:3:2 to 4:3:4 (i.e., change the proportions by doubling x and z).

14. Sketch the textbook isometric pictorial in **Figure P2.4f** (Lieu and Sorby, Chapter 2.14, Problem 4) using the isometric underlay at the end of this assignment as your guide (your sketch should be larger than the provided figure). Now re-sketch Figure P2.4f from the x:y:z ratio 2:2:3 to 4:4:3 (i.e., change the proportions by doubling x and y).

15. Sketch the isometric pictorial in Figure 2-B.13. Use the isometric underlay at the end of this assignment as a guide (your sketch should be larger than the provided figure).

16. Sketch the isometric pictorial in Figure 2-B.14. Use the isometric underlay at the end of this assignment as a guide (your sketch should be larger than the provided figure).

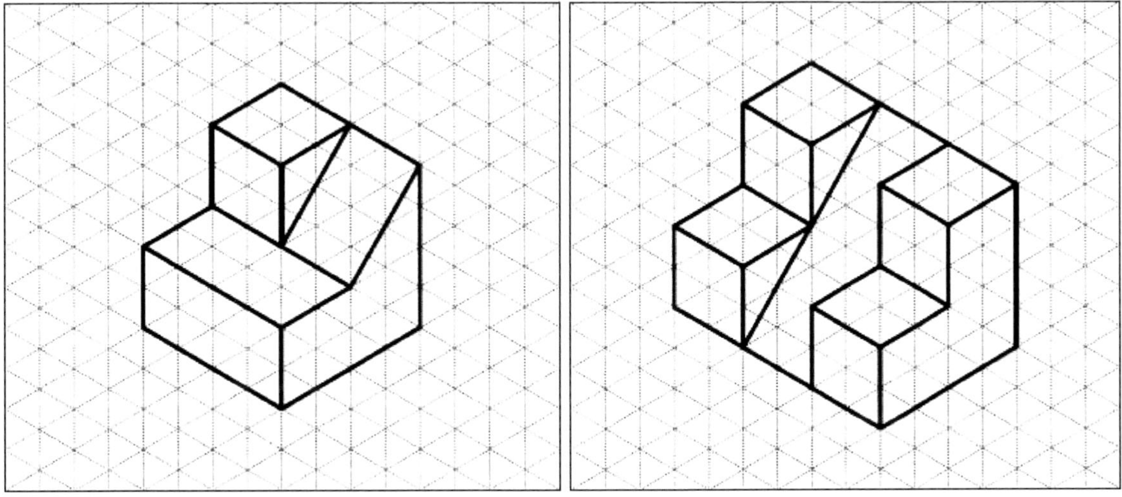

Figure 2-B.13 Figure 2-B.14

17. Sketch the isometric pictorial in Figure 2-B.15. Use the isometric underlay at the end of this assignment as a guide (your sketch should be larger than the provided figure).

18. Sketch the isometric pictorial in Figure 2-B.16. Use the isometric underlay at the end of this assignment as a guide (your sketch should be larger than the provided figure).

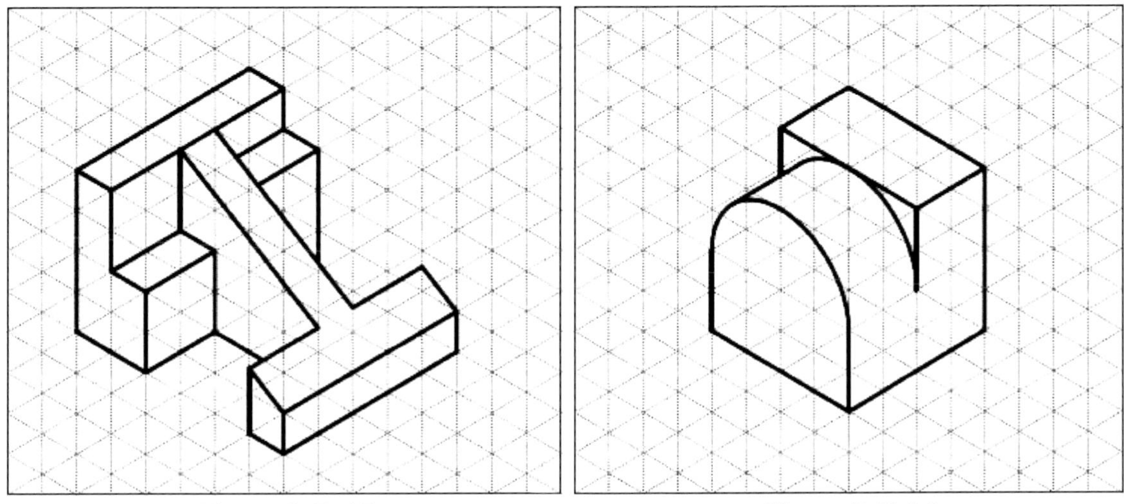

Figure 2-B.15 Figure 2-B.16

19. Sketch the isometric pictorial in Figure 2-B.17. Use the isometric underlay at the end of this assignment as a guide (your sketch should be larger than the provided figure).

20. Sketch the isometric pictorial in Figure 2-B.18. Use the isometric underlay at the end of this assignment as a guide (your sketch should be larger than the provided figure).

Figure 2-B.17 Figure 2-B.18

21. Complete Problem 4 in Chapter 3.16 from Lieu and Sorby.

22. Sketch the isometric pictorial in **Figure P3.5a** (Lieu and Sorby, Chapter 3.16, Problem 5) rotated 90 degrees clockwise about the x-axis. Use the isometric underlay at the end of this assignment as a guide (your sketch should be larger than the provided figure).

23. Sketch the isometric pictorial in **Figure P3.5b** (Lieu and Sorby, Chapter 3.16, Problem 5) rotated 90 degrees counter-clockwise about the z-axis. Use the isometric underlay at the end of this assignment as a guide (your sketch should be larger than the provided figure).

24. Sketch the isometric pictorial in **Figure P3.5f** (Lieu and Sorby, Chapter 3.16, Problem 5) rotated 90 degrees clockwise about the y-axis. Use the isometric underlay at the end of this assignment as a guide (your sketch should be larger than the provided figure).

Tutorial 2: Lesson

Tutorial 2 introduces multiview representation. In this lesson, you will learn how a multiview drawing is derived from a pictorial view using the glass box theory. You will also learn standards and conventions of multiview drawings. Finally, you will continue developing visualization skills by learning to: 1) identify missing features in certain views, 2) project a third view in a drawing, and 3) sketch isometric pictorials given only a multiview representation of an object.

Material covered in this tutorial will be assessed during your **Tutorial 2 Test**. However, subsequent tutorials will build on the material covered here, so you are strongly encouraged to attempt the assigned questions below.

TUTORIAL TOPIC: MULTIVIEW REPRESENTATION

- Standard views and the Glass Box Theory
- Multiview standards and conventions
- Alignment and arrangement of views
- Projecting a third view
- Sketching isometric pictorials given a multiview

REQUIRED READING MATERIAL:

- Lieu & Sorby, 2nd Edition: Chapter 8.02 – 8.06
- Giesecke et al., 4th Edition: Chapter 4 (Please refer to Appendix B of Custom Courseware)

SUPPLEMENTARY MATERIAL:

- AvenueToLearn Video 17 – Introduction to Orthographic Projection
- AvenueToLearn Video 18 – Layout and Spacing of Views
- AvenueToLearn Video 19 – Using Construction Lines to Create Multiviews
- AvenueToLearn Video 20 – Using Hidden Lines and Center Lines

Assignment Questions:

Attempt as many of the following questions as you can **on your own**. You should have these questions completed ahead of your Tutorial 2 Test. These questions are for practice and will not be graded. Avoid the use of a straight edge or other drawing aids (an isometric underlay is acceptable). Such drawing aids are not permitted for tests.

For your Tutorial 2 Test, you will be tested on a subset of these question and/or question types. For additional practice problems, please refer to the appropriate textbook sections.

1. Re-sketch the textbook multiview drawing in **Figure P8.3a** (Lieu and Sorby, Chapter 8.12, Problem 3) and add the missing lines. Missing lines may either be 'visible' or 'hidden', or both. You may use the rectangular grid at the end of this assignment as a guide. However, it is strongly recommended you also attempt this problem without use of a grid.

2. Re-sketch the textbook multiview drawing in **Figure P8.3e** (Lieu and Sorby, Chapter 8.12, Problem 3) and add the missing lines. Missing lines may either be 'visible' or 'hidden', or both. You may use the rectangular grid at the end of this assignment as a guide. However, it is strongly recommended you also attempt this problem without use of a grid.

3. Re-sketch the textbook multiview drawing in **Figure P8.3i** (Lieu and Sorby, Chapter 8.12, Problem 3) and add the missing lines. Missing lines may either be 'visible' or 'hidden', or both. You may use the rectangular grid at the end of this assignment as a guide. However, it is strongly recommended you also attempt this problem without use of a grid.

4. Re-sketch the textbook multiview drawing in **Figure P8.3m** (Lieu and Sorby, Chapter 8.12, Problem 3) and add the missing lines. Missing lines may either be 'visible' or 'hidden', or both. You may use the rectangular grid at the end of this assignment as a guide. It is strongly recommended you also attempt this problem without use of a grid.

5. Re-sketch the textbook multiview drawing in **Figure P8.3q** (Lieu and Sorby, Chapter 8.12, Problem 3) and add the missing lines. Missing lines may either be 'visible' or 'hidden', or both. You may use the rectangular grid at the end of this assignment as a guide. However, it is strongly recommended you also attempt this problem without use of a grid.

6. Re-sketch the textbook multiview drawing in **Figure P8.3u** (Lieu and Sorby, Chapter 8.12, Problem 3) and add the missing lines. Missing lines may either be 'visible' or 'hidden', or both. You may use the rectangular grid at the end of this assignment as a guide. However, it is strongly recommended you also attempt this problem without use of a grid.

7. Given the two views appearing in Figure 2-B.19 below, sketch the missing view.

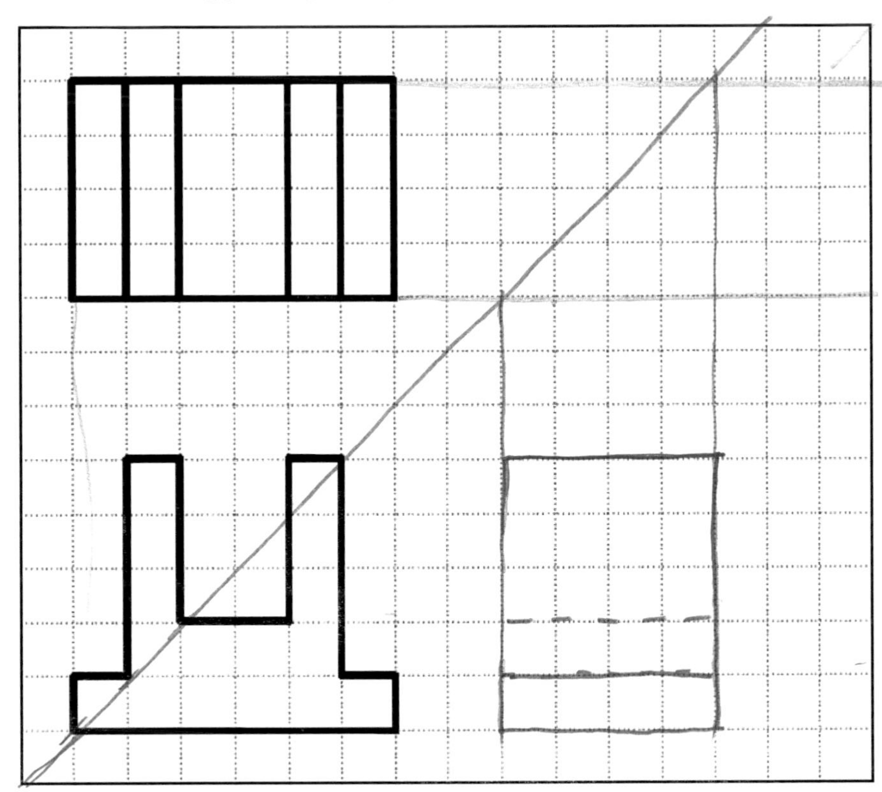

Figure 2-B.19

8. Given the two views appearing in Figure 2-B.20 below, sketch the missing view.

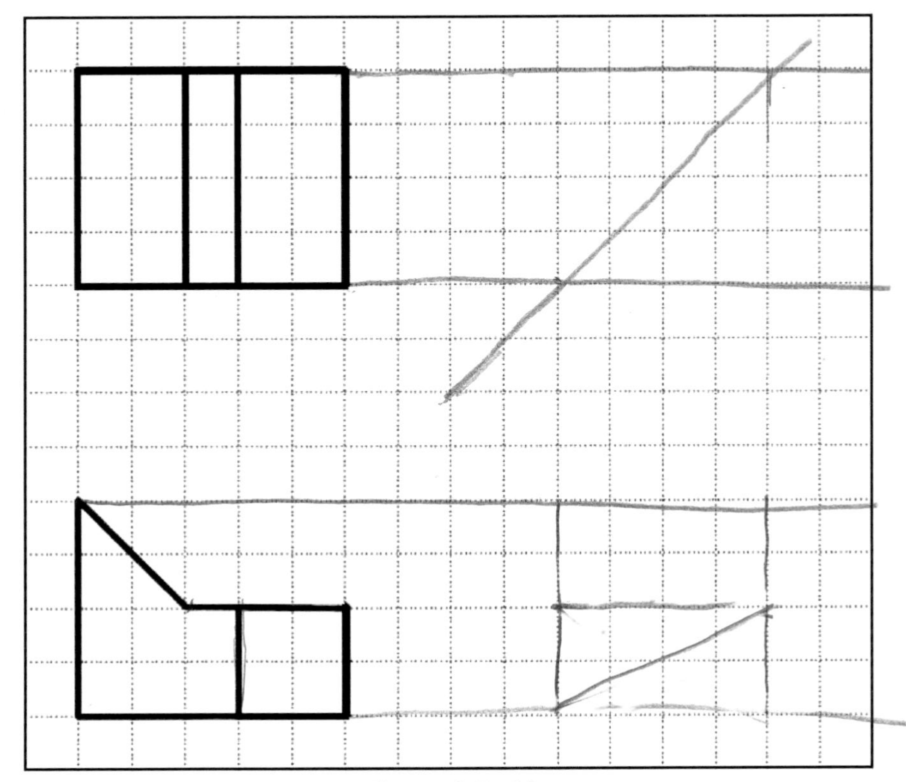

Figure 2-B.20

9. Given the two views appearing in Figure 2-B.21 below, sketch the missing view.

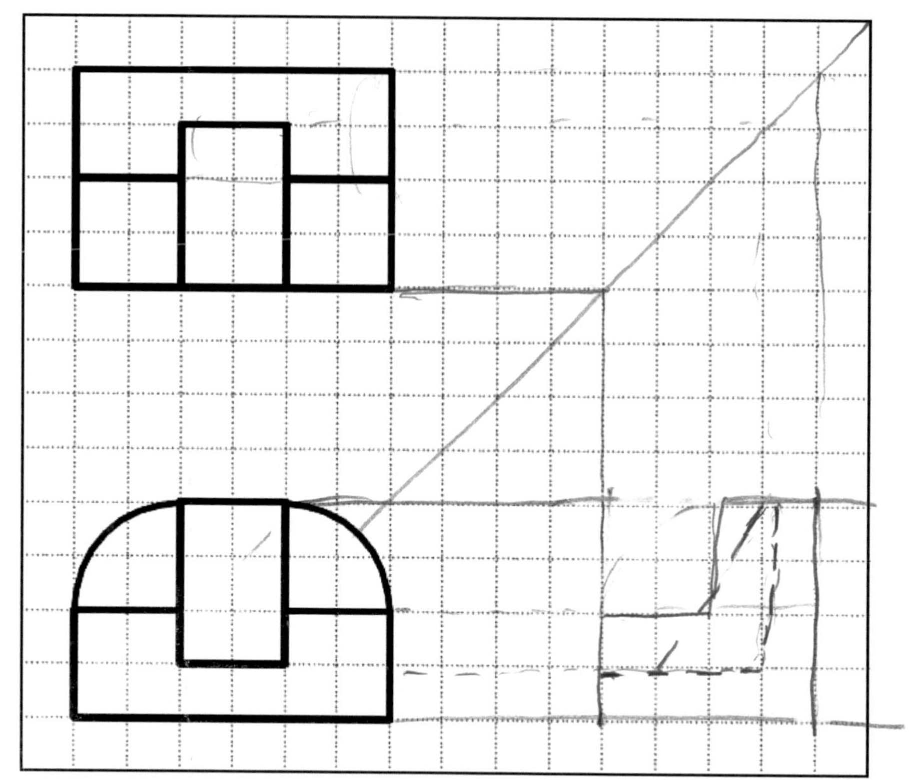

Figure 2-B.21

10. Re-sketch the two views in **Figure 11** of Giesecke et al., Chapter 4, Exercise 4.7 (please refer to Appendix B of custom courseware) and add the missing (right side) view. You may use the rectangular grid at the end of this assignment as a guide.

11. Re-sketch the two views in **Figure 15** of Giesecke et al., Chapter 4, Exercise 4.7 (please refer to Appendix B of custom courseware) and add the missing (right side) view. You may use the rectangular grid at the end of this assignment as a guide.

12. Re-sketch the two views in **Figure 23** of Giesecke et al., Chapter 4, Exercise 4.7 (please refer to Appendix B of custom courseware) and add the missing (right side) view. You may use the rectangular grid at the end of this assignment as a guide.

13. Given the multiview drawing in **Figure P9.1a** (Lieu and Sorby, Chapter 9.10, Problem 1), sketch the isometric pictorial. Use the isometric underlay at the end of this assignment as a guide (your sketch should be larger than the provided figure).

14. Given the multiview drawing in **Figure P9.1b** (Lieu and Sorby, Chapter 9.10, Problem 1), sketch the isometric pictorial. Use the isometric underlay at the end of this assignment as a guide (your sketch should be larger than the provided figure).

15. Given the multiview drawing in **Figure P9.1c** (Lieu and Sorby, Chapter 9.10, Problem 1), sketch the isometric pictorial. Use the isometric underlay at the end of this assignment as a guide (your sketch should be larger than the provided figure).

16. Given the multiview drawing in **Figure P9.1d** (Lieu and Sorby, Chapter 9.10, Problem 1), sketch the isometric pictorial. Use the isometric underlay at the end of this assignment as a guide (your sketch should be larger than the provided figure).

17. Given the multiview drawing in **Figure P9.1g** (Lieu and Sorby, Chapter 9.10, Problem 1), sketch the isometric pictorial. Use the isometric underlay at the end of this assignment as a guide (your sketch should be larger than the provided figure).

18. Given the multiview drawing in **Figure P9.1j** (Lieu and Sorby, Chapter 9.10, Problem 1), sketch the isometric pictorial. Use the isometric underlay at the end of this assignment as a guide (your sketch should be larger than the provided figure).

Rectangular Grid Underlay

Isometric Underlay

Tutorial 2: Test

For Tutorial 2 test, you will be tested on the topics introduced in the previous week. In preparation for this test, you should have attempted all the questions in the above section.

TUTORIAL TOPIC: MULTIVIEW REPRESENTATION

- This test will consist of 3 problems. The Tutorial 2 Assignment Questions provide a useful reference for the types of questions you can expect.
    - Sketch the missing lines
    - Projecting a third view
    - Sketch an isometric pictorial given the multiview
- Any questions requiring an isometric pictorial will be graded using the **Isometric Rubric** at the end of this section
- Grading breakdown for other components (e.g., sketch missing lines, project a third view) will be clearly labelled on the test sheet
- **REMINDER**: some notes on grading to keep in mind:
    - Be sure to indicate scale!
    - For sketching (i.e., visible) lines: they should be thick, solid and dark (avoid *feathering*)
    - Draw any construction and grid lines lightly so they are clearly discerned from all sketching lines
    - For isometric pictorials, the orientation (i.e., the 30-degree rule) is critical
    - Straight lines – some waviness is okay, but too much is BAD
    - Parallel lines – all parallel lines that should be parallel must be parallel
    - Be neat and tidy! It counts for marks
    - Put signature, date and title near the drawing
    - **Don't cheat!!!**

REQUIRED READING MATERIAL:

- Refer to above section

Tutorial 3: Lesson

Tutorial 3 introduces simple mechanisms. In this lesson, you will learn nomenclature of mechanisms that will be relevant to the design project. You will also learn how common gear design parameters (e.g., pitch circle diameter, number of teeth) of meshing gear teeth relate to the relative rotational speed of meshing gear teeth (i.e., the speed ratio or **gear ratio**). These principles will be applied in the design of various gear trains.

Material covered will be assessed during your **Tutorial 3 Test**. Understanding of this material is crucial to successful completion of the design project, so you are strongly encouraged to attempt the assigned questions below.

TUTORIAL TOPIC: GEAR DESIGN

- Gear design nomenclature
- The gear ratio
- Design of a gear train

REQUIRED READING MATERIAL:

- Dudley: Chapter 1 (Please refer to Appendix B of Custom Courseware)
- Uicker et al.: Chapter 7 (Please refer to Appendix B of Custom Courseware)
- Uicker et al.: Chapter 8 (Please refer to Appendix B of Custom Courseware)
- Uicker et al.: Chapter 9 (Please refer to Appendix B of Custom Courseware)

SUPPLEMENTARY MATERIAL:

- An excellent website to aid in understanding the material covered in this tutorial can be found at www.geargenerator.com

ASSIGNMENT QUESTIONS:

Attempt as many of the following questions as you can. You should have these questions completed ahead of your Tutorial 3 Test. These questions are for practice and will not be graded. It is strongly recommended, however, that you attempt these questions on your own.

1. A 17-tooth spur gear has a diametral pitch of 8 teeth/inch, runs at 1,120 RPM (revolutions per minute), and drives a second spur gear at a speed of 544 RPM. Find the number of teeth on the second gear and the theoretical center distance.

2. A 15-tooth spur gear has a module of 3-mm, runs at 1,600 RPM (revolutions per minute). The driven gear has 60 teeth. Find the speed of the driven gear, the circular pitch, and the theoretical center distance.

3. A 32-tooth gear with an 8-diametral pitch meshes with a 65-tooth gear. Determine the value of the center distance.

4. A spur gearset has a module of 6 mm and a gear ratio of 4. The pinion has 16 teeth. Find the number of teeth on the driven gear, the pitch diameters, and the theoretical center distance.

5. Two gears in a 2:1 ratio gearset and with a diametral pitch of 6 are mounted at a center distance of 5 inches. Find the number of teeth in each gear.

6. A 20-tooth pinion with a diametral pitch of 8 rotates 2,000 RPM and drives a gear at 1,000 RPM. What are the number of teeth in the gear, the center distance, and the circular pitch?

7. A 24-tooth pinion has a module of 2-mm, rotates 2,400 RPM, and drives an 800 RPM gear. Determine the number of teeth on the gear, the circular pitch, and the center distance.

8. Given only spur gears with diameters that are multiples of 10 (i.e. 10, 20, 30, 40, etc.) design a spur gear-train with an overall gear ratio of 2.5 and with the same rotational direction as the input.

9. Given only spur gears with diameters that are multiples of 10 (i.e. 10, 20, 30, 40, etc.) design a spur gear-train with an overall gear ratio of 0.4 and with the same rotational direction as the input.

10. Calculate and state the gearing ratio of the given gear train in Figure 2-B.22. If the input is 100 rpm (clockwise), what is the output speed and direction?

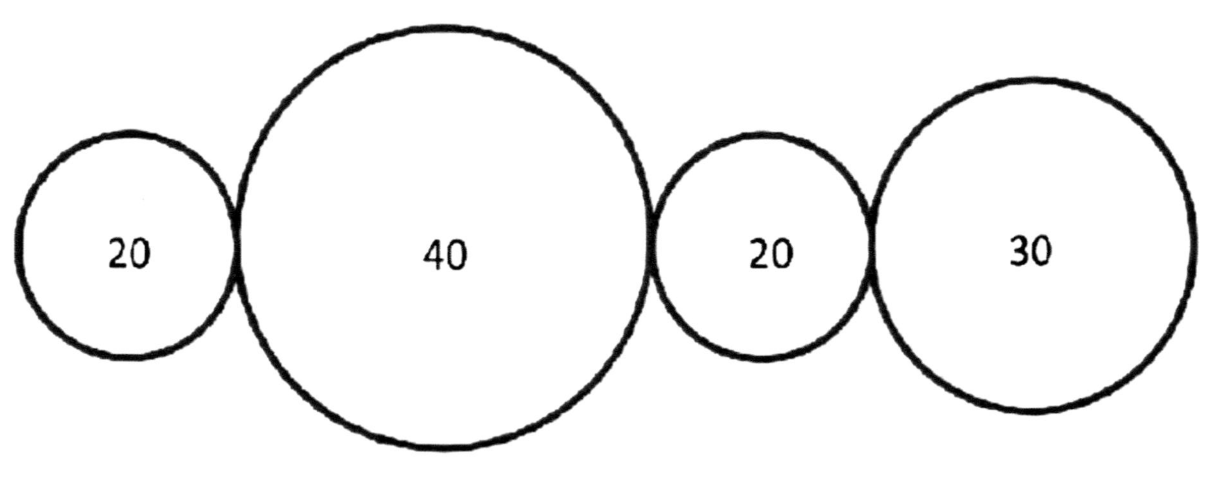

Figure 2-B.21

11. Calculate and state the gearing ratio of the given gear train in Figure 2-B.23. If the output is 1800 rpm (clockwise), what is the input speed and direction?

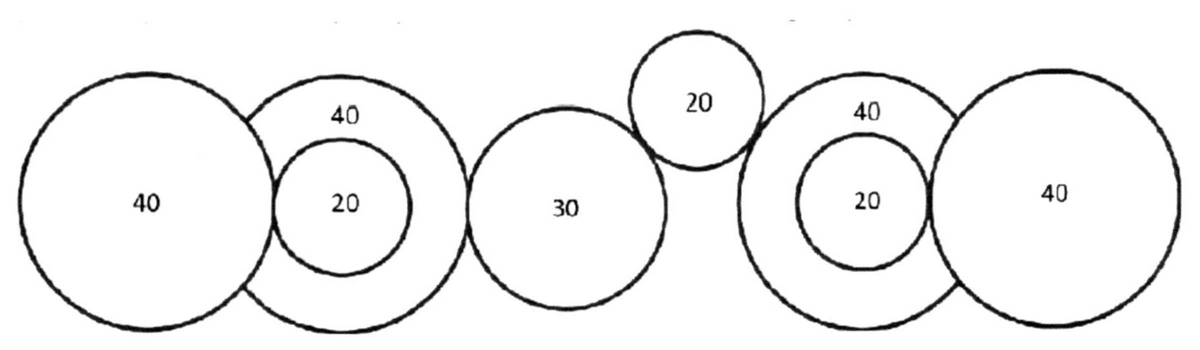

Figure 2-B.23

12. Calculate and state the gearing ratio of the given gear train in Figure 2-B.24. If the input is 25rpm (clockwise), what is the output speed and direction?

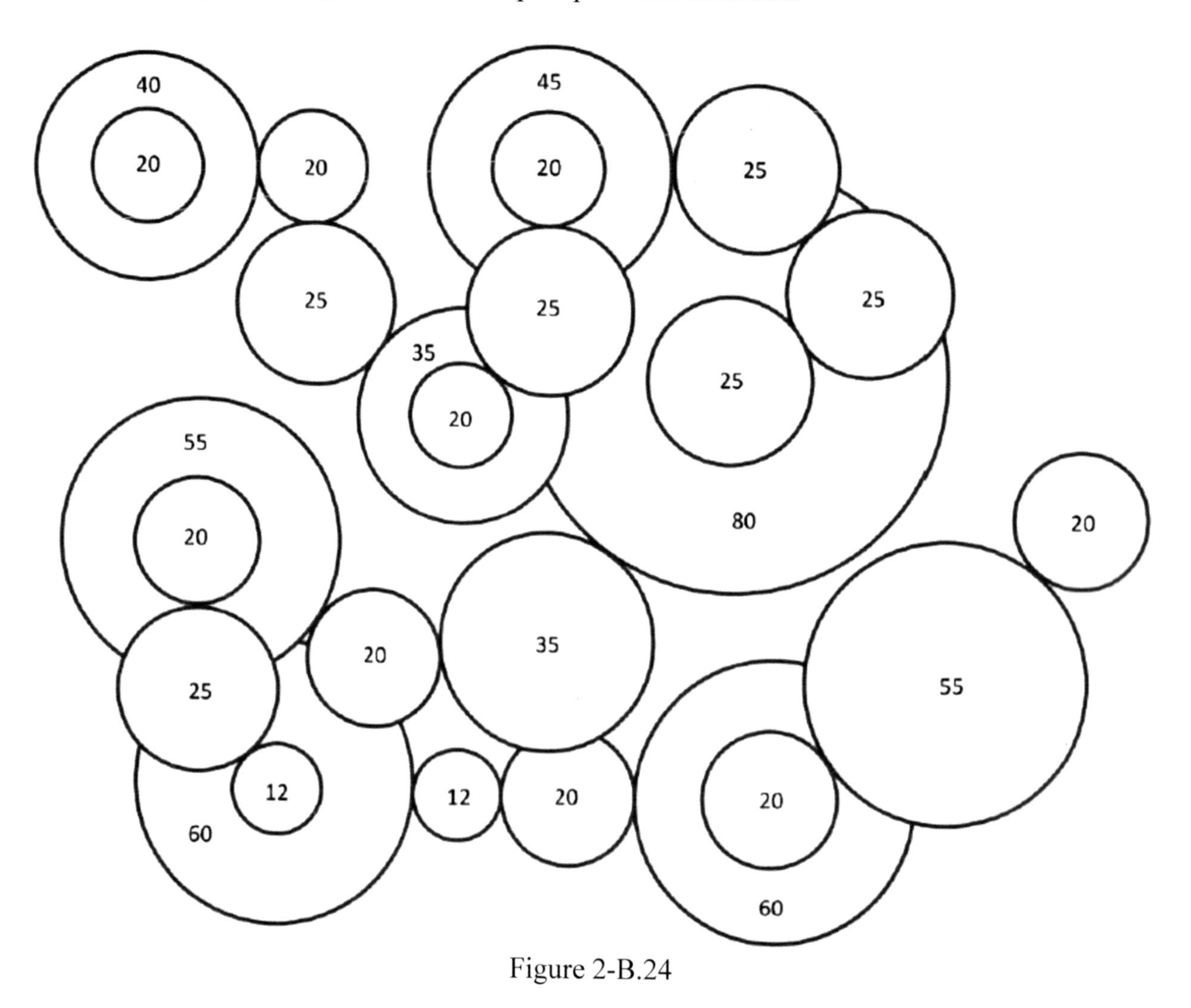

Figure 2-B.24

13. A worm gear with 50 teeth and ACP = 10 mates with a double-threaded worm. Determine (a) the gear ratio, (b) diameter of gear, and (c) lead of worm.

14. Design a mechanism that satisfies the following:
    a. Comprised of one (1) worm-worm gear pair and two (2) spur gears
    b. Worm gear and output spur gear rotate in opposite directions
    c. The angular velocity of the worm is equal to the least significant digit of your student number, plus one (1), multiplied by one hundred (100) RPM
    d. The overall gear ratio is equal to the second least significant digit of your student number, plus one (1), multiplied by ten (10) RPM

15. Design a mechanism that satisfies the following requirements:
    a. Comprised of one (1) worm-gear pair and four (4) spur gears
    b. Worm gear and output spur gear rotate in the same direction
    c. The angular velocity of the worm is equal to the least significant digit of your student number, plus one (1), multiplied by one hundred (100) RPM
    d. The angular velocity of output spur gear is equal to the sum of the second and third least significant digits of your student number plus two (2) RPM

16. Design a worm-rack that satisfies the following requirements:
    a. Rack moves at 10mm/s (Hint: Calculate necessary axis circular pitch)
    b. The angular velocity of the worm is equal to the least significant digit of your student number, plus one (1), multiplied by sixty (60) RPM

17. Design a mechanism that satisfies the following requirements:
    a. Comprised of three (3) spur gears driving a worm-gear pair that are driving a worm-rack. Gear train is set up as following: Spur Gear 1 -> Spur Gear 2 -> Spur Gear 3 -> Rod -> Worm -> Worm Gear -> Rod -> Rack.
    b. The output of the spur gear train is axially connected to the worm-gear pair
    c. The worm gear is axially connected to the worm-rack
    d. The angular velocity of the input spur gear is equal to the least significant digit of your student number, plus one (1), multiplied by thirty (30) RPM
    e. First worm must have speed that is 20 times greater than input speed
    f. Worm-gear pair must have gear ratio equal to the to the second least significant digit of your student number, plus one (1), multiplied by five (5)
    g. Rack moves at 5mm/s

Tutorial 3: Test

For Tutorial 3 Test, you will be tested on the topics introduced in the previous week. In preparation for this test, you should have attempted all the questions in the above section.

Tutorial Topic: Gear Design

- This test will consist of 2-3 problems. The Tutorial 3 Assignment Questions provide a useful reference for the types of questions you can expect.
    - Spur gear design calculations
    - Design of a compound gear train
    - Isometric pictorial sketching of a simplified mechanism
- Any questions requiring an isometric pictorial will be graded using the **Isometric Rubric** at the end of this section
- Grading breakdown for other components (e.g., correct calculations) will be clearly labelled on the test sheet

Required Reading Material:

- Refer to above section

Tutorial 4: Lesson

Tutorial 4 builds on the multiview representation topics introduced in the previous week. In this lesson, you will learn to visualize and sketch additional views in a multiview drawing (rather than the standard views covered in Tutorial 2). You will also learn to identify, visualize, and sketch special features such as fillets and rounds. Finally, you will learn to visualize and sketch a multiview drawing given only the isometric pictorial.

Material covered in this tutorial will be assessed during your **Tutorial 4 Test**. You are strongly encouraged to attempt the assigned questions below.

TUTORIAL TOPIC: ADVANCED MULTIVIEW REPRESENTATION

- Section views
- Auxiliary views
- Special features
- Sketching a multiview given an isometric pictorial

REQUIRED READING MATERIAL:

- Lieu & Sorby, 2nd Edition: Chapter 8.02 – 8.05 (**Review**)
- Giesecke et al., 4th Edition: Chapter 4 (Please refer to Appendix B of Custom Courseware) (**Review**)
- Lieu & Sorby, 2nd Edition: Chapter 10.02 – 10.12
- Lieu & Sorby, 2nd Edition: Chapter 11.02 – 11.04

SUPPLEMENTARY MATERIAL:

- None

ASSIGNMENT QUESTIONS:

Attempt as many of the following questions as you can. You should have these questions completed ahead of your Tutorial 4 Test. These questions are for practice and will not be graded. It is strongly recommended, however, that you attempt these questions on your own. Avoid the use of a straight edge or other drawing aids (an isometric underlay is not required, nor allowed). A rectangular grid underlay has been provided at the end of this assignment for practice. However, **you are strongly encouraged to attempt questions without the rectangular grid underlay, as it will *not* be permitted for your Tutorial 4 Test, nor for the final exam**.

For your Tutorial 4 Test, you will be tested on a subset of these question and/or question types. For additional practice problems, please refer to the appropriate textbook sections.

1. Sketch the multiview representation of the isometric pictorial given in Figure 2-B.25. For this problem, the front view corresponds to the front of the isometric glass box.

2. Sketch the multiview representation of the isometric pictorial given in Figure 2-B.26. For this problem, the front view corresponds to the front of the isometric glass box.

Figure 2-B.25 Figure 2-B.26

3. Sketch the multiview representation of the isometric pictorial given in Figure 2-B.27. For this problem, the front view corresponds to the front of the isometric glass box.

4. Sketch the multiview representation of the isometric pictorial given in Figure 2-B.28. For this problem, the front view corresponds to the front of the isometric glass box.

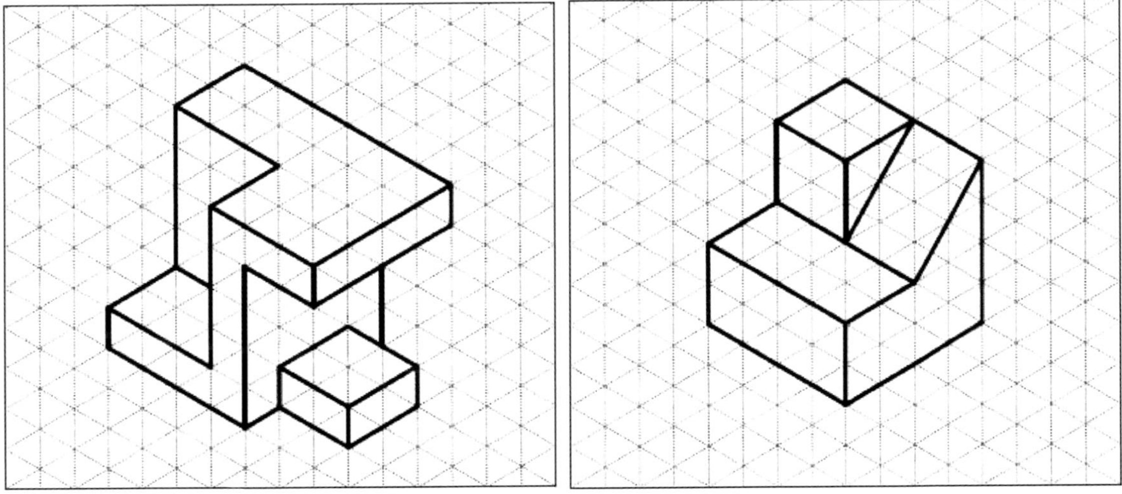

Figure 2-B.27 Figure 2-B.28

5. Sketch the multiview representation of the isometric pictorial given in Figure 2-B.29. For this problem, the front view corresponds to the front of the isometric glass box.

6. Sketch the multiview representation of the isometric pictorial given in Figure 2-B.30. For this problem, the front view corresponds to the front of the isometric glass box.

Figure 2-B.29 Figure 2-B.30

7. Sketch the multiview representation of the isometric pictorial given in Figure 2-B.31. For this problem, the front view corresponds to the front of the isometric glass box.

8. Sketch the multiview representation of the isometric pictorial given in Figure 2-B.32. For this problem, the front view corresponds to the front of the isometric glass box.

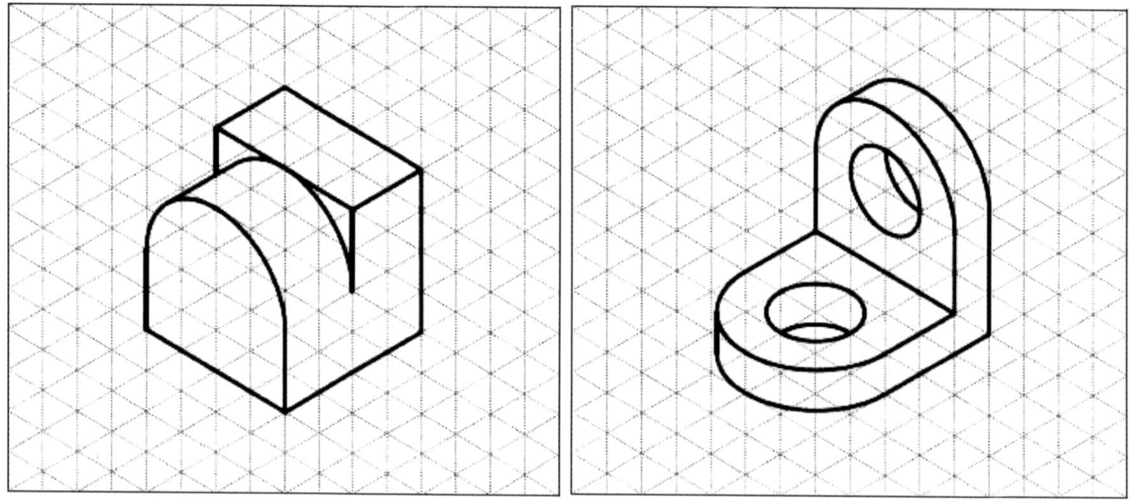

Figure 2-B.31

Figure 2-B.32

9. Sketch the multiview representation of the isometric pictorial given in Figure 2-B.33. For this problem, the front view corresponds to the front of the isometric glass box.

10. For each set of views in **Figure P10.1** (Lieu and Sorby, Chapter 10.17, Problem 1), select the correct section view from the 24 proposed views shown at the right. In each set of views, the view indicated by the balloon is to be changed to a full section view taken along the centerline in the direction indicated by the arrows in the adjacent view.

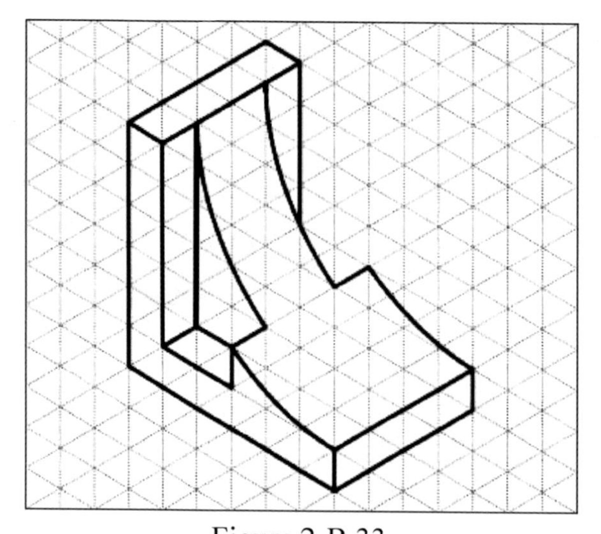

Figure 2-B.33

11. For the pictorial shown in **Figure P11.1c** (Lieu and Sorby, Chapter 11.09, Problem 1), sketch the primary views as well as an auxiliary view to present only the true shape of the inclined surface.

12. For the pictorial shown in **Figure P11.1e** (Lieu and Sorby, Chapter 11.09, Problem 1), sketch the primary views as well as an auxiliary view to present only the true shape of the inclined surface.

13. For the pictorial shown in **Figure P11.1h** (Lieu and Sorby, Chapter 11.09, Problem 1), sketch the primary views as well as an auxiliary view to present only the true shape of the inclined surface.

14. For the pictorial shown in **Figure P11.2c** (Lieu and Sorby, Chapter 11.09, Problem 2), sketch the primary views as well as an auxiliary view to present only the true shape of the inclined surface.

15. For the pictorial shown in **Figure P11.2a** (Lieu and Sorby, Chapter 11.09, Problem 2), sketch the primary views as well as an auxiliary view to present the entire object with the inclined surface shown in its true shape.

16. For the pictorial shown in **Figure P11.2b** (Lieu and Sorby, Chapter 11.09, Problem 2), sketch the primary views as well as an auxiliary view to present the entire object with the inclined surface shown in its true shape.

17. For the pictorial shown in **Figure P11.2d** (Lieu and Sorby, Chapter 11.09, Problem 2), sketch the primary views as well as an auxiliary view to present the entire object with the inclined surface shown in its true shape.

18. List two international standards organizations that specify orthographic projection conventions.

19. In your own words, explain (in *words*) the difference between first angle and third angle projection. Your explanation should be in terms of the projection/viewing planes of each.

20. Using a figure from one of the above assigned questions in this section, illustrate the difference between first angle and third angle projection (i.e., by sketching the appropriate multiview for each projection).

21. Extending on question 19 above, label each multiview using an appropriate projection symbol.

Rectangular Grid Underlay

Tutorial 4: Test

For Tutorial 4 test, you will be tested on the topics introduced in the previous week. In preparation for this test, you should have attempted all the questions in the above section.

TUTORIAL TOPIC: ADVANCED MULTIVIEW REPRESENTATION

- This test will consist of 3-4 problems. The Tutorial 4 Assignment Questions provide a useful reference for the types of questions you can expect.
  - Sketch a multiview given an isometric pictorial
  - Sketch auxiliary views
  - Explain and/or illustrate different standards and/or conventions
- Any questions requiring a multiview drawing will be graded using the **Multiview Rubric** at the end of this section
- **REMINDER**: some notes on grading to keep in mind:
  - Be sure to indicate scale!
  - For visible lines: they should be thick, solid and dark (avoid *feathering*)
  - For hidden and centre lines: they should be thin, solid and light
  - Draw any construction and grid lines lightly so they are clearly discerned from all sketching lines
  - Straight lines – some waviness is okay, but too much is BAD
  - Be neat and tidy! It counts for marks
  - Put signature, date and title near the drawing
  - **Don't cheat!!!**

REQUIRED READING MATERIAL:

- Refer to above section

Tutorial 5: Lesson

Tutorial 5 introduces the concept of design analysis and, specifically, reverse engineering. In this lesson, you will learn the importance of analysis in the design process. You will also learn the methodology of reverse engineering and be introduced to common metrology tools. Finally, you will have the opportunity to apply a part of the reverse engineering process in a simple dissection and measurement exercise. In this exercise, you must measure and document the size and features of a simple device using handheld metrology equipment.

Please note that the topics covered in this lesson will necessarily be applied to your **Lab 5 Test**. For this test, you will be required to dissect, measure, model, assemble and simulate a simple mechanical device. The dissection and measurement phases of the test will draw on this tutorial lesson (Tutorial 5) whereas the modelling, assembly and simulation phases will draw on your preceding laboratory lessons (Lab 1, 2, 3, 4 and 5). Full details of the Lab 5 Test will be posted to Avenue.

Note that there are no submission criteria for this tutorial. Attending your Tutorial 5 Lesson and completing the simple dissection and measurement exercise is sufficient preparation for your Lab 5 Test.

TUTORIAL TOPIC: DESIGN ANALYSIS

- Reverse engineering
- Metrology tools for reverse engineering
- The reverse engineering process
- Simple dissection and measurement exercise

REQUIRED READING MATERIAL:

- Lieu & Sorby, 2nd Edition: Chapter 7.02 – 7.04

SUPPLEMENTARY MATERIAL:

- None

Tutorial 5: Test

Your Tutorial 5 test will require you to design a mechanism to achieve a specified task. This **cumulative assessment** will cover the topics introduced in Tutorial 5 Lesson (Design Analysis) as well as sketching and gear design topics introduced throughout the term.

TUTORIAL TOPIC: DESIGN ANALYSIS

- This test will require you to design a mechanism by applying what you have learned over the duration of the term
  - Technical Sketching (Isometric pictorials & Multiview drawings)
  - Gear design
  - Reverse engineering
- Additional details will be presented leading up to this assessment

Notes on Tutorial Grading

Tutorial assessment will be through **four (4) tutorial tests**. Three of your tutorial tests (1, 2, and 4) will cover visualization and sketching problems, and one test (Test 3) will cover mechanism design. Two types of sketching problems that will be especially emphasized are:

1. Isometric Pictorials
2. Multiview Drawings

The tests and final exam will be graded in exactly the same way using the rubrics below.

Grading rubrics for isometric pictorials and multiview drawings is given on the following pages.

Your TA should have your work graded before your next tutorial session, but it may take slightly longer to have the marks posted on Avenue. If you wish to appeal or have your mark reviewed, you will be required to produce your original graded work; keep all of your work until the final grade is released and you have verified it. A returned grade of 0 means the work has not been submitted and/or graded.

Important Considerations for Tutorial Grading

No Straight Edges Allowed: rulers and other straight edges (including student cards) are not allowed when writing tests.

Isometric Underlay: an isometric underlay may be used for Tutorial Tests 1 and 2. Underlays are not required after Tutorial Test 2 and are not permitted for the Final Exam.

Size / Orientation Penalty: drawings of insufficient size and/or incorrect orientation will be subject to a 50% penalty. A technical sketch must take up more than one-third (1/3) of the available drawing (see below).

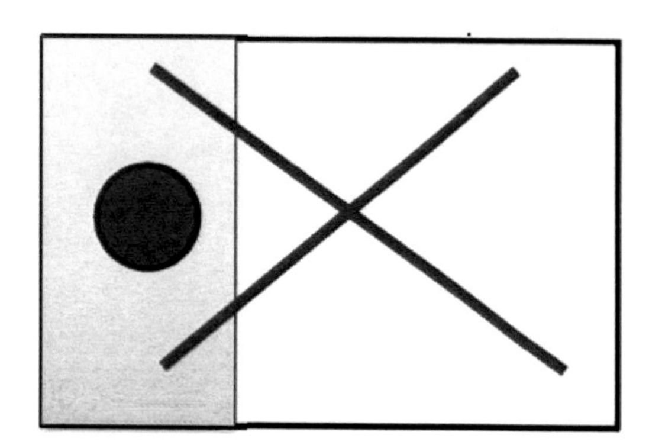

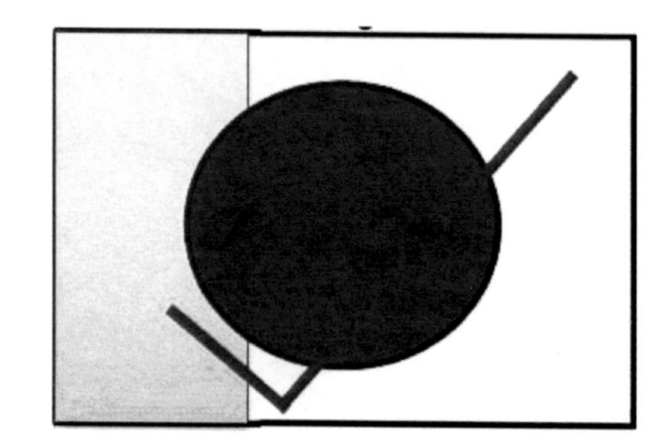

Isometric Rubric

Criteria	Below Expectations (x0)	Marginal (x1)	Meets Expectations (x2)	Exceeds Expectations (x3)	Weight	Total
Projection	Significant errors. Design cannot be recognized or not isometric projection.	Significant errors. Design cannot be visualized or angles to horizontal are not 30 degrees.	Some errors but design can be visualized with minor mental corrections. Correct isometric projection	Reader can fully visualize the shape of the design. Correct isometric projection	**4**	**/ 12**
Proportions and Scale	Sketch does not reflect correct proportions of design or scale is not indicated.	Sketch poorly reflects proportions of design. Sketch is misleading or scale is not indicated correctly.	Proportions generally correct. Scale indicated correctly.	Proportions correct. Scale indicated correctly.	**2**	**/ 6**
Presentation	Design cannot be discerned or text cannot be read.	Difficult to read; messy	Design and text are easily read but could be neater. Straight lines could be straighter or curves more correct. Sketch is centred and fills most of the allocated space.	Very neat appearance. Sketch is well spaced with sufficient white space. Sketch is centred and fills most of the allocated space on the page. Lines are straight and curves are correct.	**1**	**/ 3**
				Total:	**/ 21**	

Multiview Rubric

3. Criteria	Below Expectations (x0)	Marginal (x1)	Meets Expectations (x2)	Exceeds Expectations (x3)	Weight	Total
View Selection	Two or more necessary views are missing.	One or more necessary views are missing.	All necessary views are present. Unnecessary views present.	No unnecessary views are present. All necessary primary, auxiliary and section views are present.	**4**	**12**
Projection	Significant errors. Design cannot be recognized or not isometric projection.	Significant errors. Design cannot be visualized or angles to horizontal are not 30 degrees.	Some errors but design visualized with minor mental corrections. Correct isometric. Hidden lines missing.	Reader can fully visualize the shape of the design. Correct isometric projection.	**4**	**12**
Proportions and Scale	Sketch does not reflect correct proportions of design or scale is not indicated.	Sketch poorly reflects proportions of design, is misleading or is not indicated correctly.	Proportions generally correct. Scaled indicated correctly.	Proportions correct. Scale indicated correctly.	**2**	**6**
Centre Lines and Marks	Missing	Some missing necessary centre lines/marks.	Centre marks and lines placed correctly on holes and cylinders, but centre lines used for aligned of arc/lines are missing, centre line/ marks not aligned or unnecessary centre lines/marks present.	Centre marks with the project lines placed on all circles in true shape. Centre lines placed on all holes and cylinders and extended beyond the ends. Additional centre lines used for alignment of arc/ circle centres where necessary. Centre lines/ Marks aligned.	**1**	**3**
Presentation	Design cannot be discerned or text cannot be read	Difficult to read. Messy	Design and text are easily read but could be neater. Straight lines could be straighter or curves more correct. Sketch is centred and fills most of the allocated space.	Very neat appearance. Sketch is well spaced with sufficient white space. Sketch is centred and fills most of the allocated space on the page. Lines are straight and curves are correct.	**1**	**3**
				Total:	**/ 36**	

SECTION 2: Design Project

2.1 Design Project Overview

Working in design teams, you will incorporate all elements covered in the course (technical sketching, part modelling, assemblies, gear design, simulation, and prototyping) towards the design of a gearing mechanism that accomplishes a specific task. This project requires that you:

- Complete/submit preliminary design that includes hand calculations and technical sketches
- Model your mechanism and all its components in Autodesk Inventor
- Create an appropriately constrained assembly of your design in Autodesk Inventor
- Create complete set of working drawings based on modelled components and assembly
- Simulate the motion of your assembly to verify it meets specifications
- Create a functioning physical prototype using 3D printers available to you in the EPIC Lab
- Complete technical report summarizing design and team's progress throughout the term
- Undergo a **Project Interview** near the end of term (see details below)

The exact project specification (i.e., what your project entails) will be introduced early in the term and all project deliverables will be submitted during your project interview near the end of term. All of the project requirements outlined above will be discussed either in lecture, lab or tutorial. However, certain practical skills will be critical for success. Included among these skills are:

- **Working in a Team Environment**: this project requires you to collaborate with your design team members outside of normal lecture/lab/tutorial hours. This means that it is imperative you set time aside each week and schedule regular meetings with your group (these meetings should be properly documented).
- **Good Time Management**: many of the tools (e.g., software, hardware, hand-hold tools) necessary to complete your project will be provided to you, either on campus in one of the computer labs, or in the EPIC Lab. The EPIC Lab is where all the 3D printers reside. However, the EPIC Lab will only be available through scheduled bookings, given the large volume of students and small number of printers. As the term draws to a close and the project deadline approaches, you will likely find the EPIC Lab extremely busy, and thus access extremely limited. The sooner you print components, the less stressed you will be!
- **Communication**: in addition to completing a written technical report, you and your team will present your design near the end of term as part of an individual and team oral assessment (referred to as a **Project Interview**). As part of this assessment, each group member will be expected to discuss, using proper technical language, specific aspects of their design as well as specific elements of the course.

2.2 Guide to the EPIC Lab

The Experiential Playground and Innovation Classroom (EPIC) is located in the Engineering Technology Building (ETB). The main goal of this lab is to expose first year engineering students to hands-on learning.

The EPIC Lab serves as an Experiential Learning centre for ENG 1C03 and ENG 1D04, allowing more hands-on minds-on learning outside of the classroom. **The EPIC lab will serve as a crucial hub for completion of your design project**. There is also ample workspace for brainstorming and idea generation, and the Lab is equipped with desktop workstations and 3D printers for developing and printing physical prototype components. You will also find all the necessary tools needed for building and assembly for your final design.

The EPIC Lab will be available to you shortly after the term begins, and you are strongly encouraged to stop in and learn what equipment is available for your use as well as how this equipment works. However, **as various project deadlines draw near, demand for use of the EPIC Lab resources increases significantly!** Demand for use of the 3D printers is especially high, and student groups are required to schedule time in the Lab to use these printers (through a URL that will be provided). **To ensure everything flows smoothly, all students must read this section of the courseware very carefully!** Below you will find details on:

1. Booking Time in the EPIC Lab (for using 3D printers)
2. What you need to do before coming to the EPIC Lab
3. Preparing to print your parts
4. Using the 3D printers

Booking Time in the EPIC Lab

All 3D printer bookings will be booked through an appointment scheduler called SuperSaas (http://www.supersaas.com/) and specific links will be posed on the news feed of the Engineering 1C03 Avenue page as the term progresses. If you run into any issues using the scheduling website, please email the Engineering 1 Course Coordinator (engic@mcmaster.ca) with the details of the problem you are experiencing and don't forget to include your student number!!

The 3D printers in the EPIC Lab will be available to you in two stages. The **first stage** (Printer Demos) is where you can schedule time in the lab to learn how the printers work and get your first chance to work with them hands on, printing whatever you like (within reason, of course). The **second stage** (Group Printing) is where your design project group books time in the lab to print off all of your design components. **Each group will have a limited number of 60-minute booking sessions available to them**. Please use these bookings wisely!

PRINTER DEMOS

Before you begin your group project printing there will be printer demo's available earlier in the term for you to see how the printers work. These are made available so you can make the best use of your limited time to print your project components. It is strongly recommended you attend a demo session! You will book your demo sessions individually not as a group. Here is the process:

1. Go to the supersaas website link provided on Avenue for the demos
2. Create a login if this is your first visit using your McMaster email address as your login and create a password for yourself (if you have a login for Engineer 1D04 EPIC it will not work here you must create a new account)

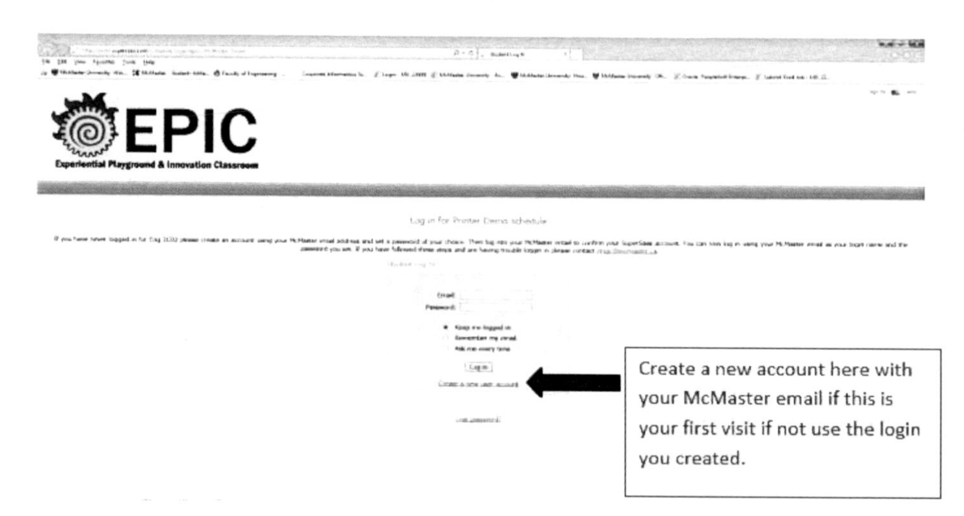

3. Once logged in you will be presented with the schedule, click on the session you want (it will show if a slot is full) and enter the required information for the booking

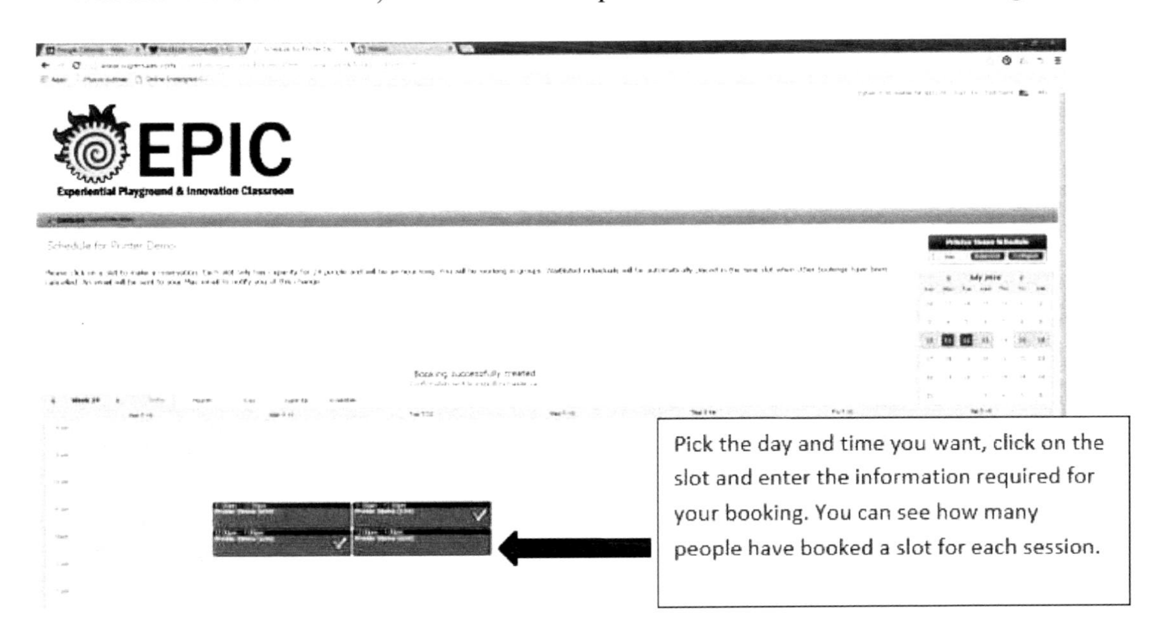

4. You will receive a confirmation email about the booking and the booking will be stored in a database
5. If you need to cancel you must do so 24 hours before the scheduled booking (in case of emergency and it is after the 24-hour mark please email engic@mcmaster.ca to have the booking canceled)

GROUP PRINTING

Once the design project has been introduced, you have signed up in a group, and have been given a group number, you will be attending the project printing sessions as a group and you will no longer need your individual login that was used for the demos.

1. Your group will be assigned a unique login and password. The login will be your group number (e.g., Group10 with no space in between the group and the number). Your password will be the sum of the student numbers of everyone in your group.
2. Go to the supersaas website link provided on Avenue. A unique URL will be provided for each group of printer bookings. Please check carefully to make sure you aren't using the wrong link!

3. Login using the appropriate credentials (i.e., correct login and password, please see above).

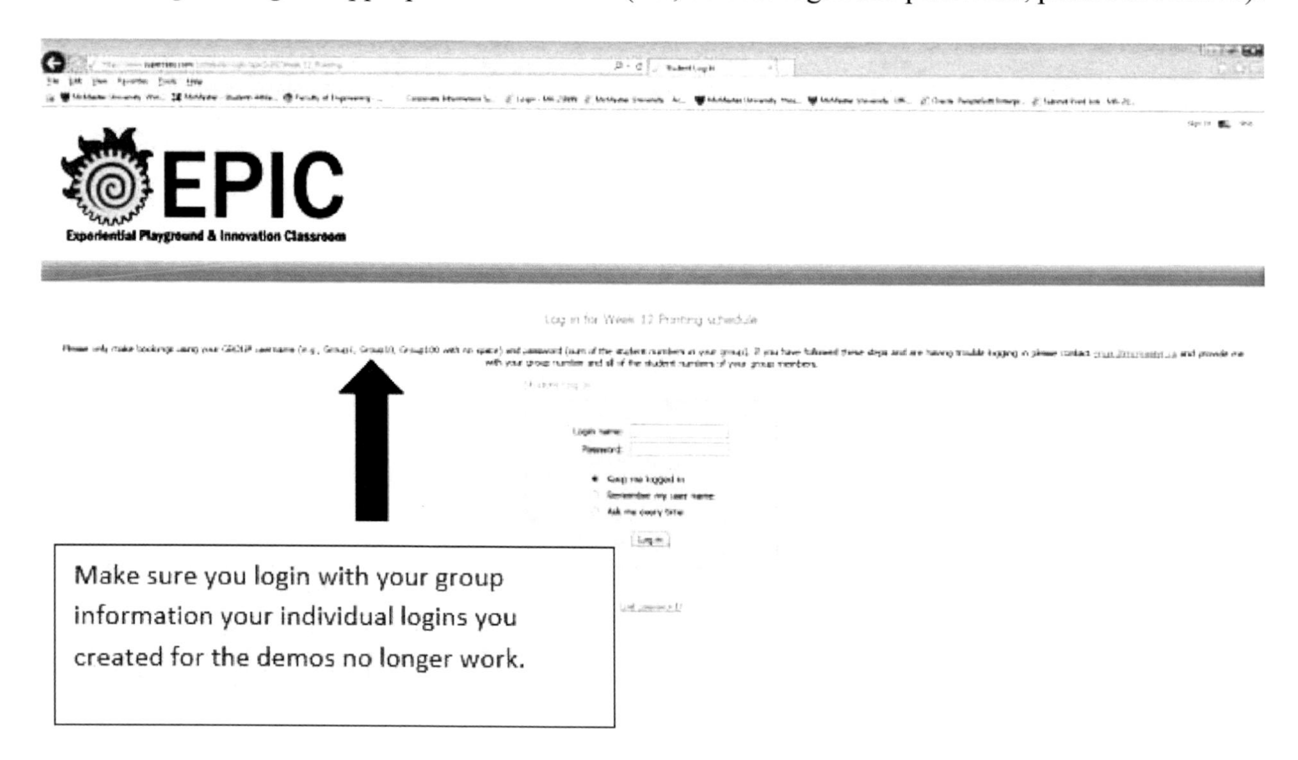

4. Once logged in you will be presented with the schedule which is the same as the screenshot of the demo schedule, click on the session you want (it will show if a slot is full) and enter the required information for the booking. The booking will be stored in a database.
5. If you need to cancel you must do so 24 hours before the scheduled booking (in case of emergency and it is after the 24 hour mark please email engic@mcmaster.ca to have the booking canceled)

Before Coming to the EPIC Lab

The 3D printers in the EPIC Lab 'print' a physical representation of a solid CAD model that you design in Autodesk Inventor. Successful use of the printers (i.e., being able to print what you intend to print) means **coming to the lab prepared**. How do you do that?

1. When modeling your parts, give thought to your final design and how everything will come together. Some parts may require assembly after they are printed. The **3D printers have limitations and may not parts exactly as they were modeled**. The ENG 1C03 team is here to help you. Don't hesitate to ask questions!
2. Your use of the 3D printers is limited! Give s
3. Create a solid model of all the parts you will be printing **before coming** to the EPIC Lab
4. Bring the files with you in the **correct file format**!
5. You can't print everything at once, but you also won't have enough bookings to print one part at a time. **Plan your use of the printers wisely**!

THE DESIGN OF YOUR MECHANISM

For your design project, you will need to print a number of different components, including, but not limited to:

1. Gears
2. A mounting bracket for housing the gears, and
3. Components for your gripper.

There are constraints given to you in terms of where certain components must be positioned, as well as required input and output parameters. However, there is also a great deal of flexibility afforded to you in terms of what your final design should look like and how it should function.

- **Fastening components together**: you may decide to attach parts together using rods, screws, bolts or other fasteners, or you may decide not to use any fasteners at all. Should you decide to use fasteners, these can be purchased from a craft or hardware store and do not need to be printed.
- **Keep final assembly in mind during modeling**: for parts that will need to be assembled together, limitations in the 3D printers may mean some parts may not fit or align exactly as you intend (e.g., holes may not align perfectly). Your design should have the necessary flexibility that you can make some 'post-processing' adjustments after printing (e.g., drilling, cutting, sanding, etc.)
- **Minimize 'post-processing' steps where you can**: if a part such as a gear connects to a rod, be sure to model the gear with a hole!

- **The bigger the part, the longer the print**: the less material your parts need, the less print time you'll require, which lowers the likelihood of something going wrong (all equipment can 'act up' from time-to-time)
    - Want to minimize print time, **don't make your parts too thick!** Not sure of an appropriate thickness? Go to the lab when it's not busy and print stuff out yourself! That's what the lab is intended for!
    - Some sample print times for various components is provided for your reference (Table 1 and Table 2)
- **Considerations for printing gears**:
    - Not too thick!
    - Create holes in the center of your gears!
    - Teeth size should be fairly big to ensure proper meshing and fewer errors in the prints. Select any size you desire, but it is YOUR responsibility to ensure that, when printed, those gears will properly mesh.

Table 1. Sample Print Times for Sample Spur Gear

Sample Spur Gears	Details	Printing Time
	Pitch Diameter: 30-mm Thickness: 3-mm	11 minutes
	Pitch Diameter: 30-mm Thickness: 6-mm	17 minutes
	Pitch Diameter: 30-mm Thickness: 9-mm	25 minutes

Table 2. Sample Print Times for Other Sample Components

Sample Mounting Bracket	Details	Printing Time
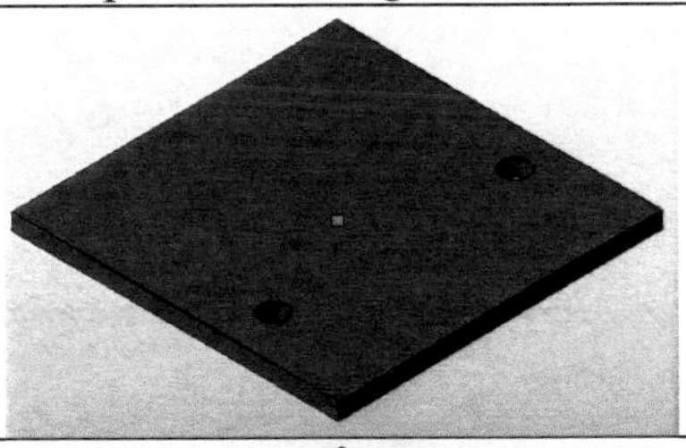	Dimensions: 60x60-mm Thickness: 3-mm	18 minutes
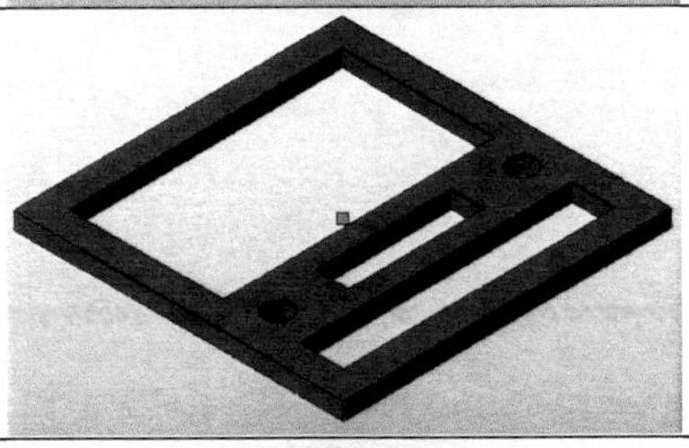		13 minutes
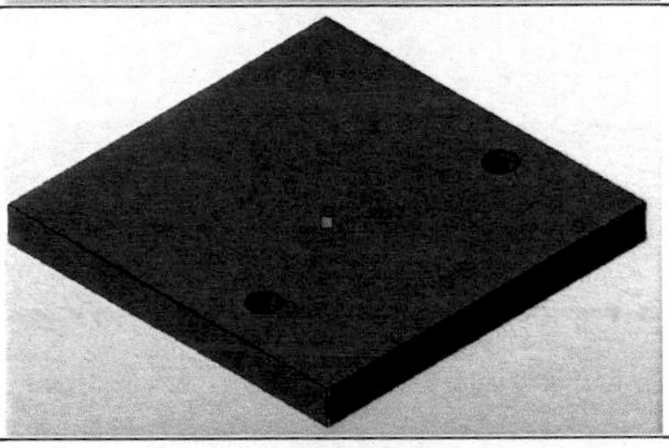	Dimensions: 60x60-mm Thickness: 6-mm	35 minutes
		21 minutes
	Dimensions: 60x60-mm Thickness: 9-mm	44 minutes
		29 minutes

Creating Solid Models in Autodesk Inventor

All of the parts you intend to print will need to be modeled in Autodesk Inventor. Part modeling will be covered in Labs 1 and 2. For some of your parts (e.g., your mounting bracket), there is a great deal of flexibility afforded to you. Be creative!

Components such as gears are created using the **Design Accelerator** feature in Autodesk Inventor. Use of the Design Accelerator will be covered in Lab 3 of your lab.

Successful use of the printers requires proper use of your time. Given the high demand of the printers and the limited number of bookings available, it is likely that you may only have time to complete ONE print per 60-minute session. Therefore, using that time to print off, for example, a single gear is not recommended. There are two options available to you:

1. Create and export individual parts and upload the files (in the correct format!) to the 3D printing software one at a time
2. Create and export a single assembly of several components (e.g., gears)

The following section (Modelling Multiple Gears in Inventor) details how to create and arrange multiple gears for printing.

Modeling Multiple Gears in Autodesk Inventor

Instead of printing gears one at a time, it is strongly recommended that you create a Gear Assembly in Inventor, where multiple gears are arranged on a common plane and spaced apart such that their teeth are not in mesh.

1. **Creation of Gears**: the first step is to create the desired gears that you will need, using the Design Accelerator tool in Inventor (Lab 3). After creation of gears, save assembly document so you will have access to the individual IPT files of the gears you created.

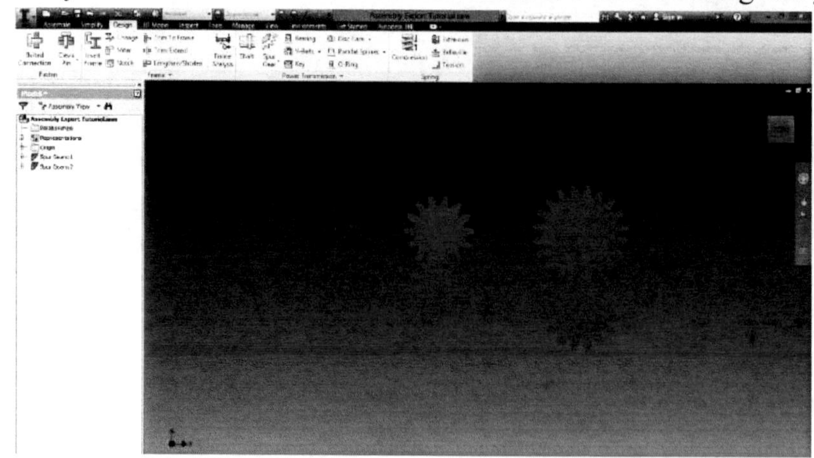

2. **Decoupling the Gears**: to arrange the gears, you will have to start a new metric assembly file in Inventor. Afterwards, place the individual gears (as IPT) into the assembly environment. This way, your gears are no longer meshed together and will print as individual gears.

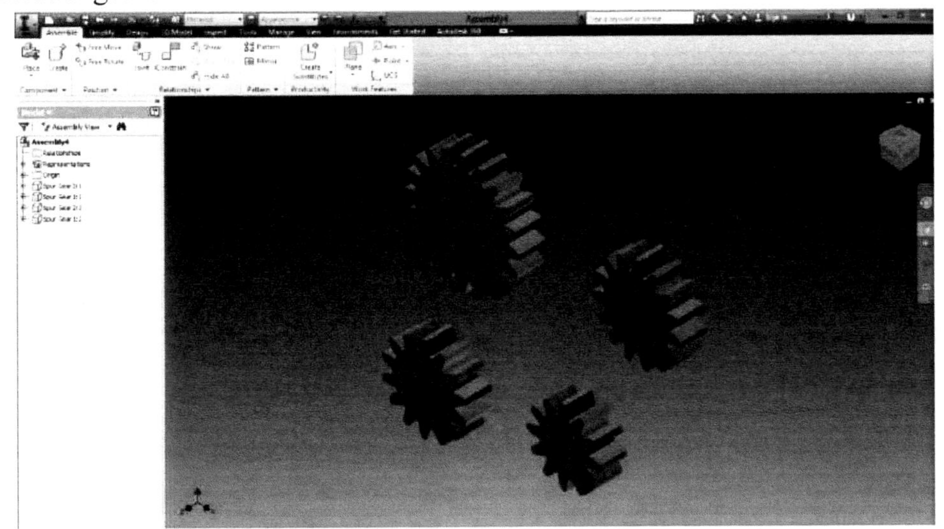

3. **Arranging the Gears**: This step important because of the restrictions of the printer and the amount time you have. You will have to rearrange the gears in such a way that they are aligned (flush) on the same surface. To do this, use the flush tool to make one side of the gears flush to each other (<u>use of the flush tool and other constraints will be covered in your Lab 3</u>).

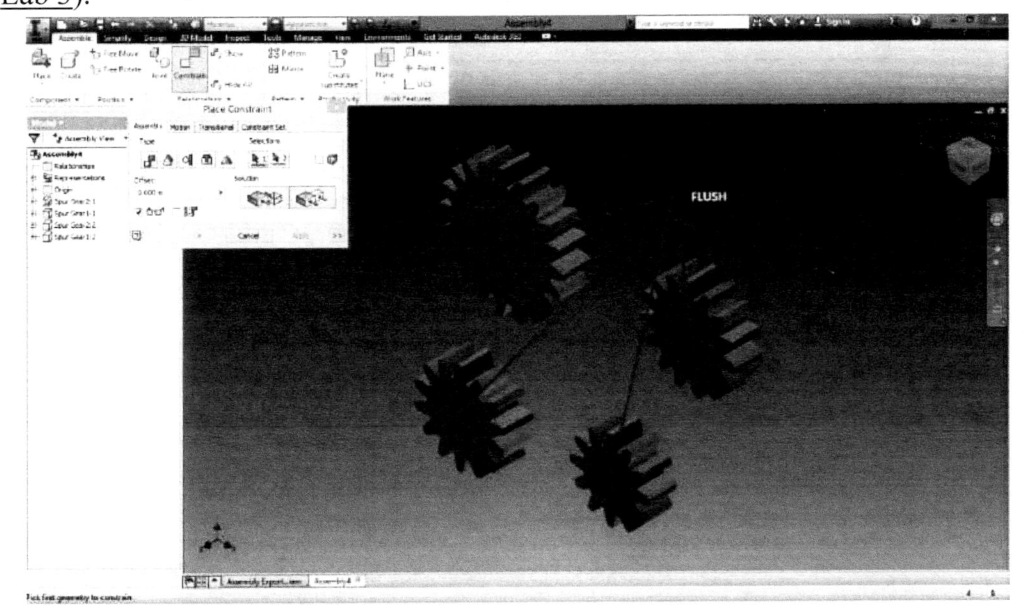

4. **Select Top View for Printing**: set the view to be top so when you export this will be considered the top view and all your gears will lie on the bed of the printer.

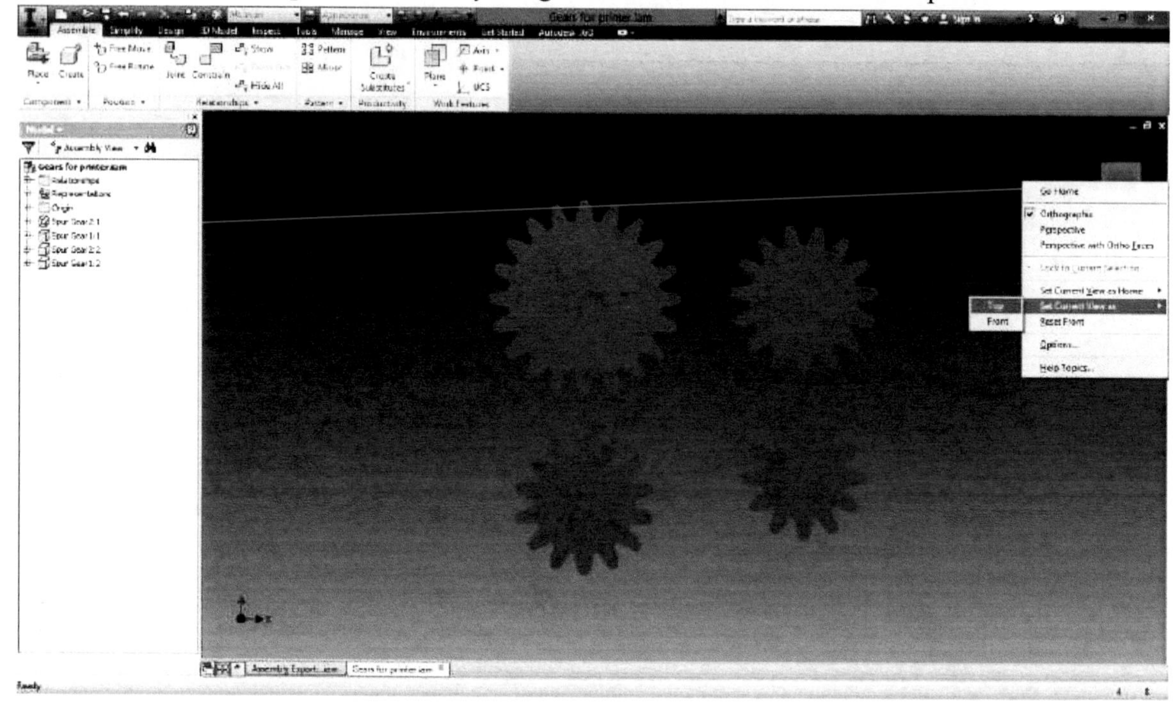

5. **Spacing your Gears**: your gears must be space apart such that they fit within the dimensions of the 3D printer bed! To determine this area, please visit the EPIC lab! To check the distance, you can use the measure tool found under the Inspect tab of Autodesk Inventor. Check the distance by clicking on the edge of the teeth of your gears, and ensure there is a 5mm gap between each of your gears.

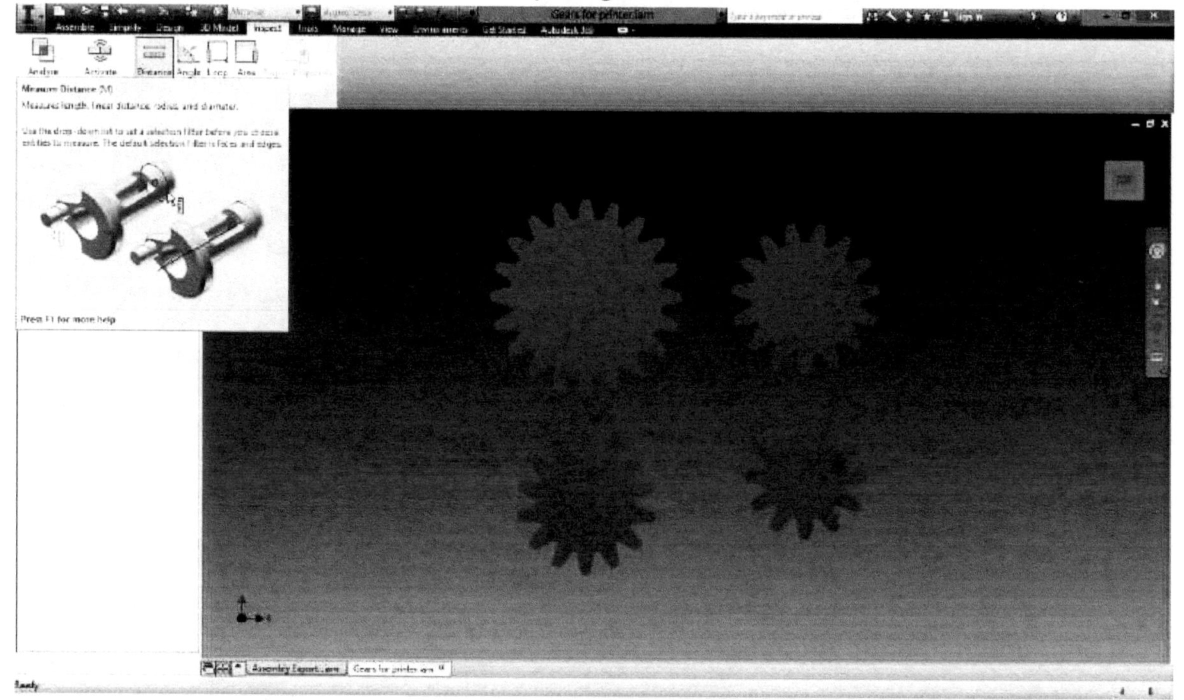

EXPORTING SOLID MODELS AND ASSEMBLIES FOR PRINTING

Before you can print your parts, they must be exported into an appropriate file format. Autodesk Inventor, by default, saves files in either *.ipt or *.iam format (native to the Inventor software). However, the program used to print your components in the EPIC lab will not recognize either of these formats!

In order to print components, **your files must be exported in the *.STL format**. To do this, select "Export" > "CAD Format".

Under "Save as type:", select "STL Files (*.stl)".

Preparing to Print your Parts

When you come to the EPIC Lab, you should have all your parts (saved as *.STL files) with you either on a USB stick or saved as an email attachment. Before printing, there's a few things to consider:

1. How many parts can I print at a time?
2. Do I need to scale any of my parts? (i.e., make them bigger or smaller)

Limitations in Size of Prints

The maximum dimensions that a part can be in the X-Y plane is limited to the size of the 3D printer bed (you will be using the Orion printers for the Design Project). Although the bed is circular, you must design your components such that they fit within a square that fits inside this circle. The best way to determine the diameter of the bed is to go to the lab and measure it! If you bring a part (or assembly) to the EPIC lab that exceeds this dimensions, you will not be allowed to print until you have corrected them!

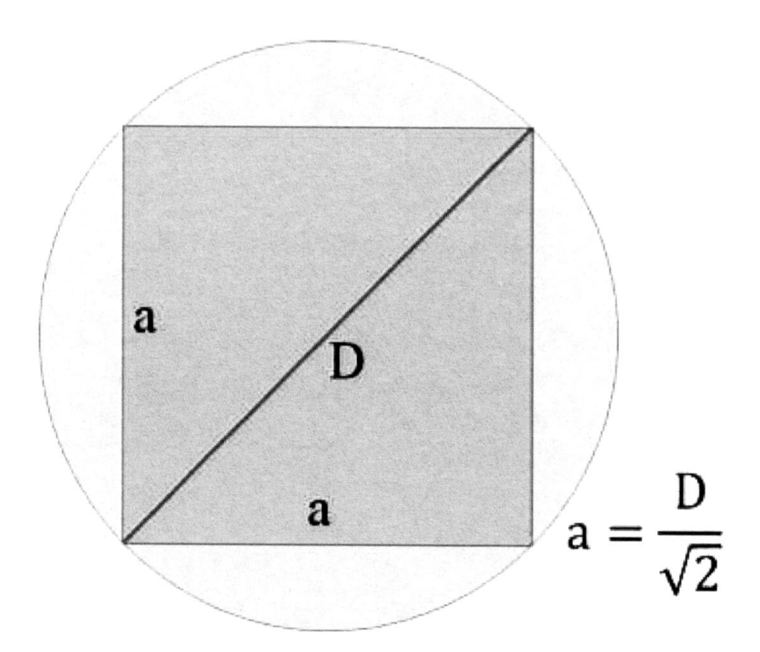

You should also be aware that a print that takes up the entire 3D printer bed will take a significant amount of time. Each printer booking is limited to 60 minutes, which includes the time to prepare and upload your files for printing. Realistically, **each of your prints should take a maximum of 30 minutes to complete**. In the next section – Using the 3D Printers, you will see how you can get an estimate of print time *before* printing your parts.

SCALING *STL FILES

The file format of your files (*.STL) is such that the file itself cannot be edited (i.e., you cannot add/delete/modify features). However, STL files can be modified in other ways. For example, you can 'cut' parts along a plane, rotate them about a particular axis, or scale them (increase/decrease part size) along one or more axes.

There are a few possibilities for why you would need to scale your parts.

1. Your part(s), as designed, is quite thick and will take too long to print. Scaling a part along a single axis allows you to make a part(s) 'thinner' in order to save print time
2. Your part(s), when exported from Autodesk Inventor as an STL file, was exported in units other than what the 3D printers require
    a. It is not uncommon for a part file to be exported as an STL file from Autodesk Inventor in different units than intended. The most common units are either 'cm' (centimetres) or inches.
    b. What does this mean? Well, if you were to model a 10mm x 10mm x 10mm cube in Inventor and export it as an STL file, it would actually be exported as 1cm x 1cm x 1cm cube (this can be changed, but is the default setting).
    c. Why is this a problem? The problem is that, when opening STL files in a different program (such as the one used in the EPIC Lab) and on a different computer, units are not carried over to that different program.
    d. So.... Why is that a problem? Well, since the units are not carried over from program-to-program, the other program used in the EPIC Lab (called netfabb) assumes these units of your part are in millimetres.
    e. What does that mean exactly? It means that your 1cm x 1cm x 1cm cube that you exported would actually be opened as a cube with dimensions 1mm x 1mm x 1mm. **This means your part is actually smaller by a factor of 10x!**
    f. You can avoid this issue by changing the default settings directly in Autodesk Inventor. But in the event your parts are indeed smaller than intended, this section will cover how best to address this conversion error.

1. **Open Netfabb Basic**: click on the Netfabb Basic icon located on the desktop of one of the EPIC Lab workstations. When prompted, select "I accept the terms of usage", and click the "Later" button when it becomes enabled after 10 seconds.

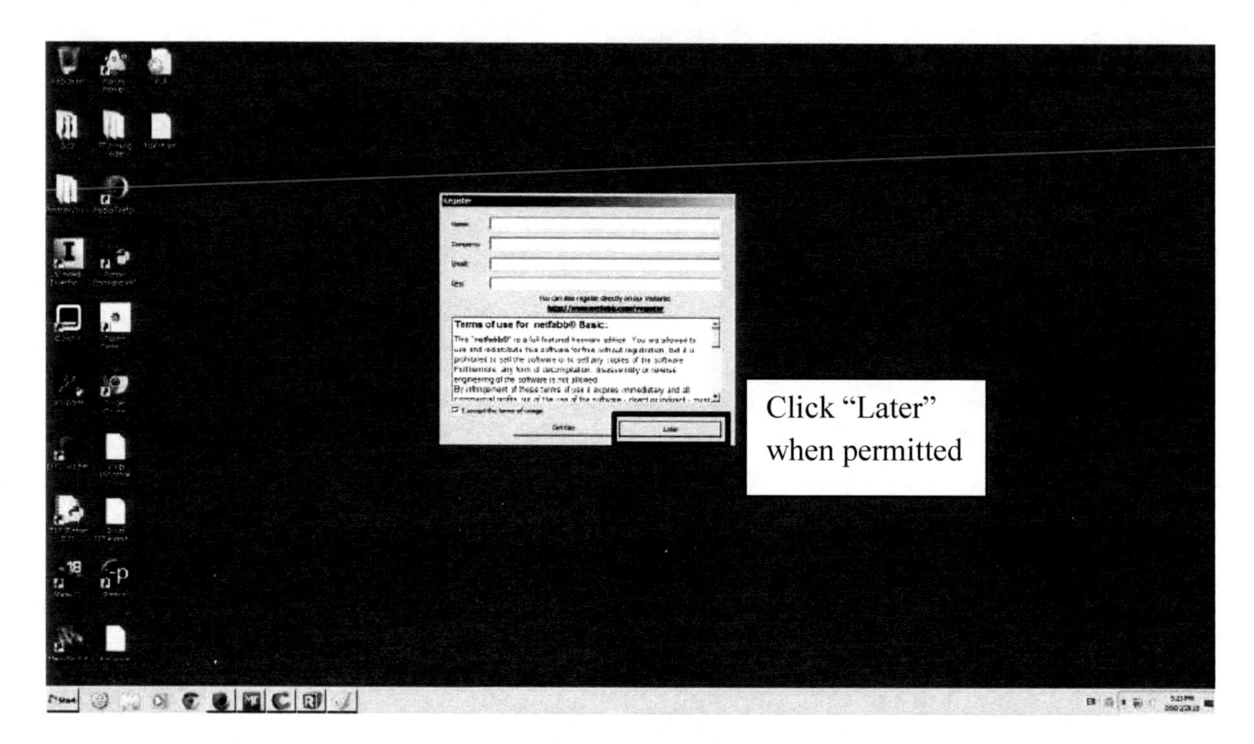

2. **Open STL File**: Click the 'Open File' icon on the top left of the screen and locate your file from the appropriate directory (ideally the Desktop)

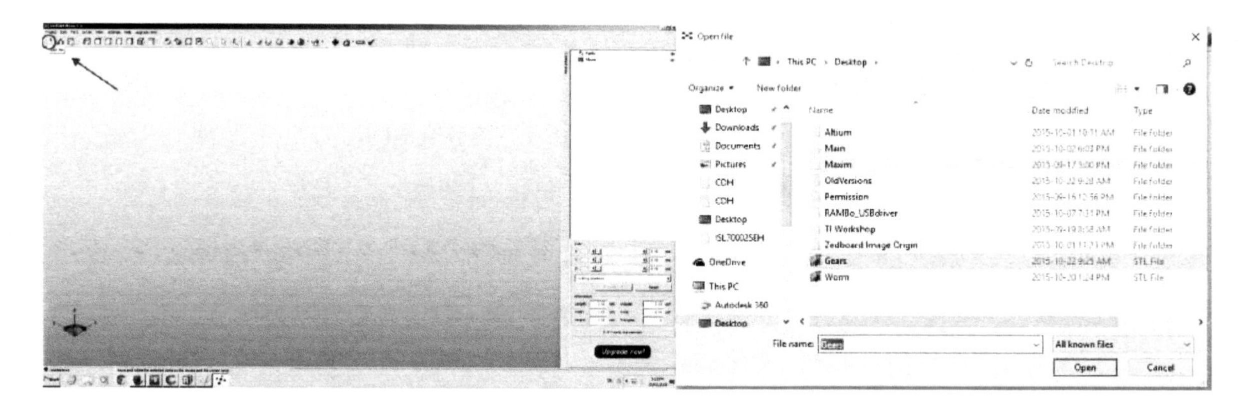

3. **Check Dimensions of the STL File**: Is it the correct (i.e., intended) size? If not, it will need to be scaled. STL dimensions can be found under 'Information' on the right side of the screen (large red box below).
4. **Scale Your STL File (if necessary)**: Right click on the part with your mouse and select 'Scale Part'. Alternatively, you can also click the Scale icon at the top (a grey sphere with an arrow pointing upward and an arrow pointing to the right – see small red box below). <u>Your part can only be scaled if it has been selected</u>! If your part appears green, it has been selected. If it appears grey, it has been 'unselected'.

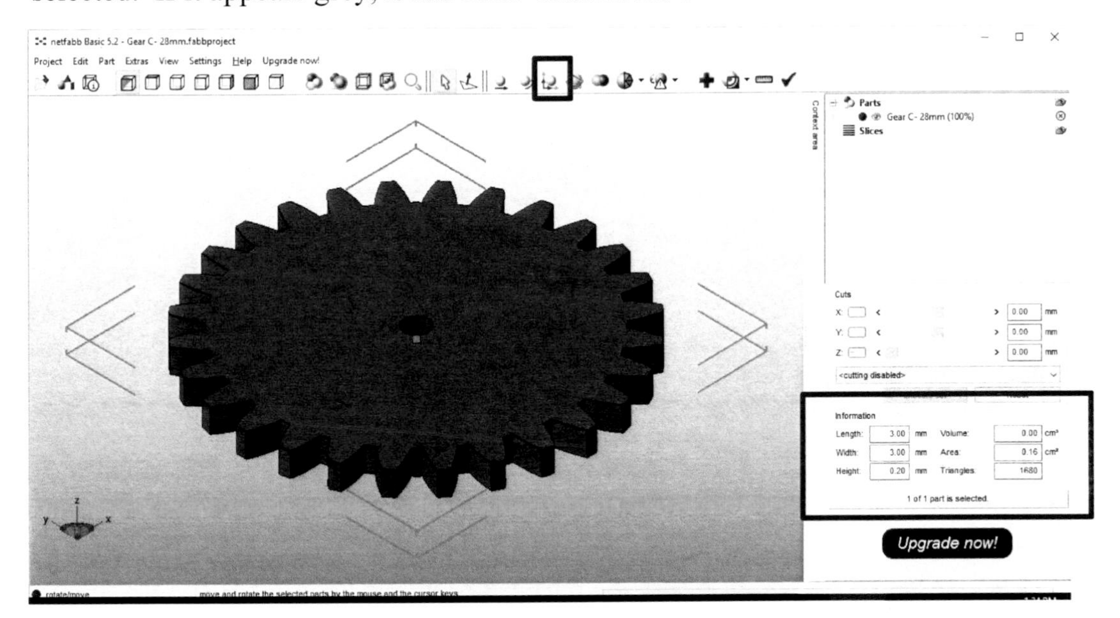

5. **Scale Parts Isometrically or Non-Isometrically**: depending on your needs, you may choose to scale isometrically (equally in X, Y and Z) or scale in one direction only (e.g., only along the Z-axis). When you click on 'Scale', the 'Scale parts' dialog box will open. For isometric scaling, leave the 'Fix scaling ratio' box **checked**. To scale in one direction, you need to '**uncheck**' this box.

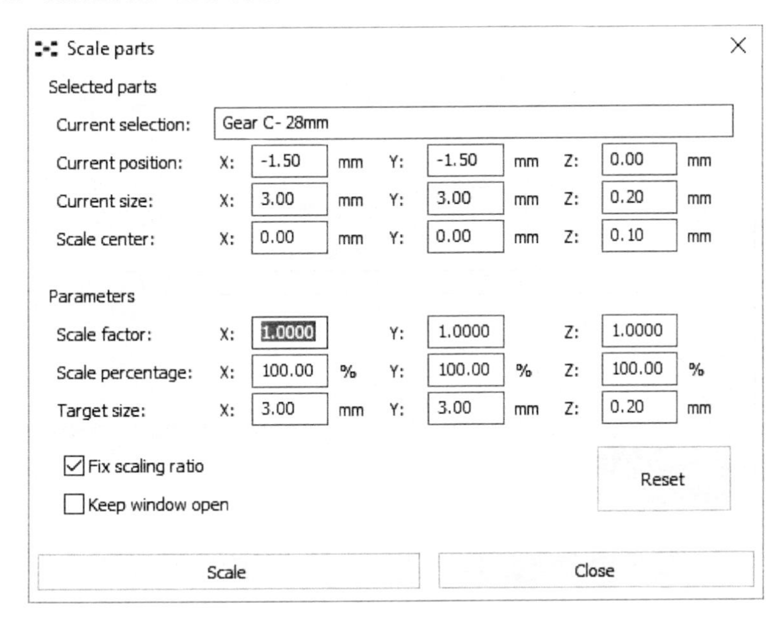

6. **Enter Scale Factor**: you have different options for resizing your parts. Entering a 'Scale Factor' of 10, for example, will increase your part size by 10x. Alternatively, you can enter the desired size as a percentage of the actual size or specify an exact size. If you left the 'Fix scaling ratio' box checked, you will see that entering a value in one direction (e.g., X) automatically enters the same value in the other directions.

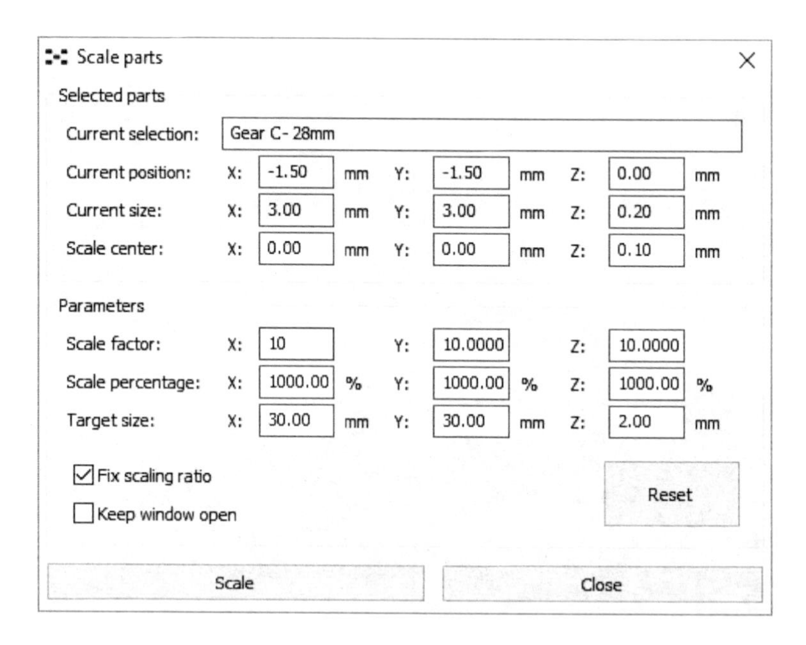

7. **Export STL for Printing**: once satisfied with scaling, right click on the part and export the *.STL file for printing. Be sure the part has been selected (i.e., it should appear green) before exporting!

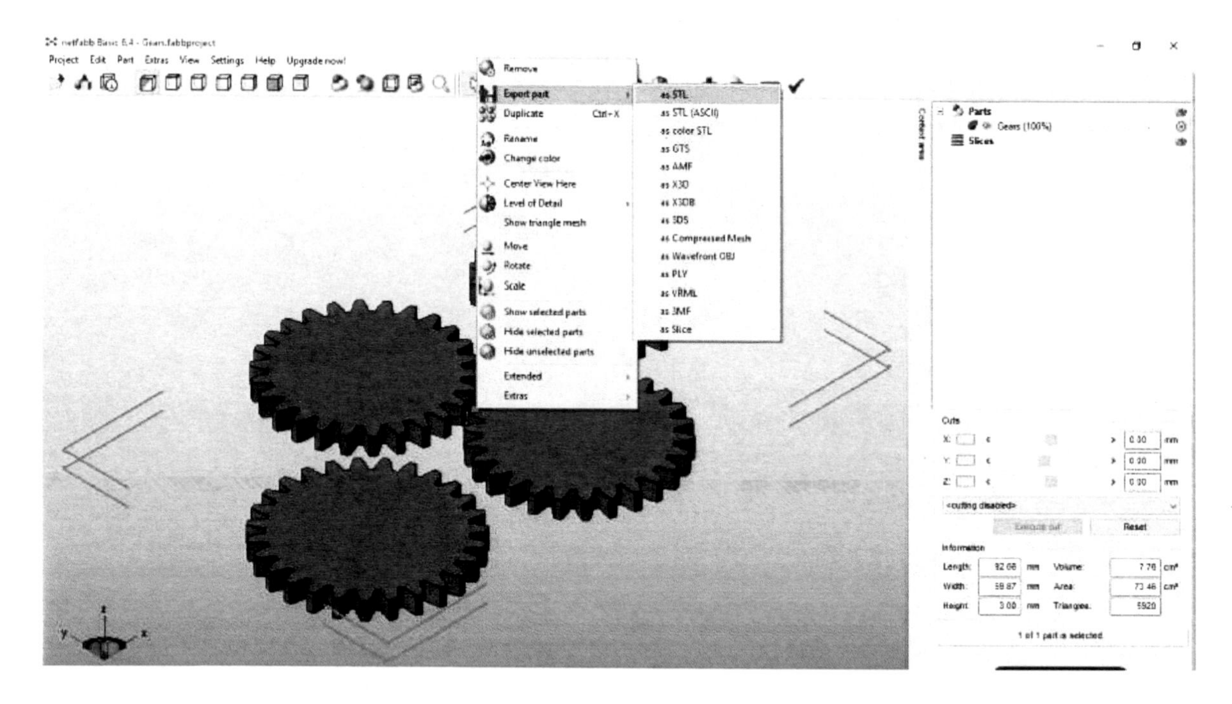

Using the 3D Printers

1. You will be printing all your project components on the SeeMeCNC Orion Delta printers. The software being used is called Repetier. Open it from the Desktop or Start Menu

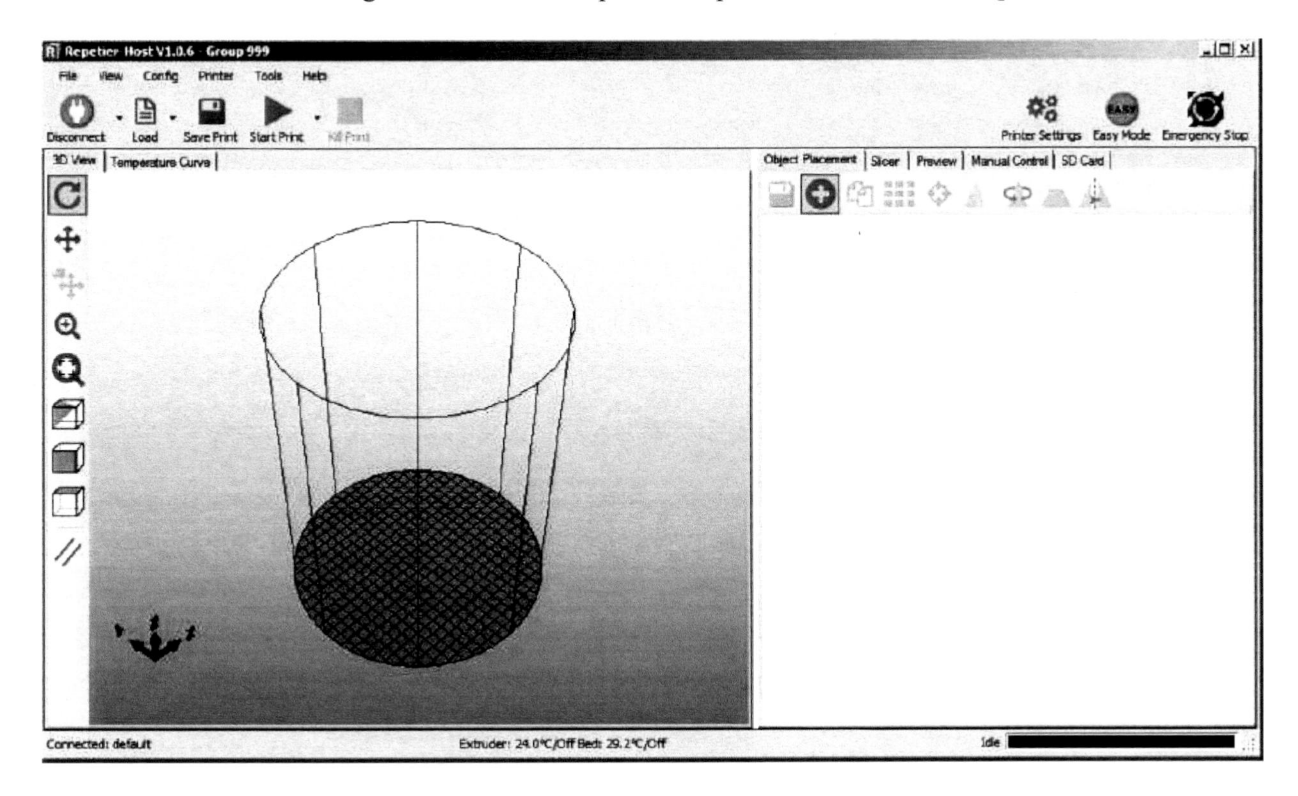

2. Add the file that was previously scaled in Netfabb Basic to the Repetier queue (On the right side under the object placement tab, the grey circle with the plus sign)
   a. Repeat this step for each file

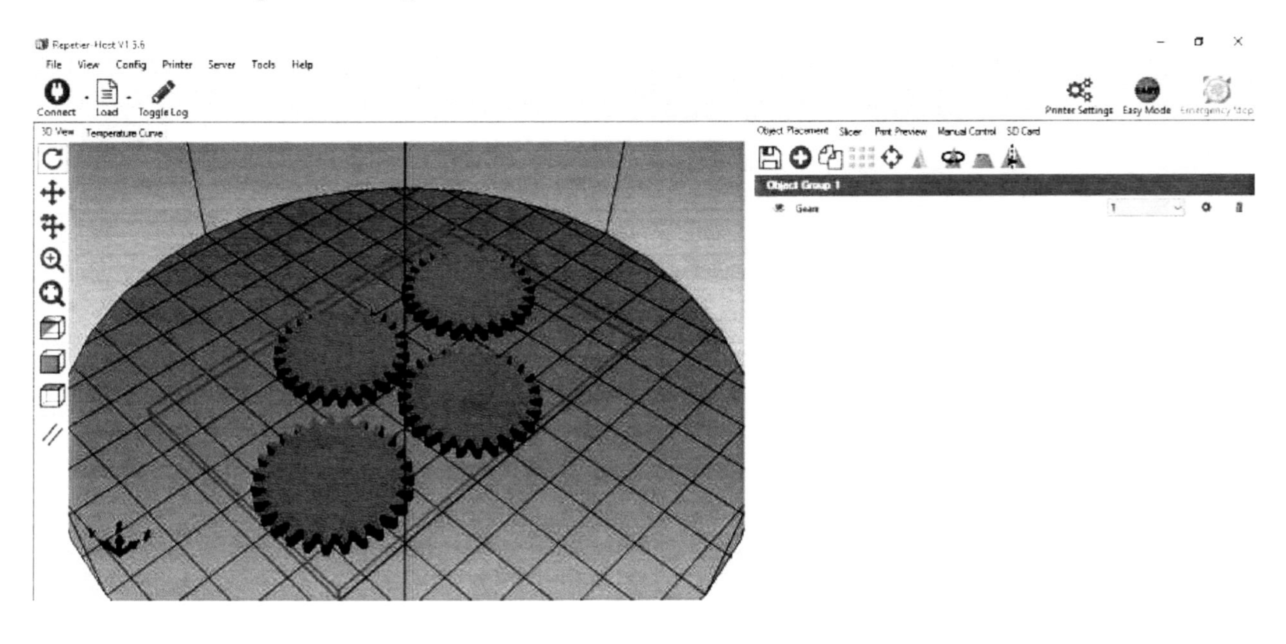

3. Navigate to the Slicer tab and click "Slice with Slic3r". This will slice the STL into many layers, and write directions that the printer needs to print them.

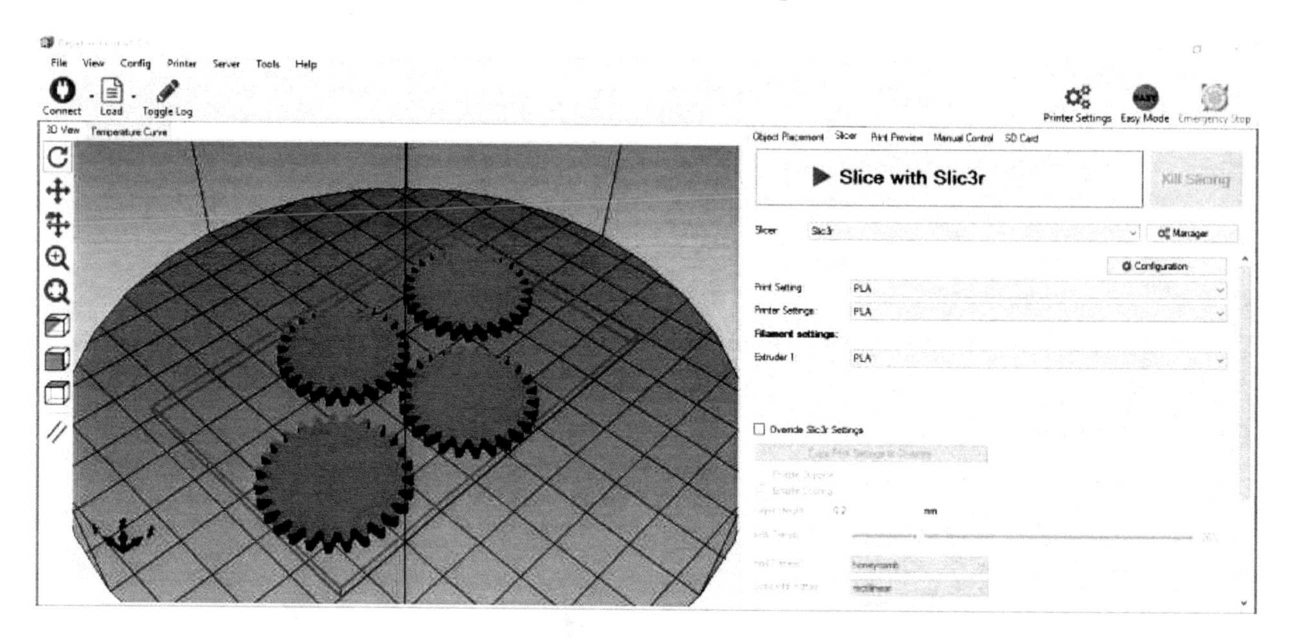

4. Check the **Estimated Print Time** on the right side of the screen and verify that it is acceptable.
    a. **REMEMBER:** Each printer booking is limited to 60 minutes, which includes the time to prepare and upload your files for printing. Realistically, **each of your prints should take a maximum of 30 minutes to complete**

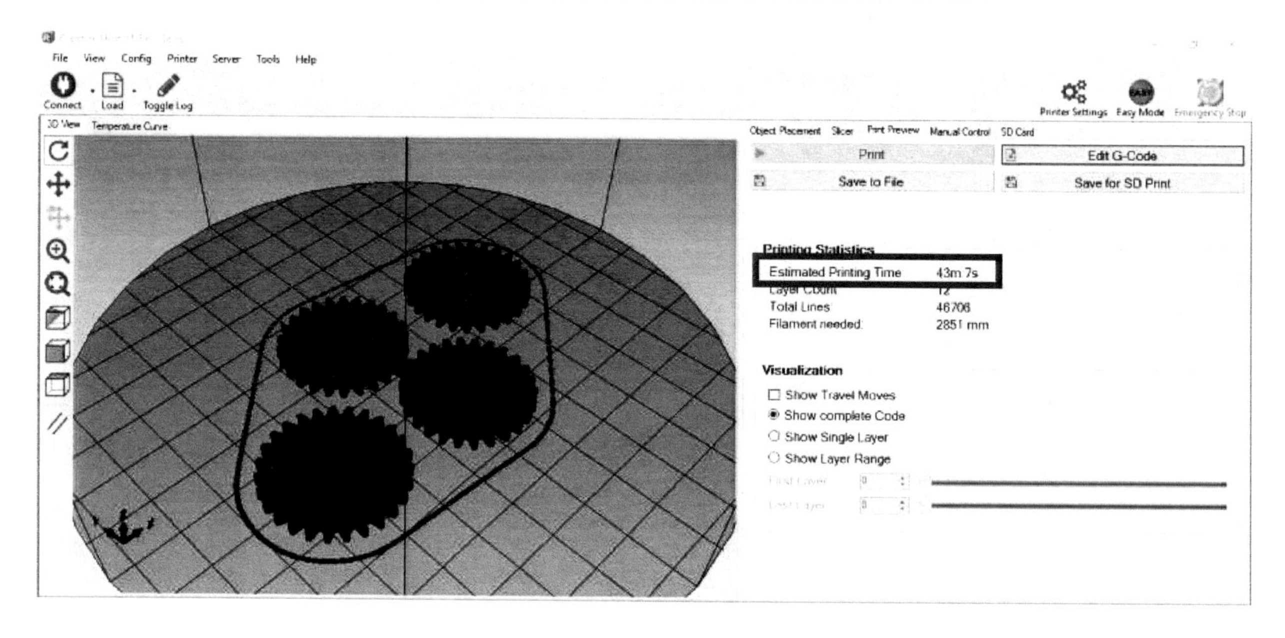

5. After the G-code is created, click the ‘Save for SD print’, which can be found under the Print Preview tab (Save this to the desktop).

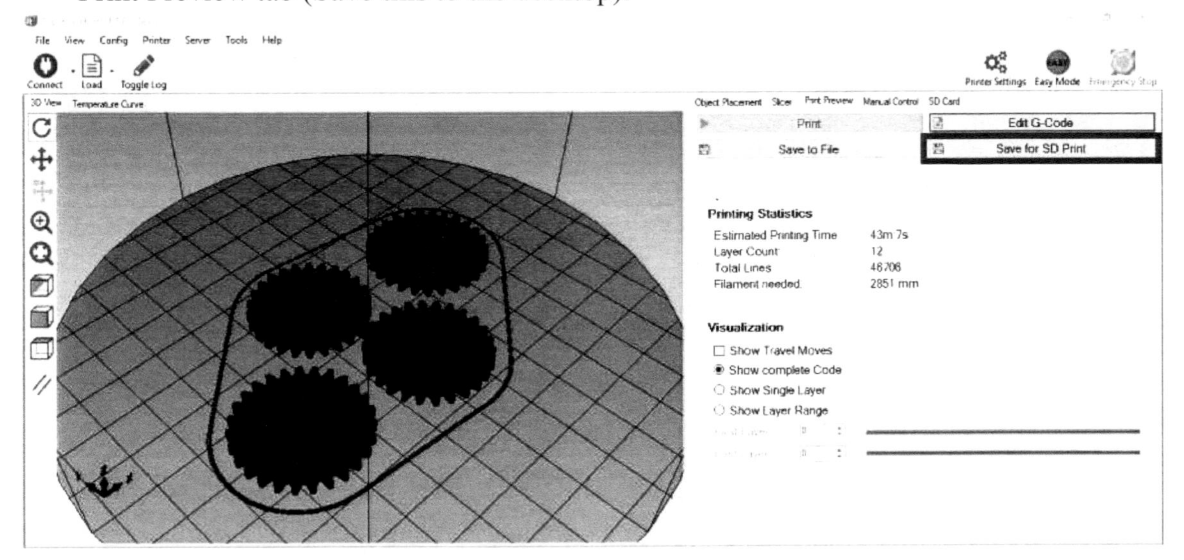

6. Make sure to leave the ‘Include “Start” and “End” code’ and ‘Include Job Finished Commands’ boxes checked. Now click save and save it to the desktop.

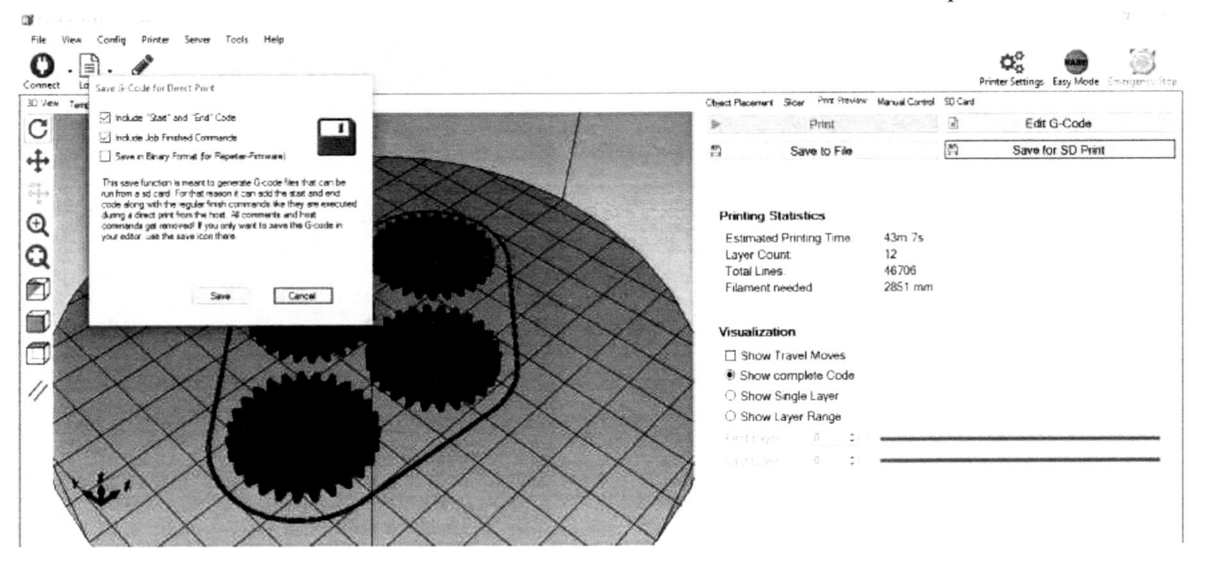

7. Once you have these files saved on the desktop transfer them to the SD card, if there is no SD card in the USB slot you will have to go get one from a printer that is not running.
8. Once the files are saved on the SD card, remove the SD card and plug it into the 3D printer. The files on the SD card should pop up on the 3D printer screen. Highlight the G-code file you want by turning the knob, press the knob in to select the file.
9. The 3D printers bed and nozzle should begin to heat up and the print will begin once the bed and nozzle hit their target temperature. Do not touch the printer once the temperature on the nozzle and bed begin to rise.

It is your job to keep an eye on your print, come get an IAI or TA if something appears wrong.

SECTION 3: Gear Design

3.1 Mixed Gear Type Design

In the design of our gear trains, it is not uncommon to use different types of gears. Additional considerations need to be given to evaluating functional requirements *and* constraints.

Typical set of functional requirements include:

1. Input speed type (i.e., linear speed or rotational speed) and magnitude
2. Output speed type and magnitude

Typical constraints include:

1. Spatial
2. Geometric
3. Practical gear design (e.g., gear ratio, gear size, etc.)
4. Standards
5. Availability

The following example endeavours to provide an approach to solving a gear train design problem where the designer must satisfy the functional requirements (i.e., input/output speeds) while also considering the spatial, geometric, and practical constraints.

A scaled sketch of the spatial and geometric constraints is necessary to get started. The sketch below illustrates the top (horizontal plane) view. A rotating motor serves as the input and is represented by a red dot (the rotational axis of this motor is perpendicular to the horizontal plane, directed out of the page, toward the reader). The output is a linear translation, and the linear output path is represented by the red arrow. The height of this pathway (i.e., out of the page) is 20 mm.

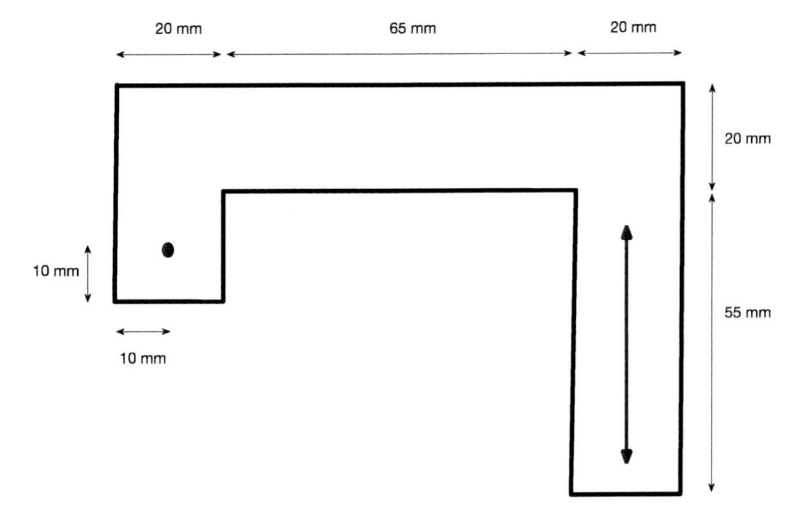

For this design we have an input rotational speed of 1000 revolutions per second (rps) and an output linear speed of 1 mm/s.

We can simplify the task of designing the entire gear train by partitioning this design into three stages: A, B, and C.

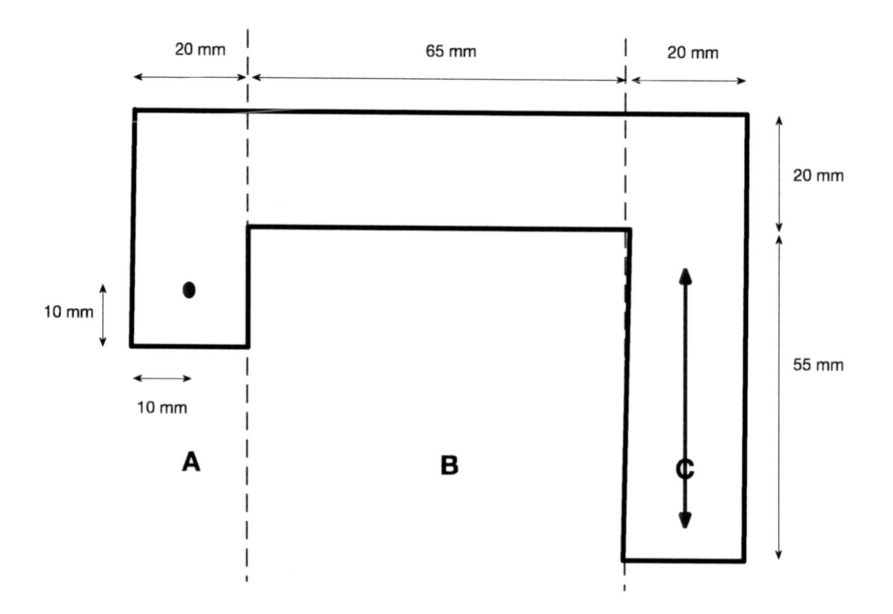

Now identify each section's gear train elements so that each stage can be assigned an input and output speed and motion requirement.

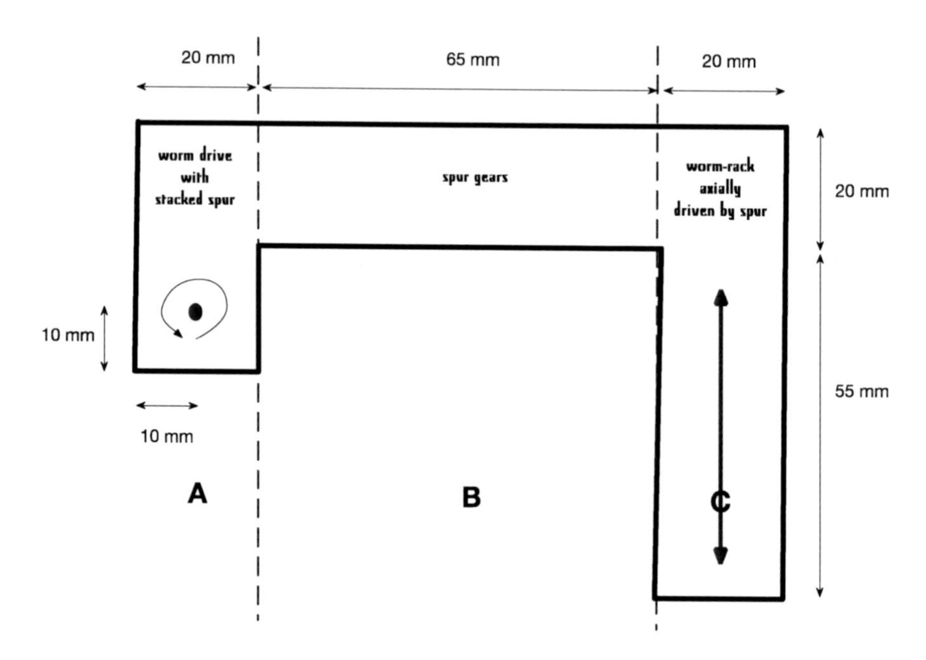

The following pages discuss how to approach each stage. Specific types of gears (e.g., spur, worm, bevel) are discussed. Some of these gears will be introduced and discussed as the term progresses. Complementary reading material can be found in Appendix B (Dudley, 1st Edition; Uicker et al., 4th Edition).

Stage-A

Given: Motor input is rotating about a vertical axis centered at 10mm by 10mm. The motor is rotating at 1000 rps (direction is not important at this time).

Observed: In order to transition from stage A to B, we have identified two geometric constraints:

1. Vertical axis of motor
2. If stage C's linear motion is to be axially driven by a spur gear then the face of the spur gears (A, B, and C) must lie in the vertical plane.

Chosen: We can visualize two solutions to stage A. The first involves a worm drive (worm to worm-gear) axially driving a spur gear. The second involves a bevel gear pair axially driving a spur gear.

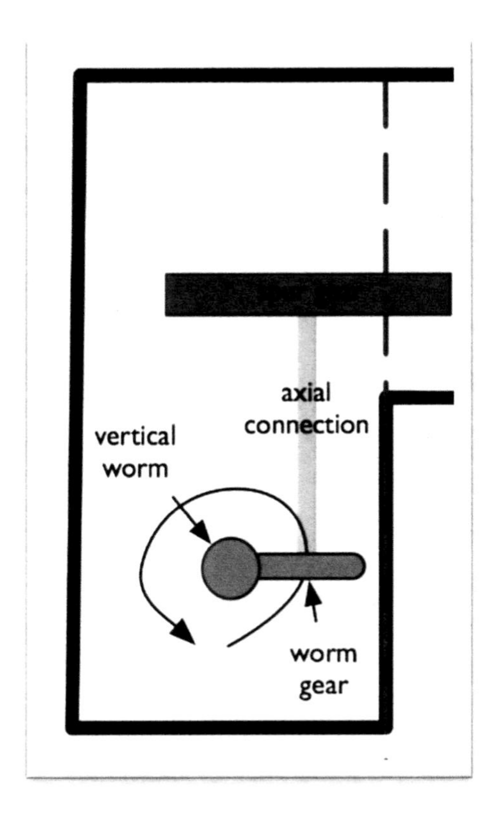

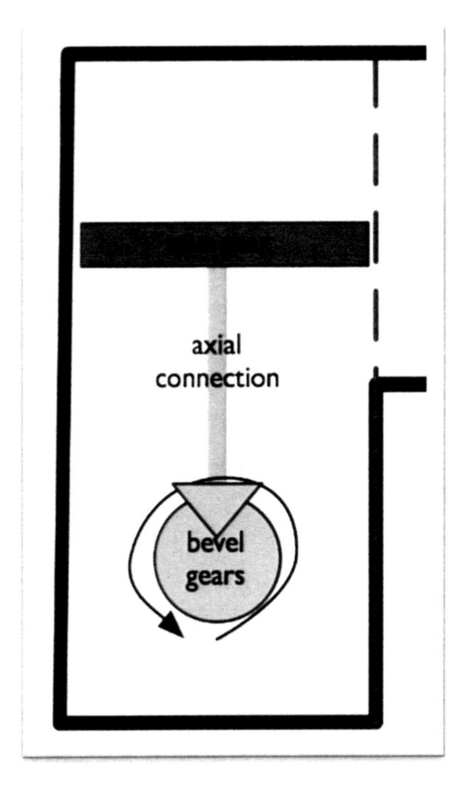

Both arrangements are suitable solutions to meet the geometric constraints. The spatial constraint (w x h : 20 mm x 20 mm) may steer the designer toward the bevel gear pair because our worm drive's worm gear will need to be less than 10 mm. The decision of which gear arrangement for stage A can be deferred with the knowledge that the smallest speed reduction that a worm drive can generate is 10x (Refer to the gear equations and practical design constraints). We can easily design a bevel gear pair with the same speed reduction of 10x.

Calculated: Given the Stage-A input speed of 1000 rps and the gear train speed reduction of 10x, the Stage-A output speed is 100 rps.

Stage-C

A designer may choose which stage to next examine; however, it is often advantageous to first work from the interface (i.e., input / output) stages.

Given: Linear output speed is 1 mm/s.

Observed: Stage C's linear motion is to be axially driven by a spur gear. The face of the spur gears (A, B, and C) must lie in the vertical plane.

Chosen: The gear arrangement is worm-rack. We have the ability to set the ACP (spacing between worm tooth). Keeping in mind that for every 1 rotation of the worm, the rack will linearly travel the distance of 1 ACP. For this example, we choose and ACP = 1 mm.

Calculated: Based upon our linear output speed requirement of 1mm/s AND A CHOSEN ACP of 1mm, our worm must be driven at 1 revolution per second (1 rps).

Stage-B

Now that we have chosen and calculated the inputs/outputs of the interface stages (see image below), the result is that we have the design requirements for the intermediate stage(s).

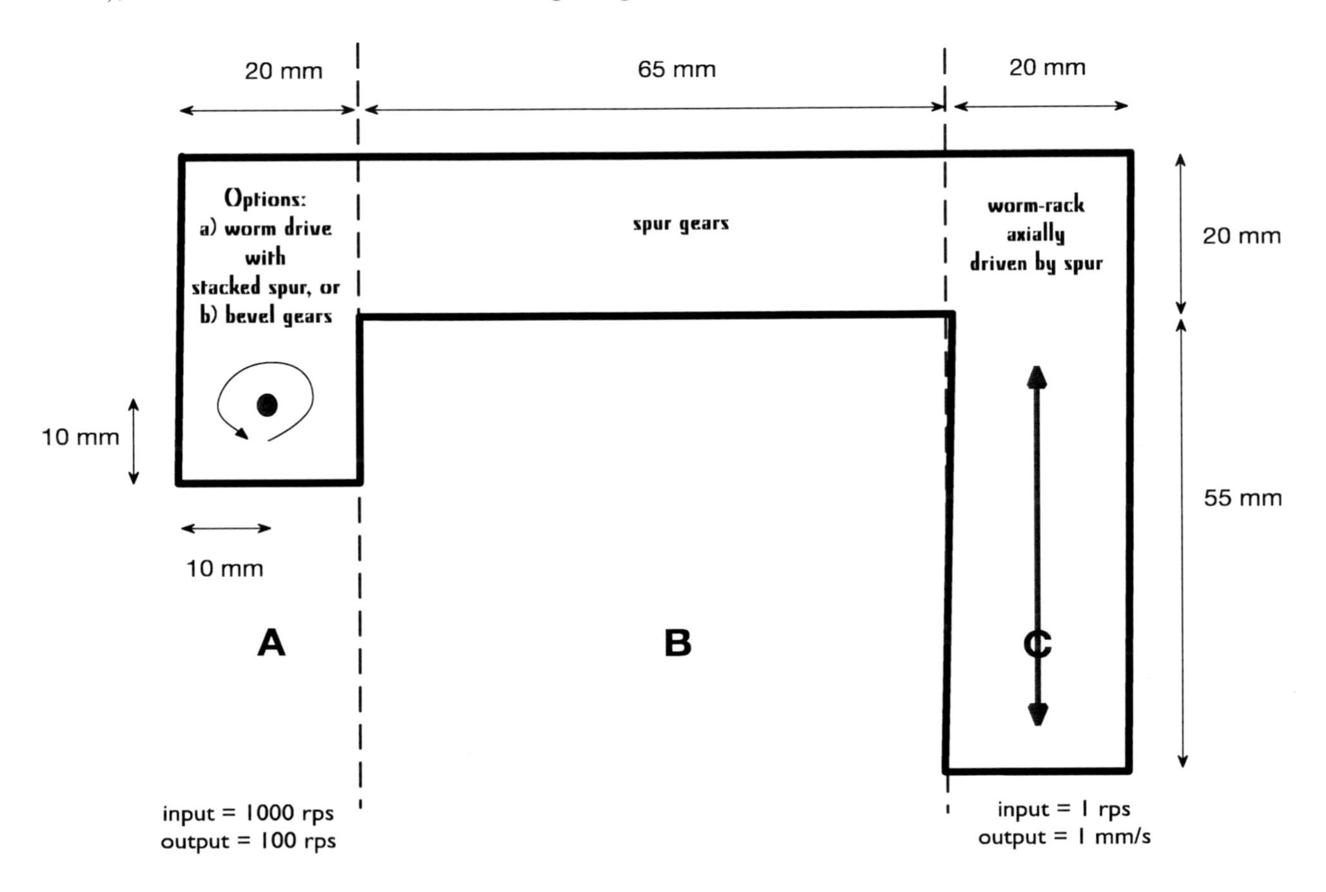

Given: Input rotational speed of 100 rps from a spur gear. Output rotational speed of 1 rps to a spur gear.

Observed: Stage A and C's rotation motion can be interfaces by meshing spur gears. The face of the spur gears (A, B, and C) must lie in the vertical plane. Gears must traverse ~70-75 mm.

Chosen: A spur gear train with an overall speed reduction of 100x.

Calculated: Speed is reduced when a smaller spur drives a meshed larger spur. This means our largest gear in the chain should be the last one. Remember the practical design considerations:

a. Maximum speed increase 5
b. Minimum speed decrease 1/5
c. Minimum number of teeth on a spur gear is 12.

The diagram below illustrates a speed reduction of 100x by employing meshing spur gears that confirm to the practical design requirements (100 = 5 x 5 x 4).

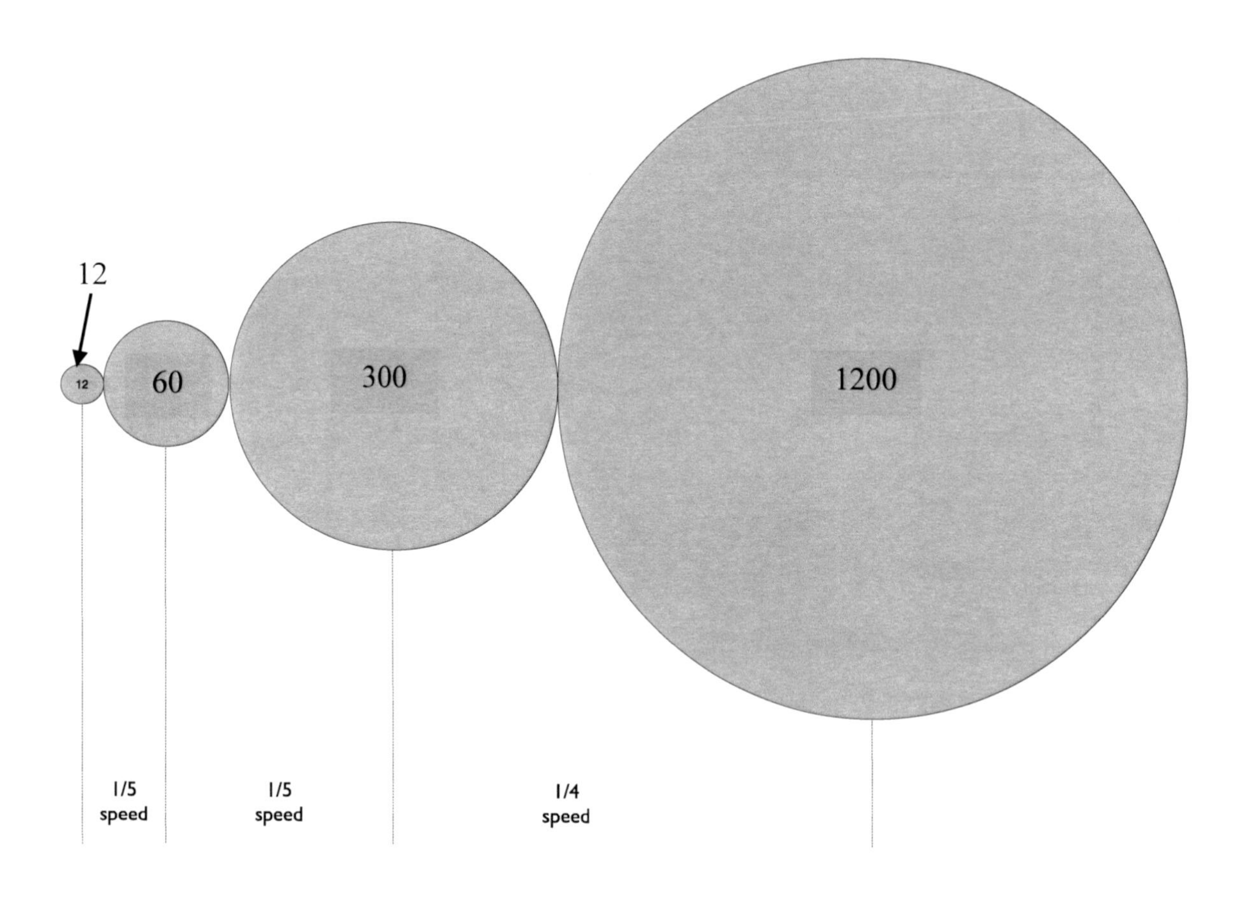

Two important questions should be asked at this point:

1. ***Will this fit my spatial constraint of 20 mm height? How?***
2. ***Does this traverse the 70-75 mm distance from stage A to C?***

To answer question 1, we need to consult our design triangle (referring to the figure on the right).
****This will be discussed in class and tutorial****

:

Pitch Circle Diameter (D) = number teeth (z) x module (m)

Keeping in mind that the Pitch Circle Diameter (D) is NOT the outer diameter of the spur gear. Pitch Circle Diameter (D) is the theoretical interface of the meshing gears and it does not account for the addendum (tooth above D).

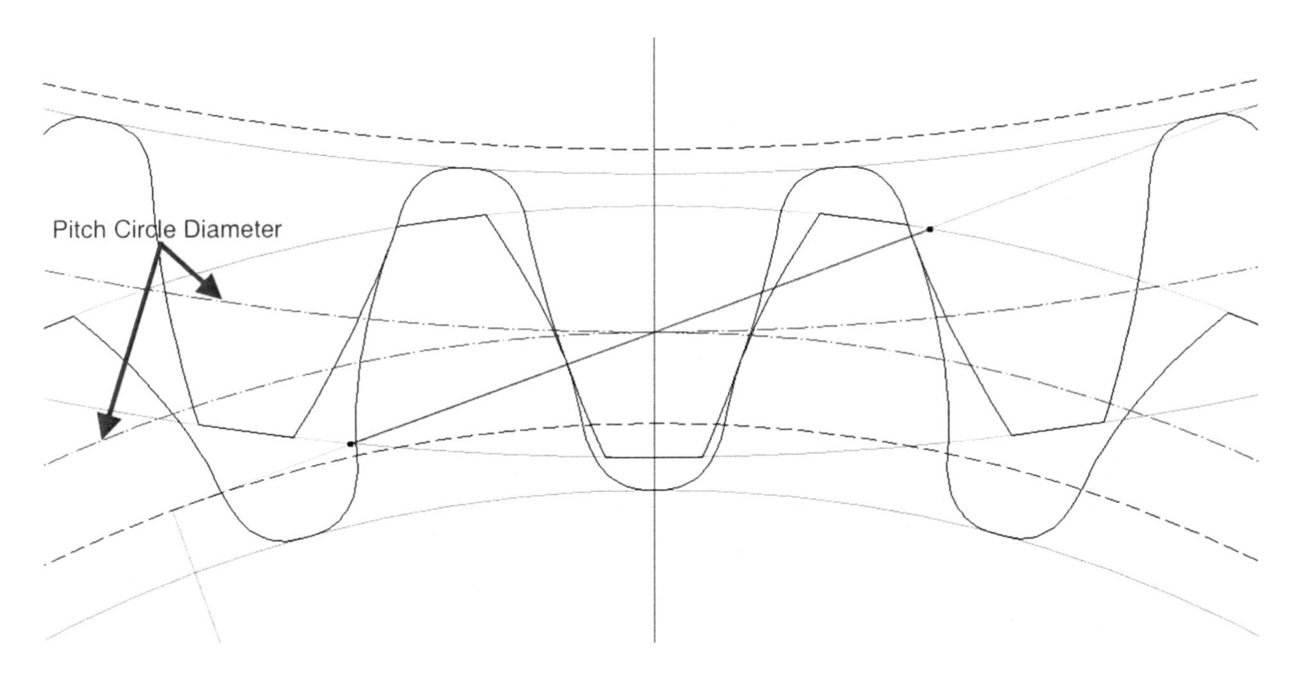

The size of the addendum depends on the specific gear design; however, we will start with a 20% accommodation for the addendum, thus we can target our largest spur gear's Pitch Circle Diameter to be 0.8 x 20mm = 16mm.

Pitch Circle Diameter of largest gear (D) = number teeth (z) x module (m)

16 mm = 1200 teeth x m

Solving for m,

m = 16 /1200 mm/tooth = 0.0133 mm/tooth

With a module of 0.0133 mm/tooth we can fit our largest gear into our spatial constraint.

However, module is defined as the size of a gear tooth (note that units are mm/tooth). We should carefully consider the impact such a small module will have on our smallest gear of 12 teeth.

Pitch Circle Diameter of smallest gear (D) = number teeth (z) x module (m)

Pitch Circle Diameter of smallest gear (D) = 12 teeth x 0.0133 mm/tooth

Pitch Circle Diameter of smallest gear (D) = **0.16 mm**

Clearly this diameter is too small to be practical.

Question - How can we meet the competing requirements of meet the gear ratio(s) vs spatial limitation?

Answer - rather than meshing all spur gears in a linear chain as we have done above, stack gear pairs by axially connecting gear sets. For example

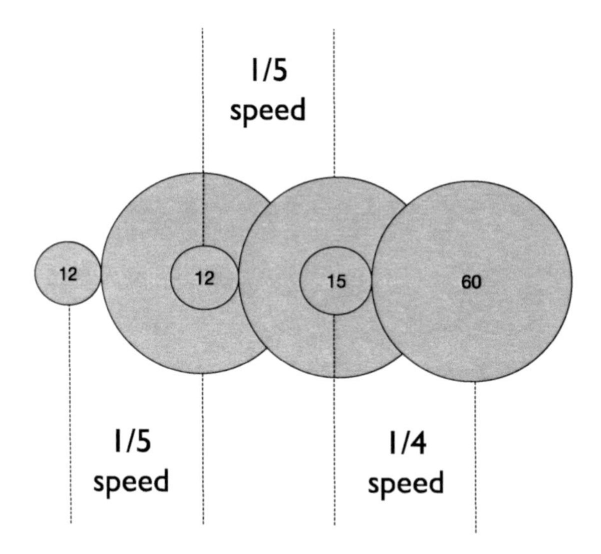

With this new stacked arrangement, instead of meshing all of the gears, each pair is axially connected to the next pair. Our largest gear now has 60 teeth (vs 1200!)

Pitch Circle Diameter of largest gear (D) = number teeth (z) x module (m)

16 mm = 60 teeth x m

Solving for m,

m = 16 /60 mm/tooth = 0.267 mm/tooth

With a module of 0.267 mm/tooth we can fit our largest gear into our spatial constraint.

Again, we must consider how the module impact our smallest gear of 12 teeth.

Pitch Circle Diameter of smallest gear (D) = number teeth (z) x module (m)

Pitch Circle Diameter of smallest gear (D) = 12 teeth x 0.267 mm/tooth

Pitch Circle Diameter of smallest gear (D) = **3.2 mm**

This diameter is also too small to be practical, but we're on the right track.

We clearly need to stack our gears by using axial connectors to keep them small, but we do this also to meet our multiplier/divisor requirements. To make our work easiest we should repeat the smallest multiplier possible to meet our overall gear ratio. If we used a speed divisor of 2, we could stack seven such pairs (2^7) to slow our speed 128 times. Working backwards 100 = x^7 we find x = 1.9307. Such a number may seem impractical; however, our only restriction is that each gear have a whole number of teeth. Consider the combination pair of a 15 tooth spur gear meshed with 29 tooth spur gear (1.933x).

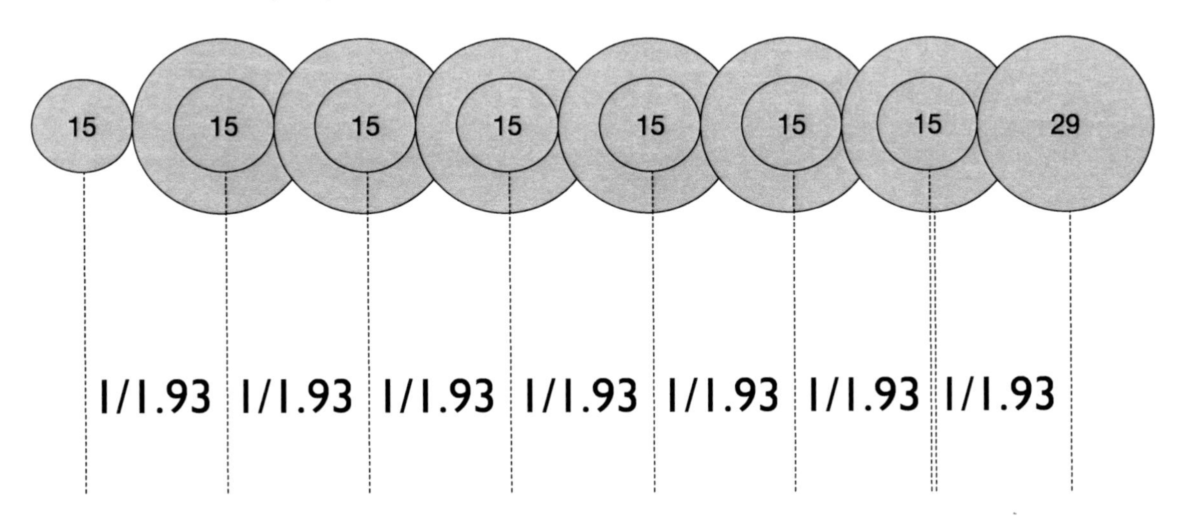

With this new stacked arrangement, instead of linearly meshing all of the gears, each pair is axially connected to the next pair. Our largest gear now has 29 teeth (vs 1200!)

Pitch Circle Diameter of largest gear (D) = number teeth (z) x module (m)

16 mm = 29 teeth x m

Solving for m,

m = 16 / 29 mm/tooth = 0.55 mm/tooth

With a module of 0.55 mm/tooth we can fit our largest gear into our spatial constraint.

Again, we must consider how the module impact our smallest gear of 15 teeth.

Pitch Circle Diameter of smallest gear (D) = number teeth (z) x module (m)

Pitch Circle Diameter of smallest gear (D) = 15 teeth x 0.55 mm/tooth

Pitch Circle Diameter of smallest gear (D) = **8.25 mm**

This diameter now manageable; however, it should be considered against 3D printer capability.

If the required *overall* gear train ratio cannot be achieved with similar accuracy, the designer must revisit either the assumptions. For example, the linear ACP (stage C) could be modified to accommodate a different input speed, or the gear pair from the motor (stage A) could be designed for a different speed reduction.

Once the specifications have been met, we are ready to now move into the solid model. As the designer can see, starting the design in the solid model can be a poor choice in time management. By segmenting the design into stages (e.g., A, B, C) and agreeing on the constraints, a team can divide and conquer the design problem.

SECTION 4: Appendices

Appendix A. Supplementary Reading Material

In addition to the course textbook, there are a number of supplementary texts that nicely complement the course content (lectures, labs and tutorials). Select chapters from these texts have been included in the following Appendix. Please note that all references to this section in the custom courseware are by author, edition and chapter. **You are expected to read each chapter as indicated in the course schedule in Appendix A**, just as you will be expected to read specified chapters in the text (Lieu and Sorby, 2nd Edition). The source of each supplementary chapter is provided below:

- Projections and Visualization
    - Giesecke et al., 4th Edition: Chapter 4 (Orthographic Projection)
    - Lieu & Sorby, 2nd Edition: Chapter 17 (Advanced Visualization Techniques)
- Gears and Mechanisms
    - Dudley, 1st Edition: Chapter 1 (Gear Design Trends)
    - Uicker et al., 4th Edition: Chapter 7 (Spur Gears)
    - Uicker et al., 4th Edition: Chapter 8 (Helical Gears, Bevel Gears, Worms and Worm Gears)
    - Uicker et al. 4th Edition: Chapter 9 (Mechanism Trains)
- Group Work
    - Lieu & Sorby, 2nd Edition: Chapter 15 (Working in a Team Environment)

Giesecke et al., 4th Edition – Chapter 4

Textbook: Modern Graphics Communication

Chapter Title: Orthographic Projection

CHAPTER FOUR

ORTHOGRAPHIC PROJECTION

OBJECTIVES

After studying the material in this chapter, you should be able to:

1. Recognize and sketch the symbol for third-angle projection.
2. List the six principal views of projection.
3. Sketch the top, front, and right-side views of an object with normal, inclined, and oblique surfaces.
4. Understand which views show depth in a drawing that shows top, front, and right-side views.
5. Know the meaning of normal, inclined, and oblique surfaces.
6. Compare and contrast using a CAD program to sketching on a sheet of paper to create 2D drawing geometry.
7. List the dimensions that transfer between top, front, and right-side views.
8. Transfer depth between the top and right-side views.
9. Label points where surfaces intersect.

Refer to the following standard:

- ANSI/ASME Y14.3—2003 Multiview and Sectional View Drawings

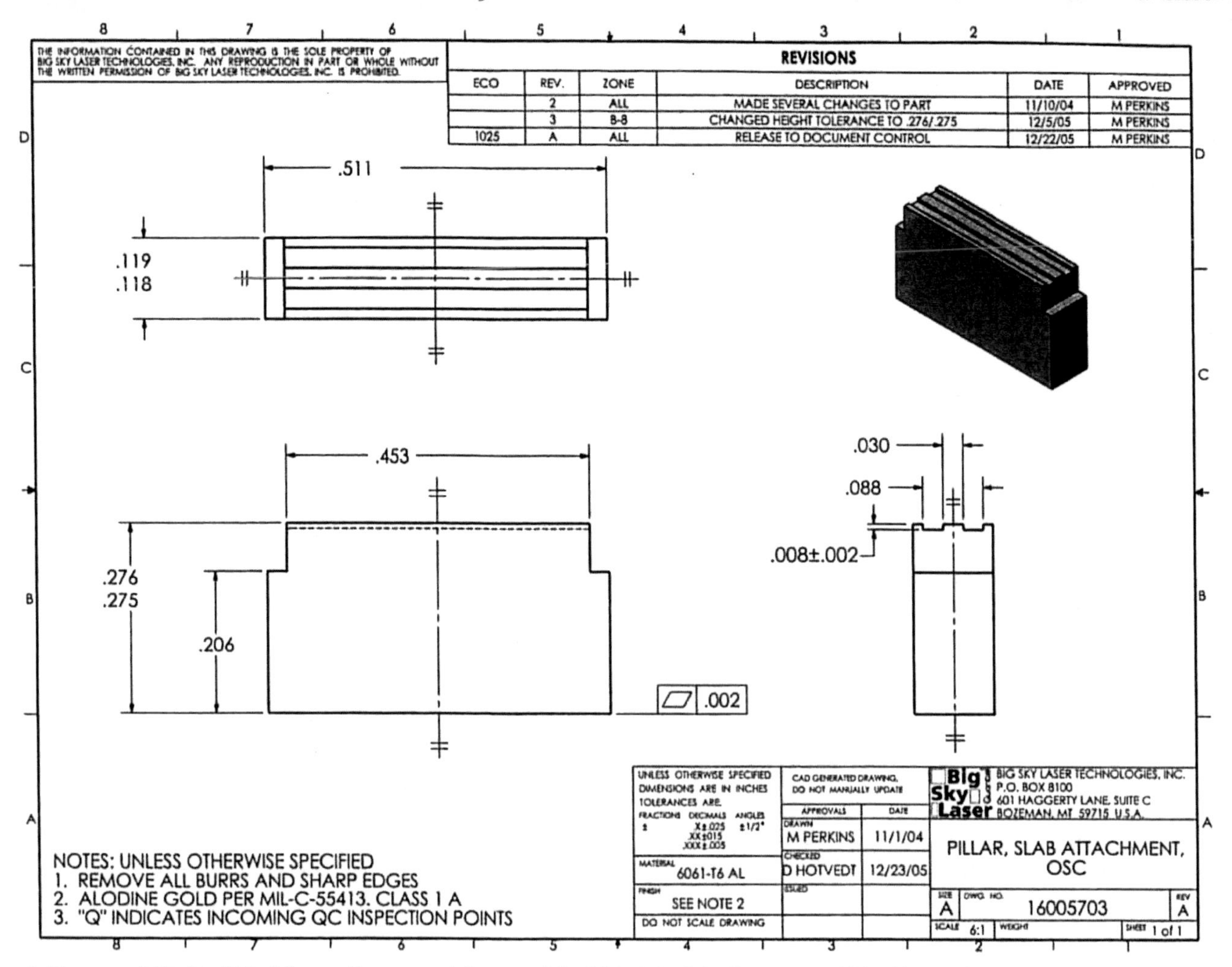

Front, Top, and Right-Side Views Generated from a 3D CAD Model. *Courtesy of Big Sky Laser.*

OVERVIEW

A view of an object is called a projection. By projecting multiple views from different directions in a systematic way, you can completely describe the shape of 3D objects.

There are certain standard practices that you must know in order to create sketches and drawings that can be accurately interpreted. For example, you need to know which views to show, how they should be oriented in your drawing, and how to represent key information such as edges, surfaces, vertices, hidden lines, centerlines, and other crucial details.

The standard published in ANSI/ASME Y14 3M-1994 is common in the United States where third-angle projection is used. Europe, Asia, and many other places use the first-angle projection system.

Search the following Web sites to learn more about orthographic projections (geomancy) and a biography of Gaspard Mongl (bib math).

- http://www.geomancy.org
- http://www.bibmath.net

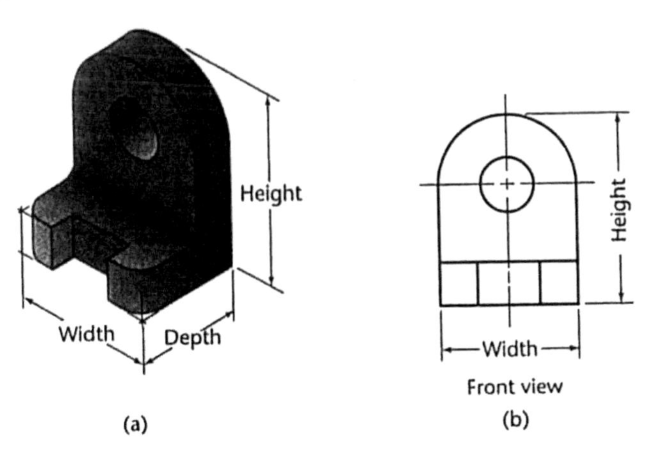

4.1 Front View of an Object

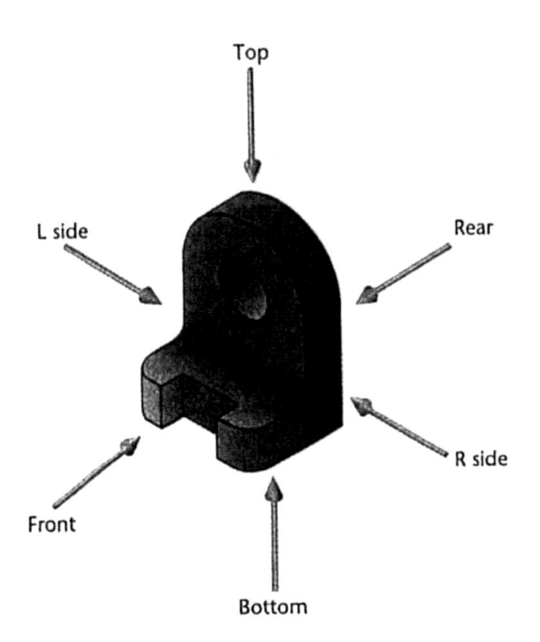

4.2 The Six Principal Views

UNDERSTANDING PROJECTIONS

In order to make and interpret drawings you need to know how to create projections and understand the standard arrangement of views. You also need to be familiar with the geometry of solid objects and be able to visualize a 3D object that is represented in a 2D sketch or drawing. The ability to identify whether surfaces are normal, inclined, or oblique in orientation can help you to visualize objects. Common features such as vertices, edges, contours, fillets, holes, and rounds are shown in a standard way, which makes drawings simpler to create and helps to prevent them from being misinterpreted.

Views of Objects

A photograph shows an object as it appears to the observer, but not necessarily as it is. It cannot describe the object accurately, no matter what distance or which direction it is taken from, because it does not show the exact shapes and sizes of the parts. It would be impossible to create an accurate three-dimensional model of an object using only a photograph for reference because it shows only one view. It is a 2D representation of a 3D object.

Drawings are two dimensional representations as well, but unlike photos, they allow you to record sizes and shapes precisely. In engineering and other fields, a complete and clear description of the shape and size of an object is necessary to be sure that it is manufactured exactly as the designer intended. To provide this information about a 3D object, a number of systematically arranged views are used.

The system of views is called **multiview projection.** Each view provides certain definite information. For example, a front view shows the true shape and size of surfaces that are parallel to the front of the object. An example showing the direction of sight and the resulting front view projection is shown in Figure 4.1. Figure 4.2 shows the same part and the six principal viewing directions, as will be discussed in the next section. Figure 4.3 shows the same six views of a house.

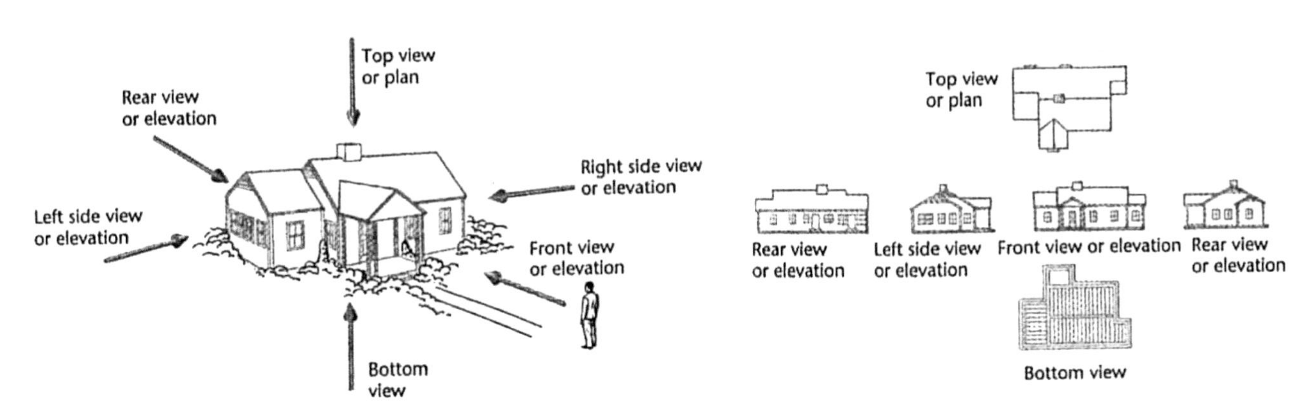

4.3 Six Views of a House

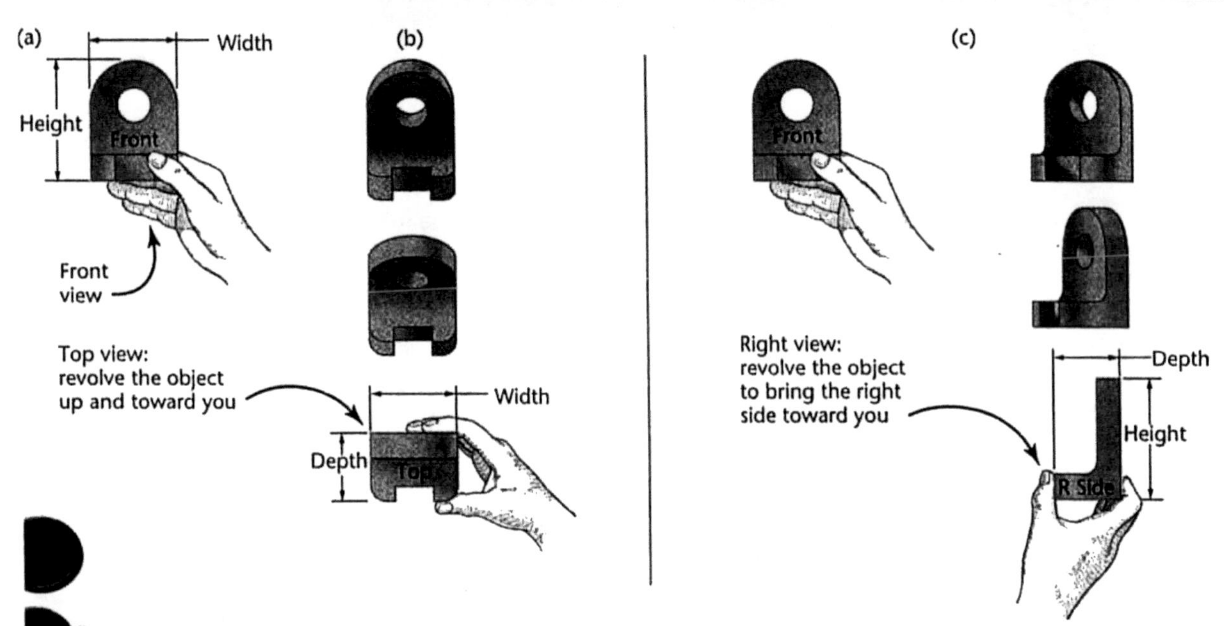

4.4 Revolving the Object to Produce Views. You can experience different views by revolving an object, as shown. (a) First, hold the object in the front view position. (b) To get the top view, tilt the object toward you to bring the top of the object into your view. (c) To get the right-side view, begin with the object's front view facing you and revolve it to bring the right side toward you. To see views of the rear, bottom, or right side, you would simply turn the object to bring those sides toward you.

The Six Standard Views

Any object can be viewed from six mutually perpendicular directions, as shown in Figure 4.2. These are called the six **principal views.**

You can think of the six views as what an observer would see by moving around the object. As shown in Figure 4.3, the observer can walk around a house and view its front, sides, and rear. You can imagine the top view as seen by an observer from an airplane and the bottom, or "worm's-eye view," as seen from underneath. The term "plan" may also be used for the top view. The term "elevation" is used for all views showing the height of the building. These terms are regularly used in architectural drawing and occasionally in other fields.

To make drawings easier to read, the views are arranged on the paper in a standard way. The views in Figure 4.3 show the American National Standard arrangement. The top, front, and bottom views align vertically. The rear, left-side, front, and right-side views align horizontally. To draw a view out of place is a serious error and is generally regarded as one of the worst mistakes in drawing. See Figure 4.4 for a demonstration of how to visualize the different views.

Principal Dimensions

The three principal dimensions of an object are **width, height,** and **depth** (Figure 4.5). In technical drawing, these fixed terms are used for dimensions shown in certain views, regardless of the shape of the object. The terms "length" and "thickness" are not used because they cannot be applied in all cases.

The front view shows only the height and width of the object and not the depth. In fact, any principal view of a 3D object shows only two of the three principal dimensions; the third is found in an adjacent view. Height is shown in the rear, left-side, front, and right-side views. Width is shown in the rear, top, front, and bottom views. Depth is shown in the left-side, top, right-side, and bottom views.

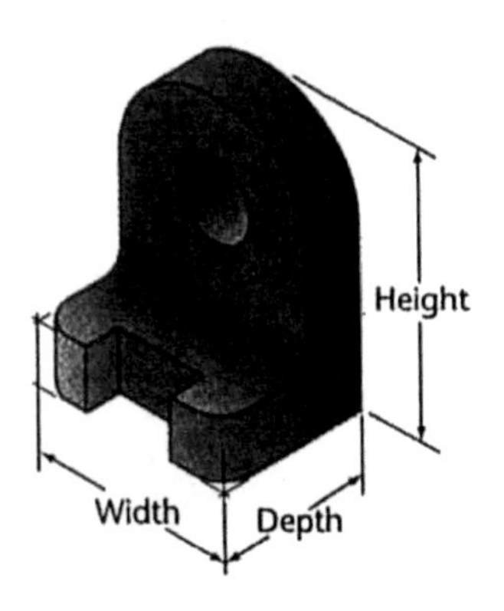

4.5 The Principal Dimensions of an Object

Projection Method

Figure 4.6 shows how to understand the front view of an object drawn using an orthographic projection. Imagine a sheet of glass parallel to the front surfaces of the object. This represents the **plane of projection.** The outline on the plane of projection shows how the object appears to the observer. In orthographic projection, rays (or projectors) from all points on the edges or contours of the object extend parallel to each other and perpendicular to the plane of projection. The word **orthographic** essentially means to draw at right angles.

Examples of top and side views are shown in Figure 4.7. The plane on which the front view is projected is called the **frontal plane.** The plane upon which the top view is projected is the **horizontal plane.** The plane upon which the side view is projected is called the **profile plane.**

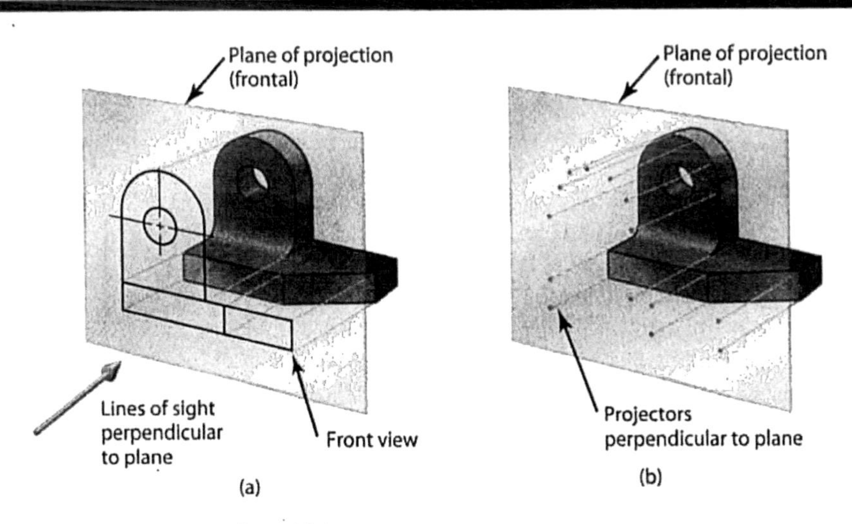

4.6 Projection of an Object

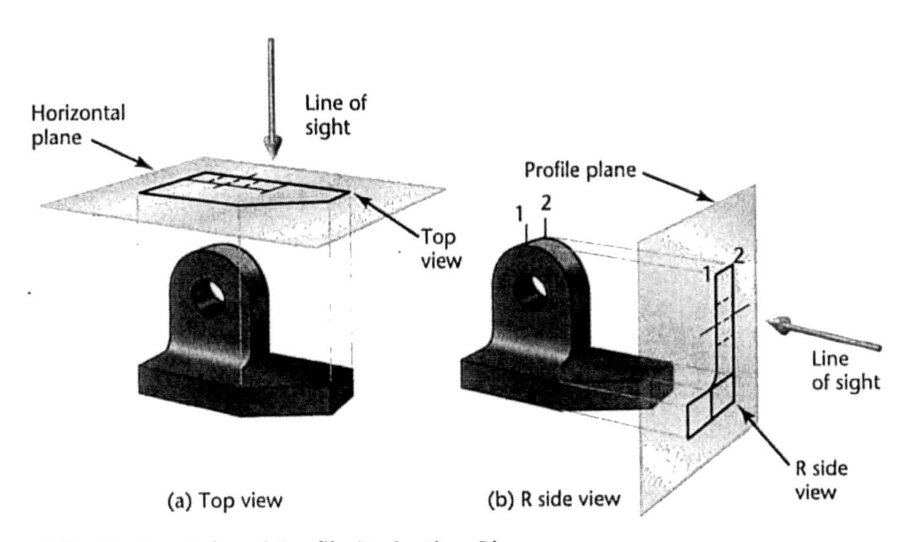

4.7 Horizontal and Profile Projection Planes

The Glass Box

One way to understand the standard arrangement of views on the sheet of paper is to envision a **glass box.** If planes of projection were placed parallel to each principal face of the object, they would form a box, as shown in Figure 4.8. The outside observer would see six standard views (front, rear, top, bottom, right side, left side) of the object through the sides of this imaginary glass box.

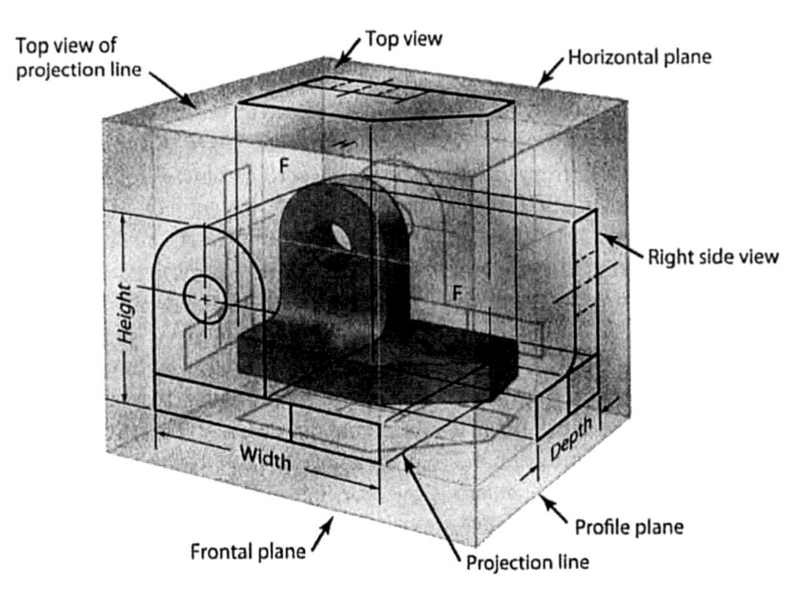

4.8 The Glass Box

To organize the views of a 3D object on a flat sheet of paper, imagine the six planes of the glass box being unfolded to lie flat, as shown in Figure 4.9. Think of all planes except the rear plane as hinged to the frontal plane. The rear plane is usually hinged to the left-side plane. Each plane folds out away from the frontal plane. The representation of the hinge lines of the glass box in a drawing are known as **folding lines.** The positions of these six planes after they have been unfolded are shown in Figure 4.10.

Carefully identify each of these planes and corresponding views with the planes' original position in the glass box.

In Figure 4.10, lines extend around the glass box from one view to another on the planes of projection. These are the projectors from a point in one view to the same point in another view. The size and position of the object in the glass box does not change. This explains why the top view is the same width as the front view and why it is placed directly above the front view. The same relation exists between the front and bottom views. Therefore, the front, top, and bottom views all line up vertically and are the same width. The rear, left-side, front, and right-side views all line up horizontally and are the same height.

Objects do not change position in the box, so the top view must be the same distance from the folding line OZ as the right side view is from the folding line OY. The bottom and left-side views are the same distance from their respective folding lines as are the right-side and the top views. The top, right-side, bottom, and left-side views are all the same distance from the respective folding lines and show the same depth.

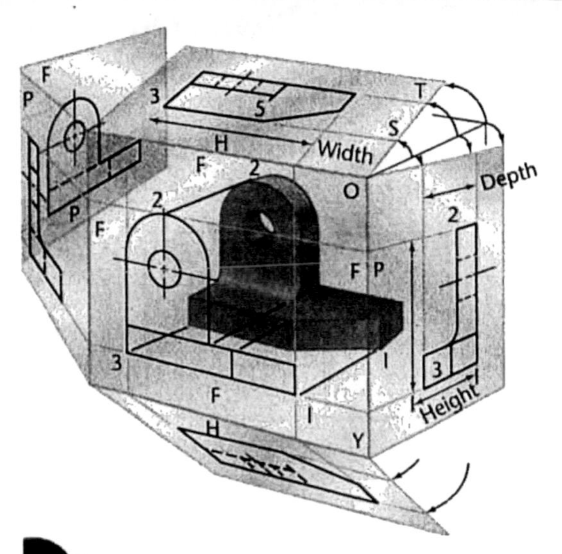

4.9 Unfolding the Glass Box

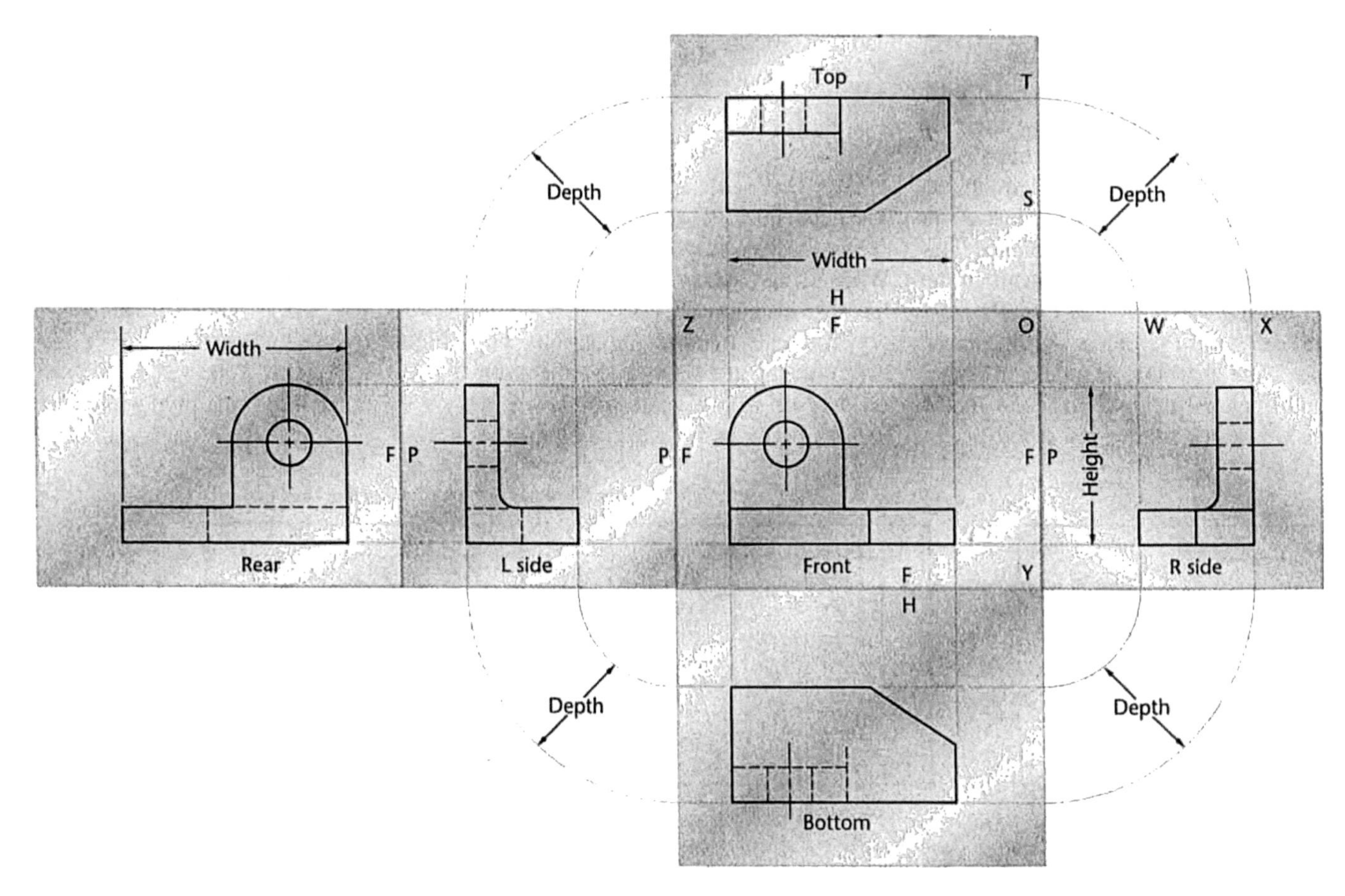

4.10 The Glass Box Unfolded

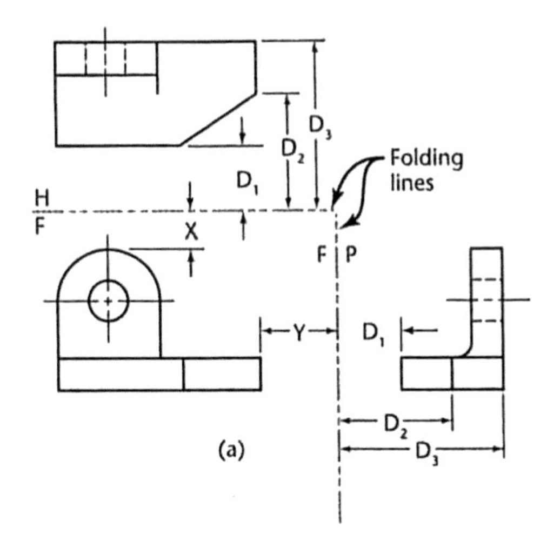

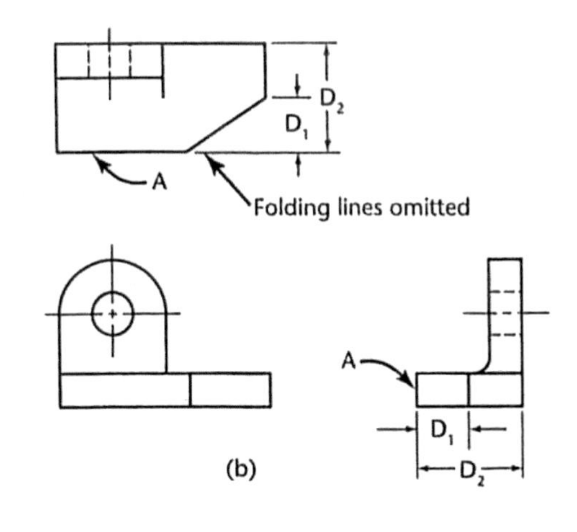

4.11 Views Shown with and without Folding Lines

The front, top, and right-side views of the object shown in the previous figures are shown in Figure 4.11a, but instead of a glass box, folding lines are shown between the views. These folding lines correspond to the hinge lines of the glass box.

The H/F folding line, between the top and front views, is the intersection of the horizontal and frontal planes. The F/P folding line, between the front and side views, is the intersection of the frontal and profile planes.

While you should understand folding lines, particularly because they are useful in solving problems in descriptive geometry, they are usually left off the drawing, as in Figure 4.11b. Instead of using the folding lines as reference lines for marking depth measurements in the top and side views, you may use the front surface (A) of the object as a reference line. Note that D1, D2, and all other depth measurements correspond in the two views as if folding lines were used.

Use Worksheet 4.1 to practice transferring depth dimensions.

Spacing between Views

Spacing between views is mainly a matter of appearance. Views should be spaced well apart, but close enough to appear related to each other. You may need to leave space between the views to add dimensions.

Transferring Depth Dimensions

The depth dimensions in the top and side views must correspond point-for-point. When using CAD or instruments, transfer these distances accurately.

You can transfer dimensions between the top and side views either with dividers or with a scale, as shown in Figures 4.12a and 4.12b. Marking the distances on a scrap of paper and using it like a scale to transfer the distance to the other view is another method that works well when sketching.

You may find it convenient to use a 45° miter line to project dimensions between top and side views, as shown in Figure 4.12c. Because the miter line is drawn at 45°, depths shown vertically in the top view Y can be transferred to be shown as horizontal depths in the side view X and vice versa.

Measuring from a Reference Surface

To transfer a dimension from one view to a related view (a view that shares that dimension) you can think of measuring from the edge view of a plane which shows on edge in both views as in Figure 4.13.

Use Worksheet 4.2 to practice measuring from a reference surface.

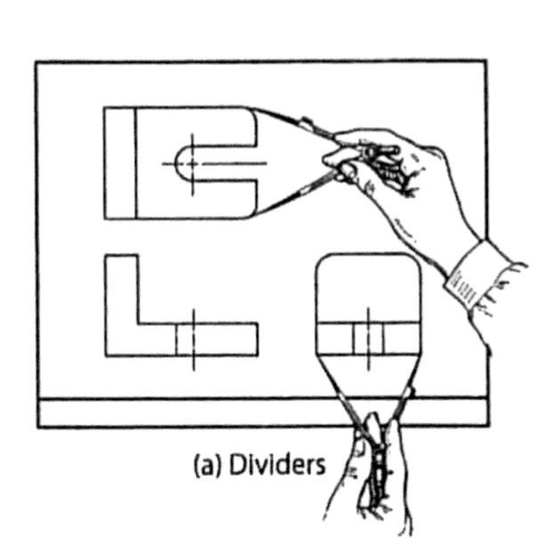

(a) Dividers

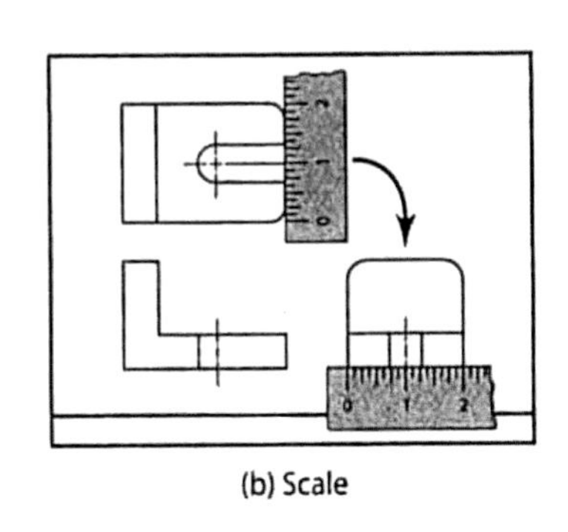

(b) Scale

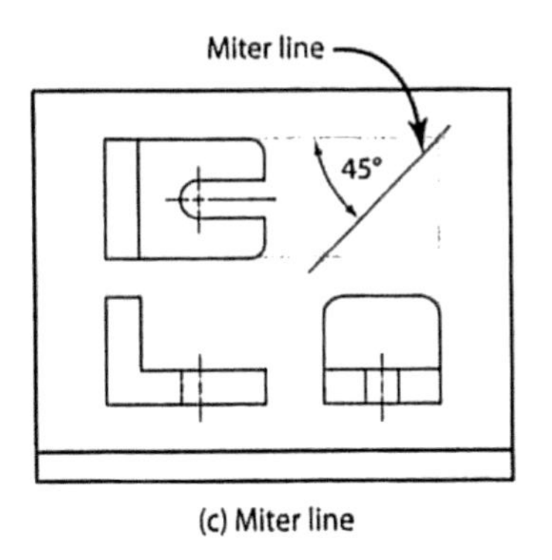

(c) Miter line

4.12 Transferring Depth Dimensions

Necessary Views

Figure 4.14 shows that right- and left-side views are essentially mirror images of each other, only with different lines appearing hidden. Hidden lines use a dashed-line pattern to represent portions of the object that are not directly visible from that direction of sight. Both the right and left views do not need to be shown, so usually the right-side view is drawn. This is also true of the top and bottom views, and of the front and rear views. The top, front, and right-side views, arranged together, are shown in Figure 4.15. These are called the **three regular views** because they are the views most frequently used.

A sketch or drawing should only contain the views needed to clearly and completely describe the object. These minimally required views are referred to as the **necessary views.** Choose the views that have the fewest hidden lines and show essential contours or shapes most clearly. Complicated objects may require more than three views or special views such as partial views.

Many objects need only two views to clearly describe their shape. If an object requires only two views and the left-side and right-side views show the object equally well, use the right-side view. If an object requires only two views and the top and bottom views show the object equally well, choose the top view. If only two views are necessary and the top view and right-side view show the object equally well, choose the combination that fits best on your paper. Some examples are shown in Figure 4.16.

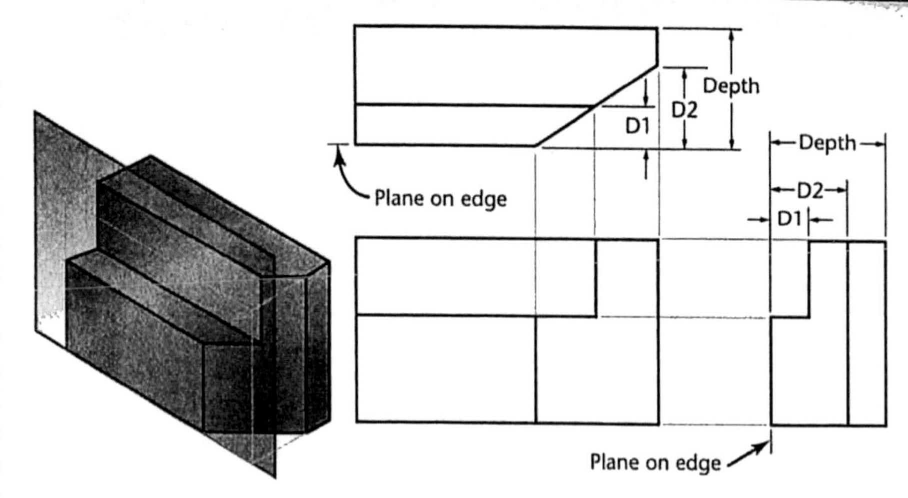

4.13 Transferring Depth Dimensions from a Reference Surface

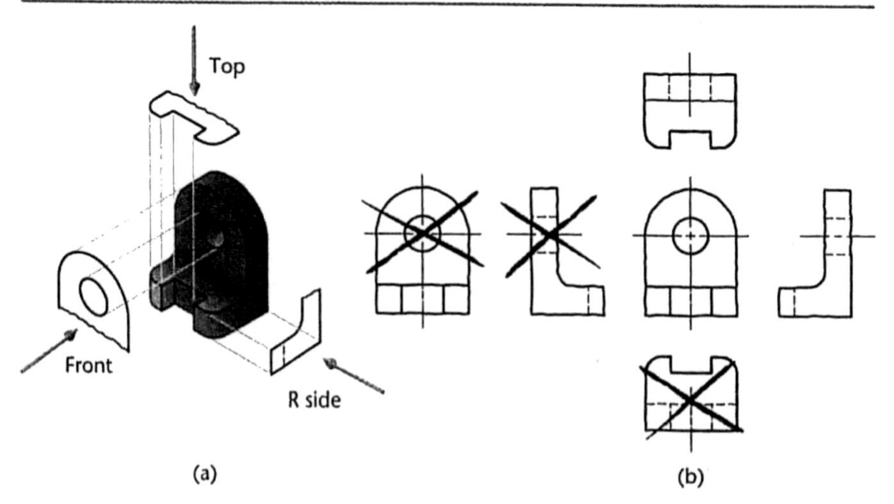

4.14 Opposite Views Are Nearly Identical

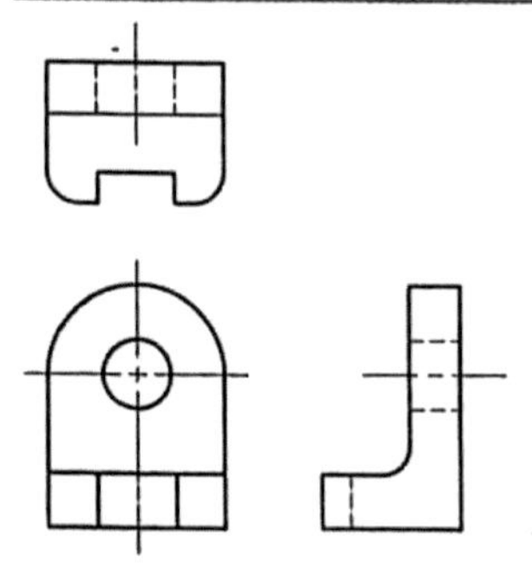

4.15 The Three Regular Views

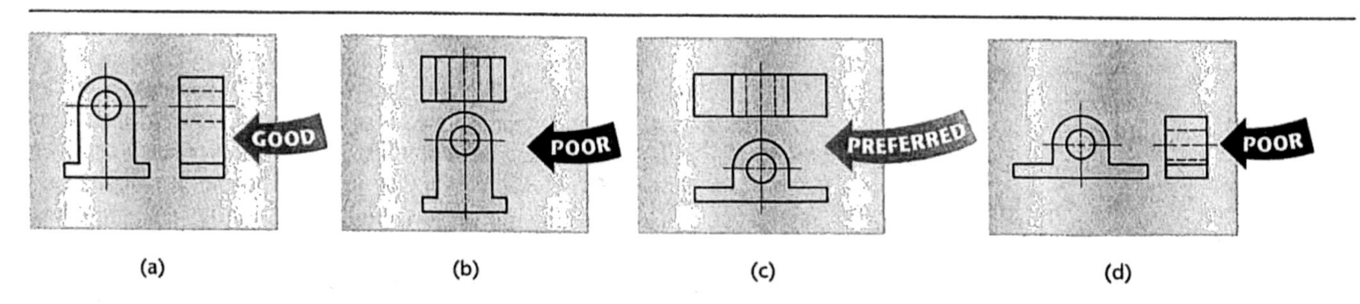

4.16 Choice of Views to Fit Paper

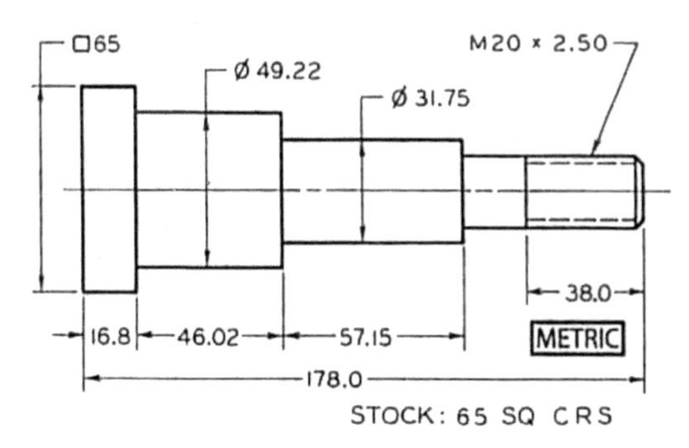

4.17 One-View Drawing of a Connecting Rod

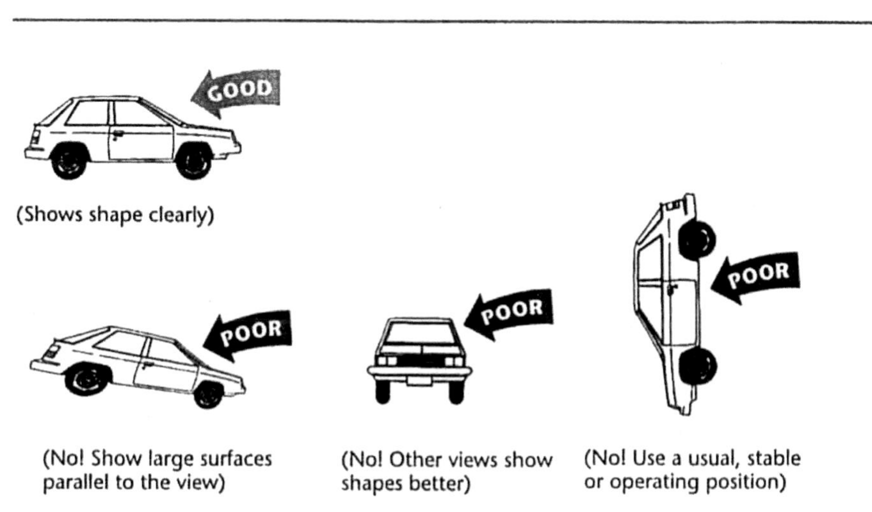

4.18 Choice of Front View

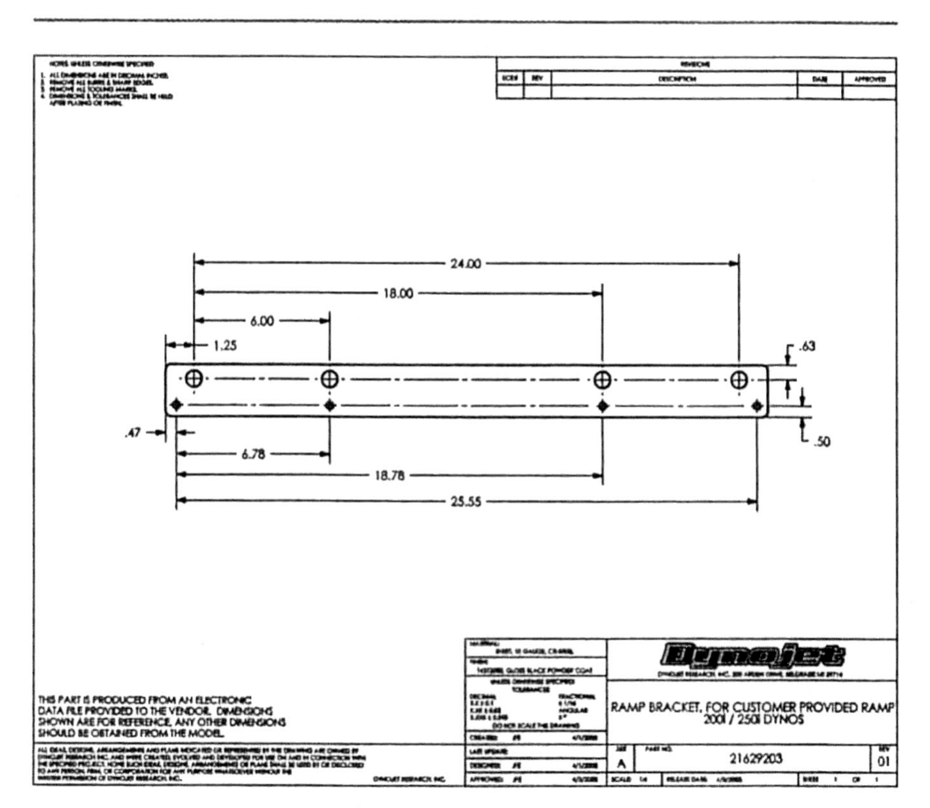

4.19 A long part looks best oriented with the long axis horizontal on the sheet. *Courtesy of Dynojet Research, Inc.*

Often, a single view supplemented by a note or by lettered symbols is enough, as shown in Figure 4.17. Objects that can be shown using a single view usually have a uniform thickness. This connecting rod is an exception. It is possible to show it in a single view due to the way it is dimensioned.

Orientation of the Front View

Four views of a compact automobile are shown in Figure 4.18. The view chosen for the front view in this case is the side, not the front, of the automobile.

- The front view should show a large surface of the part parallel to the front viewing plane.
- The front view should show the shape of the object clearly.
- The front view should show the object in a usual, stable, or operating position, particularly for familiar objects.
- When possible, a machine part is drawn in the orientation it occupies in the assembly.
- Usually screws, bolts, shafts, tubes, and other elongated parts are drawn in a horizontal position as shown in Figure 4.19.

CAD software can be used to generate orthographic views directly from a 3D model as shown in Figure 4.20. The pictorial view of this model is shown in Figure 4.21. When using CAD you still need to select a good orientation so that the part shows clearly in the front view. The standard arrangement of views shown in Figure 4.15 should be used. Do not be tempted to rearrange the views of your CAD drawing to fit the sheet better, unless you follow the practices outlined in Chapter 5 for using removed views.

First- and Third-Angle Projection

As you saw earlier in this chapter, you can think of the system of projecting views as unfolding a glass box made from the viewing planes. There are two main systems used for projecting and unfolding the views: **third-angle projection,** which is used in the United States, Canada, and some other countries, and **first-angle projection,** which is primarily used in

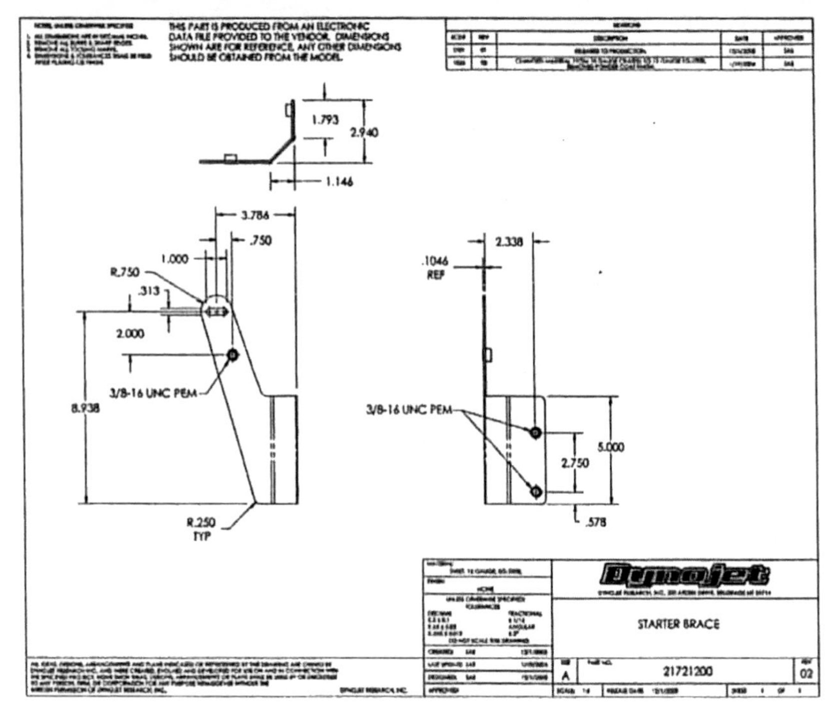

4.20 Computer-Generated Multiview Drawing from a CAD Model.
Courtesy of Dynojet Research, Inc.

4.21 Pictorial View of the CAD Model Shown in Figure 4.20.
Courtesy of Dynojet Research, Inc.

Europe and Asia. Difficulty in interpreting the drawing and manufacturing errors can result when a first-angle drawing is confused with a third-angle drawing.

Because of the global nature of technical drawings, you should thoroughly understand both methods. However, since it can be confusing to try to learn both methods intermixed, this text presents third-angle projection throughout. When you are comfortable with creating third-angle projection drawings, revisit this section. You will see that the two drawing methods are very similar and you should be able to extend the same skills to either type of drawing.

Third-Angle Projection

Figure 4.22a shows the concept of third-angle orthographic projection. To avoid misunderstanding, international **projection symbols** have been developed to distinguish between first-angle and third-angle projections on drawings. The symbol in Figure 4.22b shows two views of a truncated cone. You can examine the arrangement of the views in the symbol to determine whether first- or third-angle projection was used. On international drawings you should be sure to include this symbol.

To understand the two systems, think of the vertical and horizontal planes of projection, shown in Figure 4.22a, as indefinite in extent and intersecting at 90° with each other; the four angles produced are called the first, second, third, and fourth angles (similar to naming quadrants on a graph.) If the object to be drawn is placed below the horizontal plane and behind the vertical plane, as in the glass box you saw earlier, the object is said to be in the third angle. In third-angle projection, the views are produced as if the observer is outside, looking in.

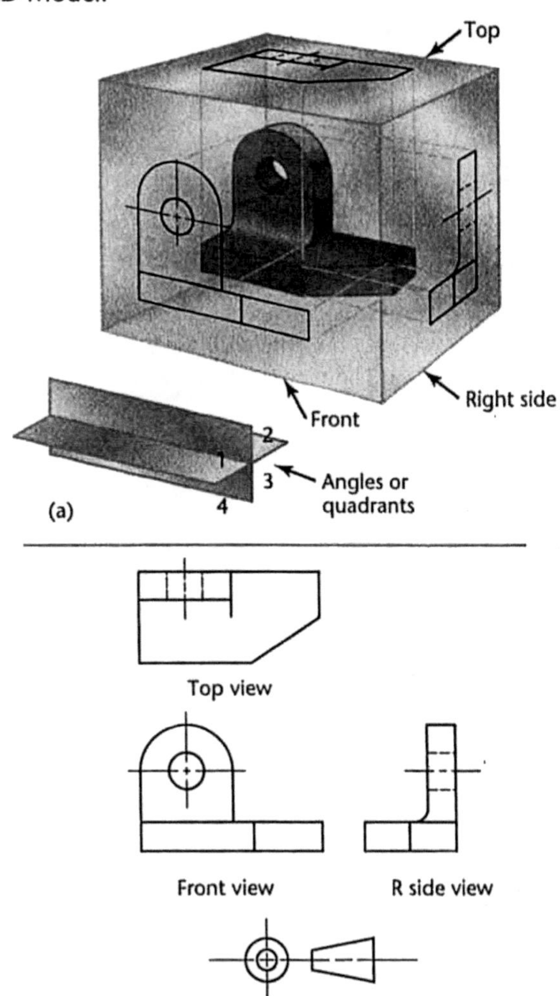

4.22 Third-Angle Projection

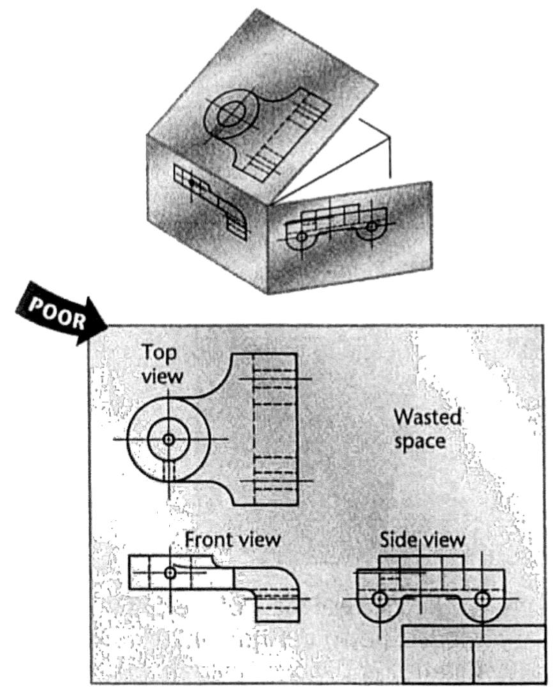

(a) Crowded arrangement of views

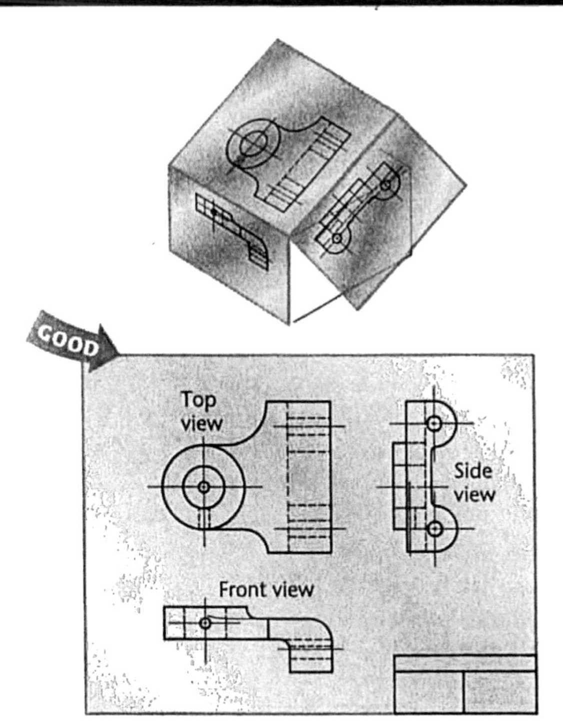

(b) Approved alternate arrangement of views

4.23 Position of Side View

Alternative Arrangements for Third-Angle Projection

Sometimes drawing three views using the conventional arrangement wastes space. (For example, see the wide flat object in Figure 4.23a.) Using the space on the paper efficiently may prevent the need to use a reduced scale.

For these cases, there is another acceptable arrangement of third-angle projection views. Imagine unfolding the glass box as shown in Figure 4.23b. The views are arranged differently, with the right-side view aligned with the top view, but these views are still using third-angle projection.

In this case, think of the profile (side view) hinged to the horizontal plane (top view) instead of to the frontal plane (front view) so that the side view is beside the top view when unfolded, as shown in Figure 4.23b. Notice the side view is rotated 90° from the orientation shown in the side view in Figure 4.23a when it is in this placement. Note also that you can now directly project the depth dimension from the top view into the side view.

If necessary, you may place the side view horizontally across from the bottom view (so the profile plane is hinged to the bottom plane of the projection).

Similarly, the rear view may be placed directly above the top view or under the bottom view. In this case, the rear plane is considered hinged to the horizontal or bottom plane and rotated to coincide with the frontal plane.

Use Worksheets 4.3 and 4.4 to practice projecting views.

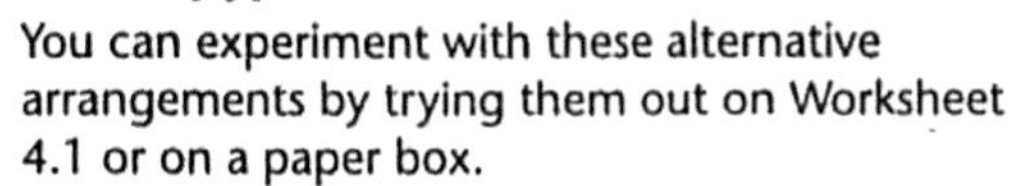

TIP

You can experiment with these alternative arrangements by trying them out on Worksheet 4.1 or on a paper box.

First-Angle Projection

If the object is placed above the horizontal plane and in front of the vertical plane, the object is in the first angle. In first-angle projection the observer looks through the object to the planes of projection. The right-side view is still obtained by looking toward the right side of the object, the front by looking toward the front, and the top by looking down toward the top; but the views are projected from the object onto a plane in each case.

The biggest difference between third-angle projection and first-angle projection is in how the planes of the glass box are unfolded, as shown in Figure 4.24. In first-angle projection, the right-side view is to the left of the front view, and the top view is below the front view, as shown.

You should understand the difference between the two systems and know the symbol that is placed on drawings to indicate which has been used. Keep in mind that you will use third-angle projection throughout this book.

Projection System Drawing Symbol

The symbol shown in Figure 4.25 is used on drawings to indicate which system of projection is used. Whenever drawings will be used internationally you should include this symbol in the title block area.

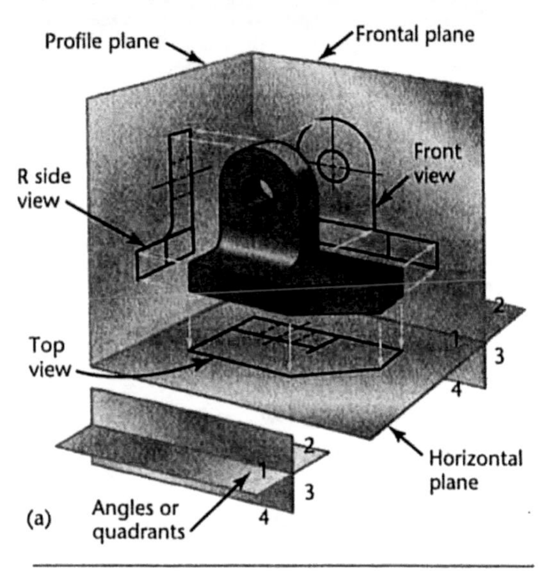

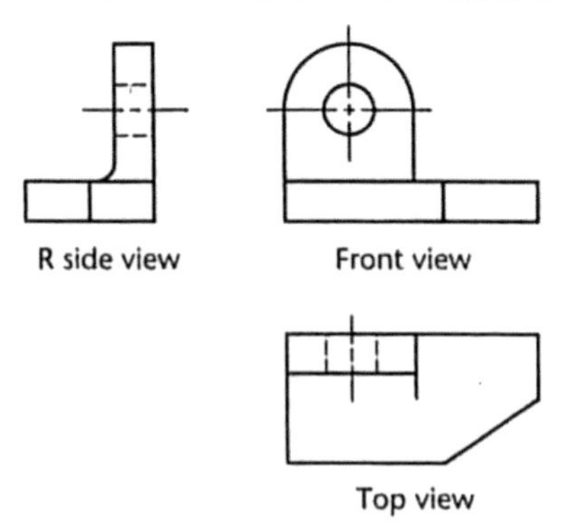

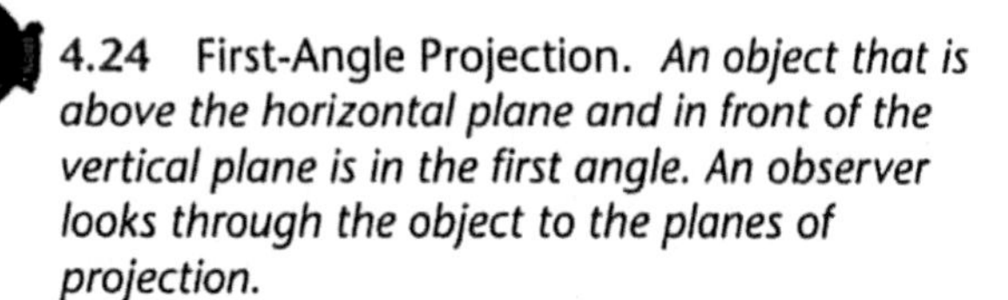

4.24 First-Angle Projection. *An object that is above the horizontal plane and in front of the vertical plane is in the first angle. An observer looks through the object to the planes of projection.*

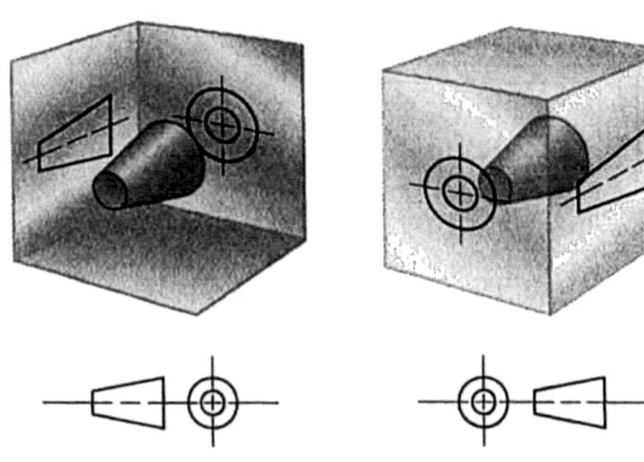

4.25 Drawing Symbols for First- and Third-Angle Projection

Hidden Lines

One advantage of orthographic views over photographs is that each view can show the entire object from that viewing direction. A photograph shows only the visible surface of an object, but an orthographic view shows the object all the way through, as if it were transparent.

Thick, dark lines represent features of the object that are directly visible. Dashed lines represent features that would be hidden behind other surfaces.

Figure 4.26 shows a part that has internal features. When a 3D view of this model is rendered using a transparent material, as shown in Figure 4.27, you can see the internal features. Figure 4.28 shows this part from the front as it would be oriented in an orthographic drawing. The features that are hidden from view are shown in orthographic views using the hidden line pattern as shown in Figure 4.29.

Whenever possible, choose views that show features with visible lines. Use hidden lines where they are needed to make the drawing clear.

Some practices for representing intersections of hidden lines with other lines may be difficult to follow when using CAD. In CAD, adjust the line patterns so that the hidden lines in your drawing have the best appearance possible.

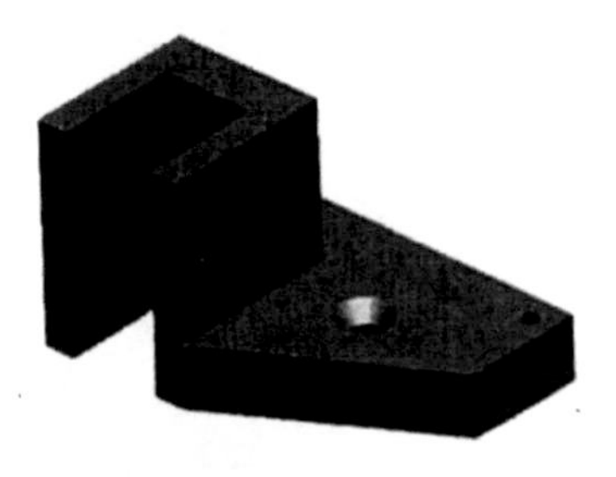

4.26 Shaded Model with Hidden Features

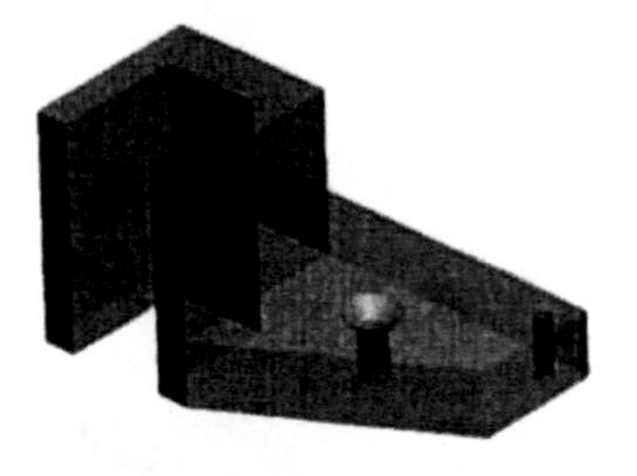

4.27 Transparent Model Showing Hidden Features

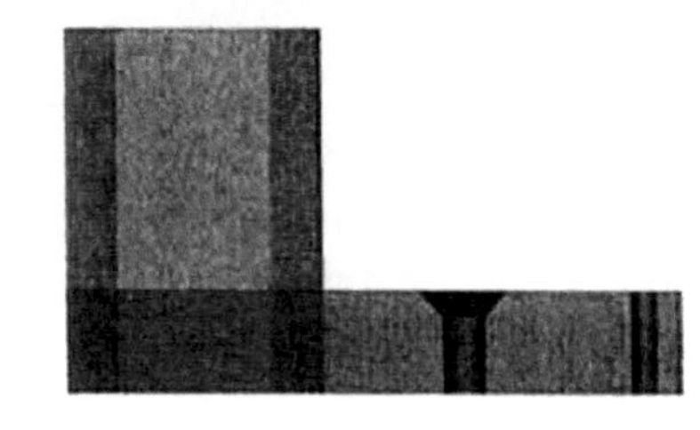

4.28 Front View of Transparent Model

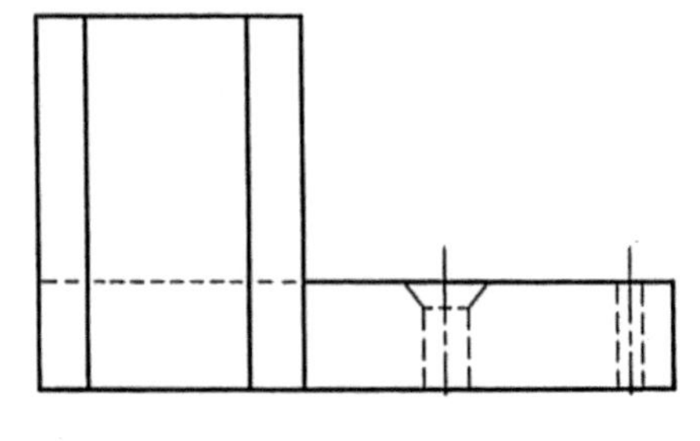

4.29 Front View Projection

CENTERLINES

The centerline pattern is used to:

- show the axis of symmetry for a feature or part
- indicate a path of motion
- show the location for bolt circles and other circular patterns

The centerline pattern is composed of three dashes: one long dash on each end with a short dash in the middle. In the drawing, centerlines are shown as thin and black. Because a centerline is not an actual part of the object, it extends beyond the symmetric feature as shown in Figure 4.30.

The most common shape that needs a centerline is a cylindrical hole. Figure 4.31 shows centerlines in a drawing. In the circular view of a hole, the centerline should form a cross to mark the center location. When a feature is too small for the centerline pattern to be shown with the long-short-long dash pattern, it is acceptable to use a straight line. You will learn more about showing hidden and centerlines in the technique sections.

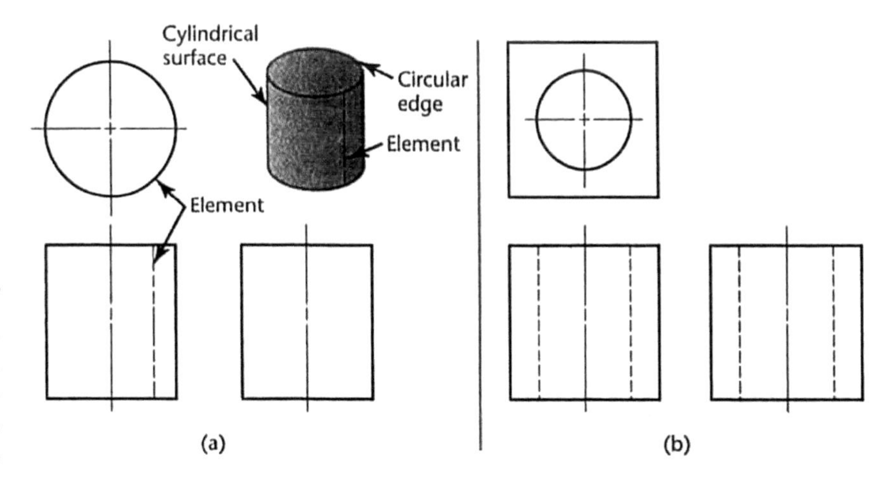

4.30 Cylindrical Surfaces

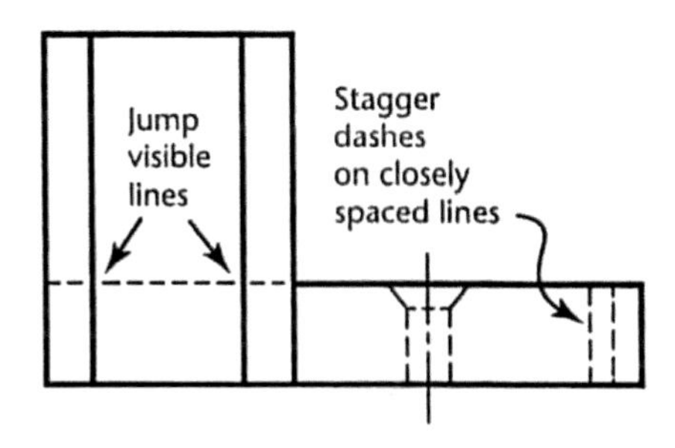

4.31 Hidden Lines

4.1 HIDDEN LINE TECHNIQUE

You can save time and reduce clutter by leaving out hidden lines that aren't necessary as long as you are certain that the remaining lines describe the object clearly and completely. If you omit unnecessary hidden lines, add a note to let the reader know that the lines were left out intentionally and that it is not an error in the drawing.

Sketch hidden lines by eye, using thin dark dashes about 5 mm long and spaced about 1 mm apart. Hidden lines should be as dark as other lines in the drawing, but should be thin.

When hidden lines intersect each other in the drawing, their dashes should meet. In general, hidden lines should intersect neatly with visible lines at the edge of an object. Leave a gap when a hidden line aligns with a visible line, so that the visible line's length remains clear.

Use Worksheet 4.5 for practice hidden line technique.

4.2 PRECEDENCE OF LINES

Visible lines, hidden lines, and centerlines (which are used to show the axis of symmetry for contoured shapes, like holes) often coincide on a drawing. There are rules for deciding which line to show. A visible line always takes precedence over and covers up a centerline or a hidden line when they coincide in a view, as shown at A and B in Figure 4.32. A hidden line takes precedence over a centerline, as shown at C. At A and C the ends of the centerline are shown separated from the view by short gaps, but the centerline can be left off entirely. Figure 4.33 shows examples of correct and incorrect hidden lines.

Use Worksheet 4.6 for practice line precedence.

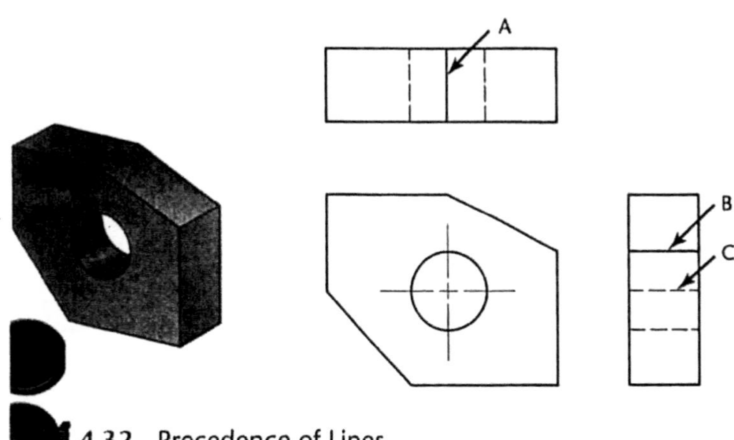

4.32 Precedence of Lines

Correct and incorrect practices for hidden lines

Make a hidden line join a visible line, except when it causes the visible line to extend too far, as shown here.

Leave a gap whenever a hidden line is a continuation of a visible line.

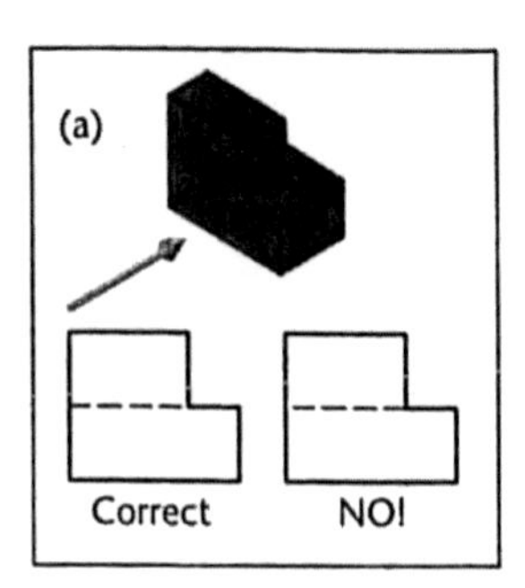

When two or three hidden lines meet at a point, join the dashes, as shown for the bottom of this drilled hole.

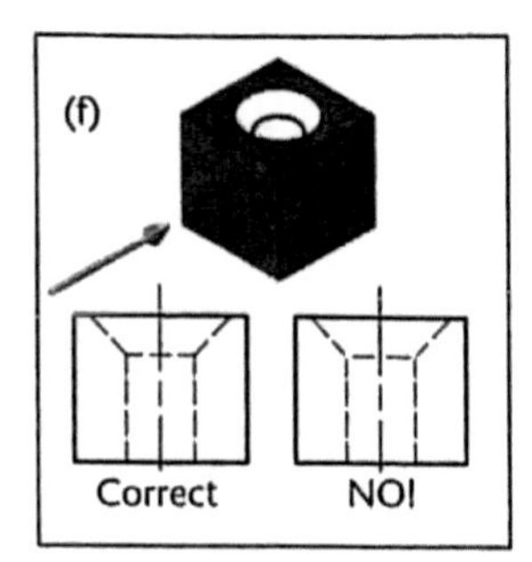

Make hidden lines intersect at L and T corners.

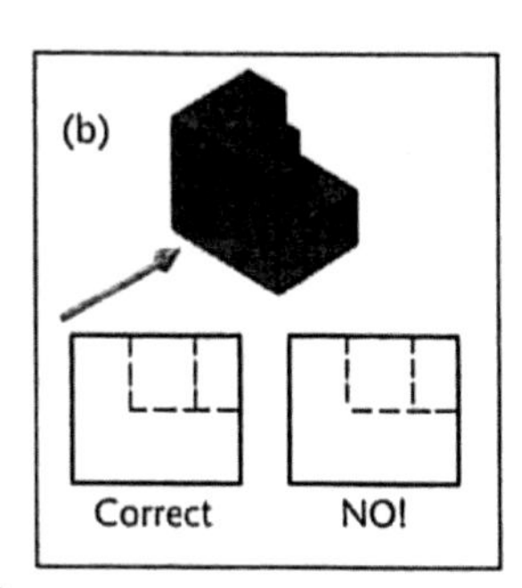

The same rule of joining the dashes when two or three hidden lines meet at a point applies for the top of this countersunk hole.

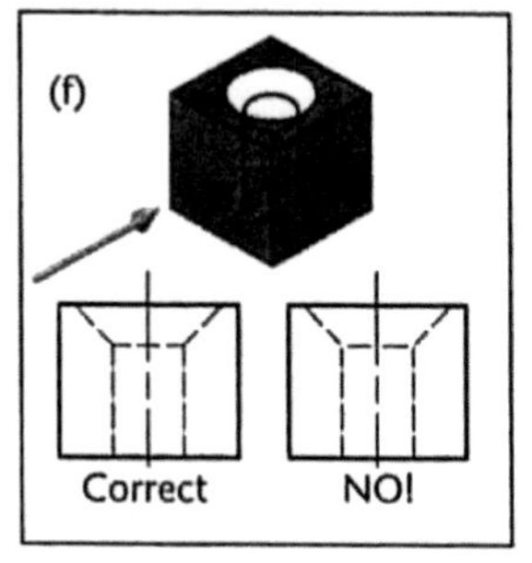

Make a hidden line "jump" a visible line when possible.

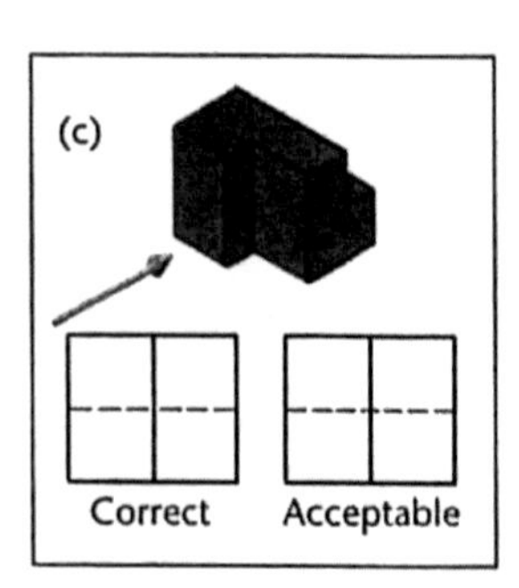

Hidden lines should not join visible lines when this makes the visible line extend too far.

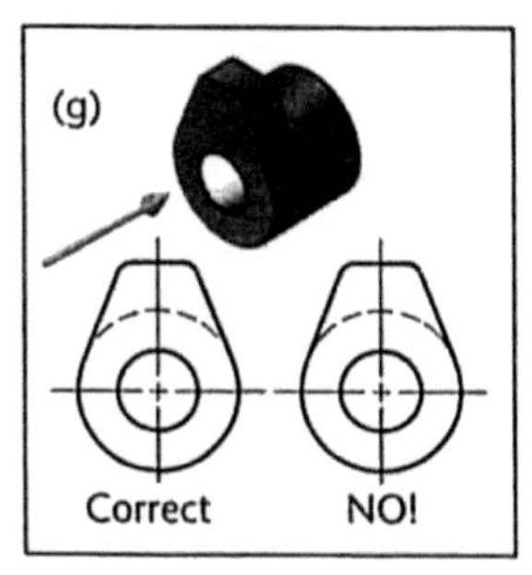

Draw parallel hidden lines so that the dashes are staggered, as in bricklaying.

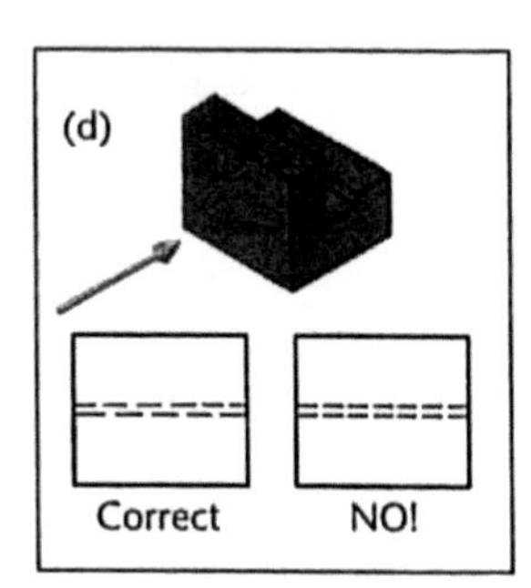

Draw hidden arcs with the arc joining the centerline, as in upper example. There should not be a gap between the arc and the centerline, as in the example below with the straightaway joining the centerline.

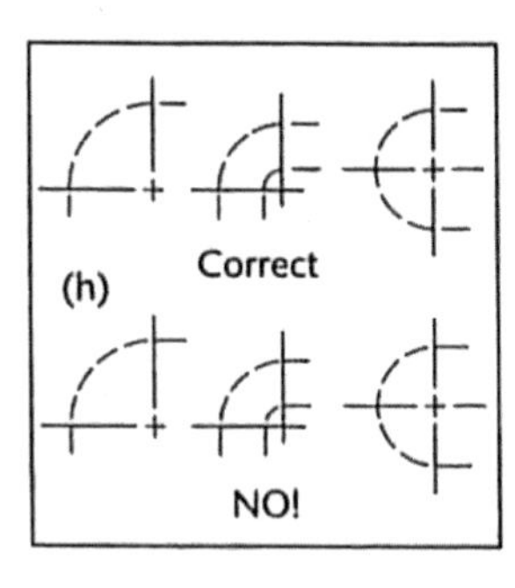

4.33 Correct and Incorrect Practices for Hidden Lines

TIP

Accent the beginning and end of each dash by pressing down on the pencil. Make hidden lines as tidy as you can so they are easy to interpret. Be sure to make hidden line dashes longer than gaps so they clearly represent lines.

4.3 CENTERLINES

Centerlines (symbol: ℄) are used to indicate symmetrical axes of objects or features, bolt circles, and paths of motion as shown in Figure 4.34. Centerlines are useful in dimensioning. They are not needed on unimportant rounded or filleted corners or on other shapes that are self-locating.

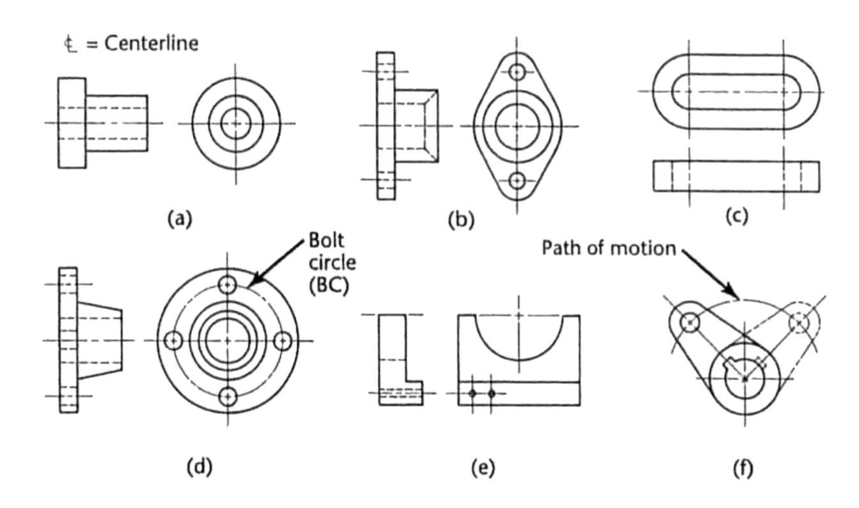

4.34 Centerlines

4.4 LAYING OUT A DRAWING

If you use 2D CAD, you can move the views later, keeping them in alignment, so you do not need to give as much attention to placement of the views in the beginning as if you were laying them out by hand. When using 3D CAD to generate views, you should still plan how the sheet will show the information clearly and select the necessary views to best represent the shape of the part. While you can easily change the scale of a CAD drawing after it is created, placing the dimensions and views on the sheet requires some planning. If you consider the purpose of the drawing, the planned scale, and the space that will be required for adding notes and dimensions, you will save the time of having to rearrange their placement later.

STEP by STEP

LAYING OUT A METRIC THREE-VIEW DRAWING

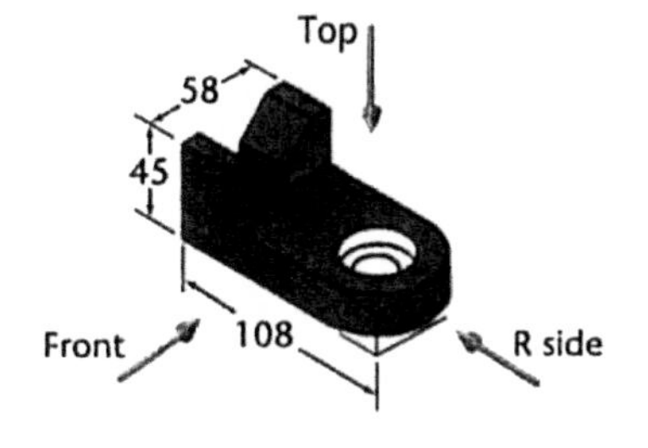

1 Determine space desired between the front and right side views, say 32 mm, C. Add this space to the sum of the length of the views that will be aligned along the long edge of the sheet. (108 + 58 + 32 = 198) To set equal distances to the paper edge, subtract this total from the sheet width, then divide the remaining number by two (266 − 198 = 70, and 70 ÷ 2 = 35). Do the same for the views to be aligned along the short side of the paper, selecting a desired space between the views. Space D need not match C. Remember to leave space for dimensions as you plan your sheet.

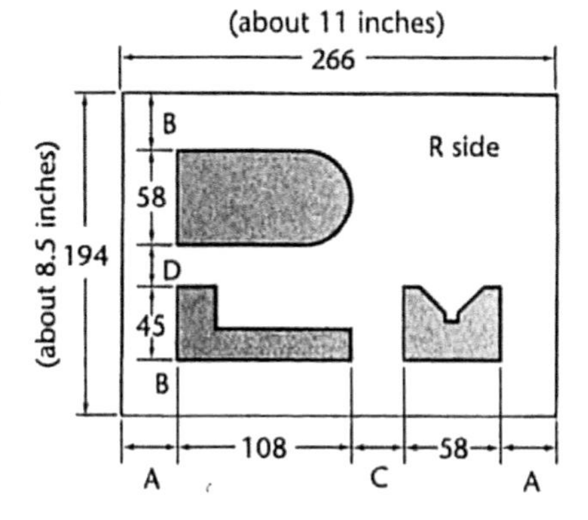

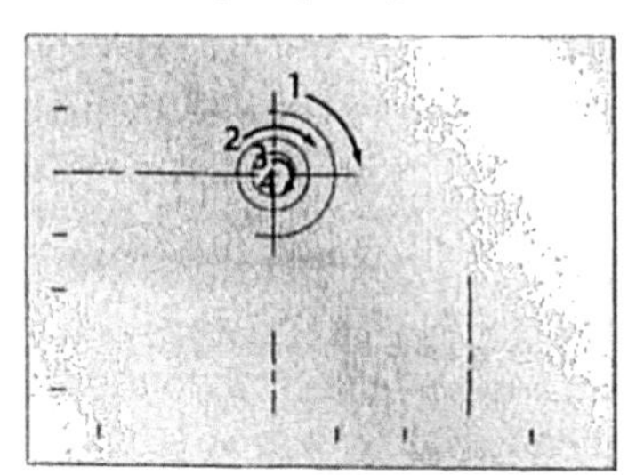

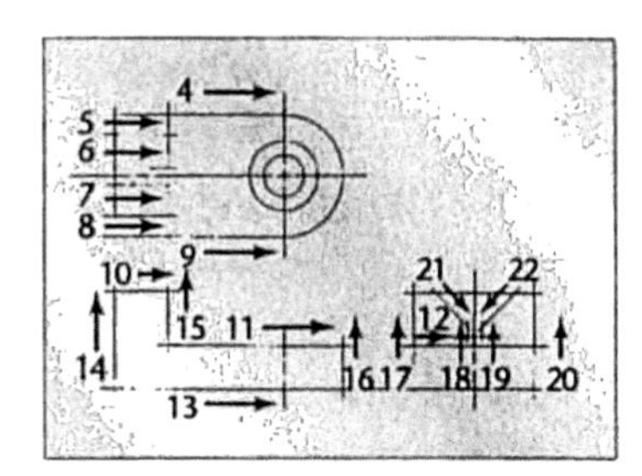

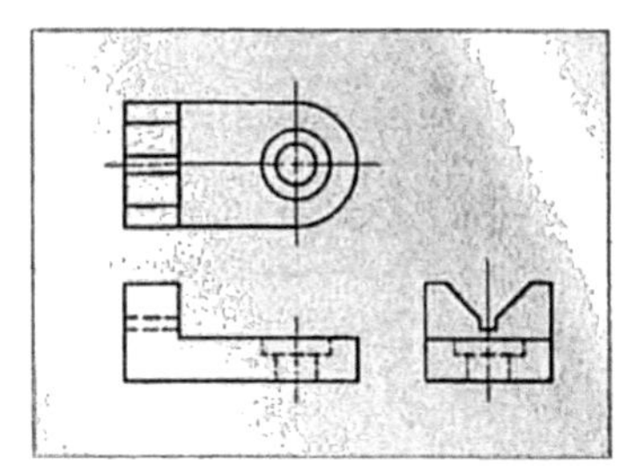

2 Set off vertical and horizontal spacing measurements with light tick marks along the edge of the sheet as shown. Locate centerlines from these spacing marks and construct arcs and circles.

3 Construct the views, drawing horizontal, vertical, and then inclined construction lines in the order shown above.

4 Add hidden lines and darken final lines.

4.5 VISUALIZATION

Along with a basic understanding of the system for projecting views, you must be able interpret multiple views to picture the object that they show. In addition to being an indispensable skill to help you capture and communicate your ideas, technical sketching is also a way for others to present their ideas to you.

Even experienced engineers, technicians, and designers can't always look at a multiview sketch and instantly visualize the object represented. You will learn to study the sketch and interpret the lines in a logical way in order to piece together a clear idea of the whole. This process is sometimes called visualization.

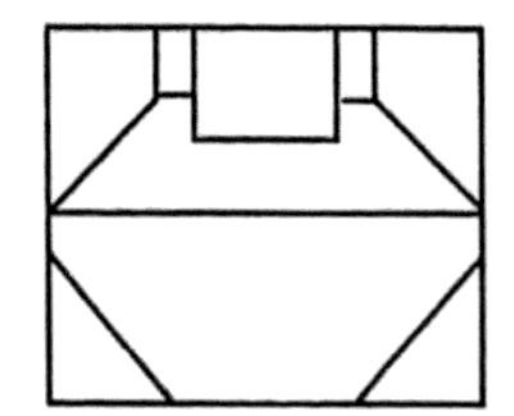

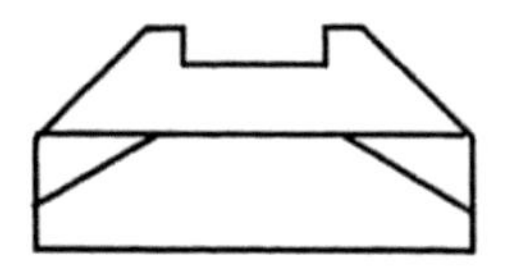

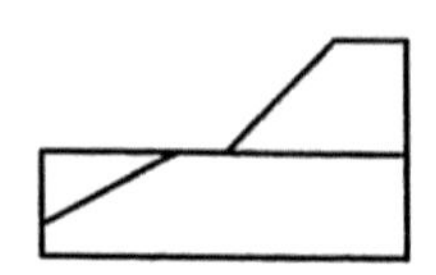

4.35 Three Views of an Object

Surfaces, Edges, and Corners

To effectively create and interpret multiview projections, you have to consider the elements that make up most solids. **Surfaces** form the boundaries of solid objects. A **plane** (flat) surface may be bounded by straight lines, curves, or a combination of the two. It takes practice to envision flat representations as 3D objects. Take a moment to examine the views shown in Figure 4.35 and try to picture the object. (See solution on page 164).

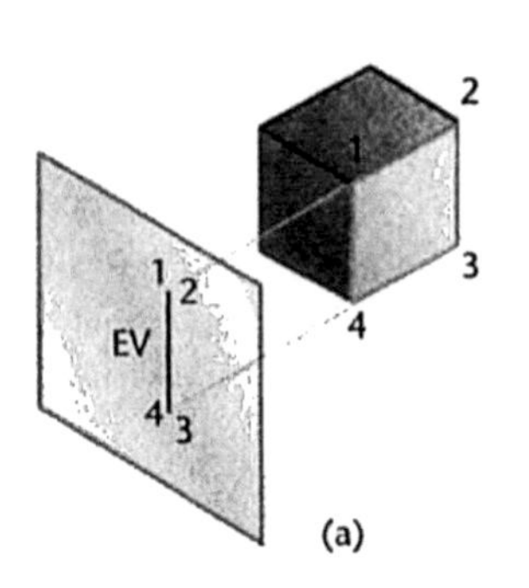

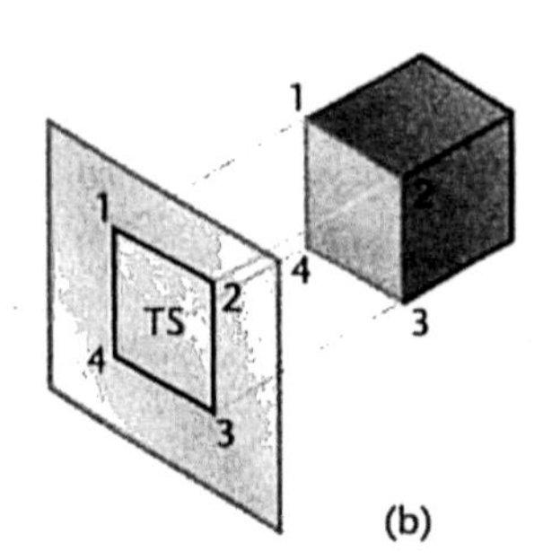

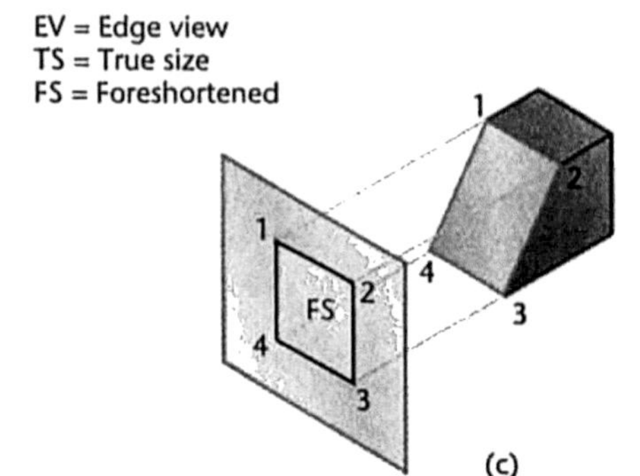

4.36 Projections of Surfaces

> **TIP**
>
> **Using Numbers to Identify Vertices**
>
> Add lightly drawn numbers to your sketches to keep track of each vertex on the surface you are considering. Each vertex is unique on the part, so each numbered vertex will appear only once in each view. Sometimes two vertices will line up one behind the other as in 4.36a. When this happens you can list them in order with the closest first, as in 1, 2, or sometimes it is useful to put numbers for the closest visible vertex outside the shape, and the farthest hidden vertex inside the shape outline.

4.6 VIEWS OF SURFACES

A plane surface that is perpendicular to a plane of projection appears on edge as a straight line (Figure 4.36a). If it is parallel to the plane of projection, it appears true size (Figure 4.36b). If it is angled to the plane of projection, it appears foreshortened or smaller than its actual size (Figure 4.36c). A plane surface always projects either on edge (appearing as a single line) or as a surface (showing its characteristic shape) in any view. It can appear foreshortened, but it can never appear larger than its true size in any view.

There are terms used for describing a surface's orientation to the plane of projection. The three orientations that a plane surface can have to the plane of projection are *normal*, *inclined*, and *oblique*. Understanding these terms will help you picture and describe objects.

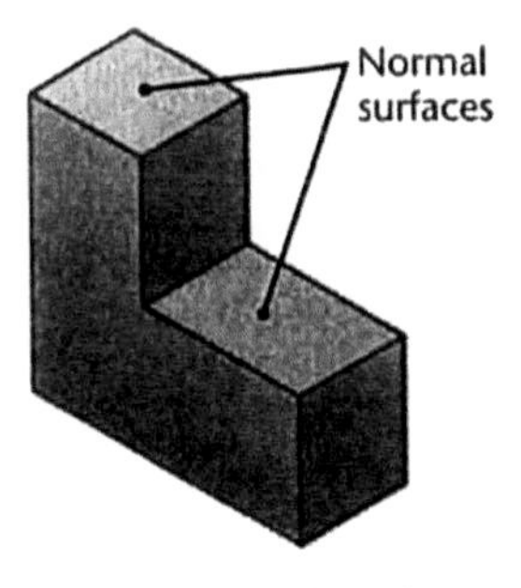

4.37 Normal Surfaces

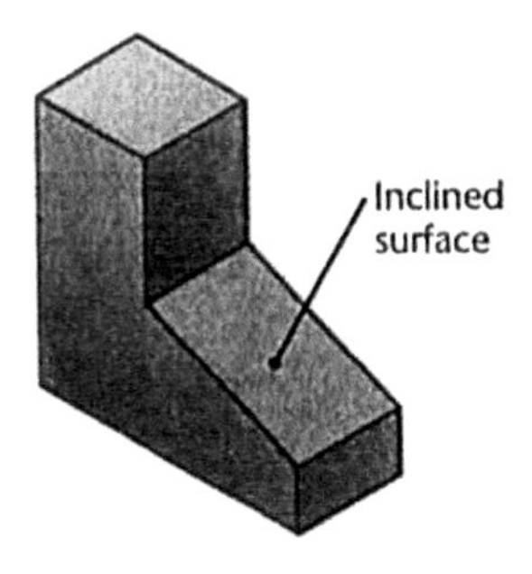

4.38 Inclined Surface

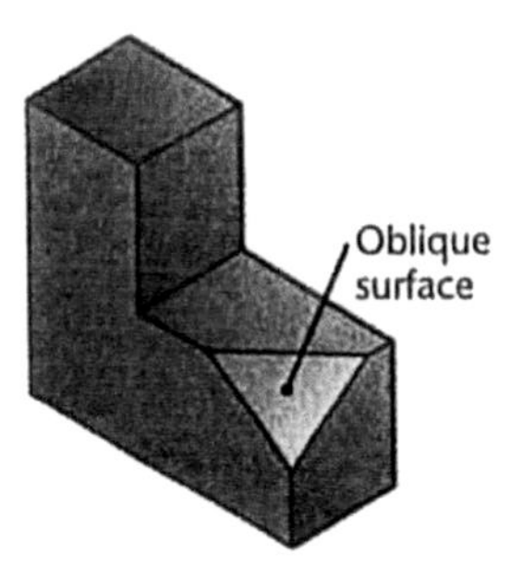

4.39 Oblique Surface

4.7 NORMAL SURFACES

A **normal surface** is parallel to a plane of projection. It appears true size and true shape on the plane to which it is parallel, and it appears as a true-length vertical or a horizontal line on adjacent planes of projection. Figure 4.37 shows an illustration of normal surfaces.

Practice identifying normal surfaces on CAD drawings. You can download orthographic views of subjects that show many normal surfaces at the following Web sites:

- http://www.constructionsite.come/harlen/8001-81.htm
- http://www.user.mc.net/hawk/cad.htm

4.8 INCLINED SURFACES

An **inclined surface** is perpendicular to one plane of projection, but inclined (or tipped) to adjacent planes. An inclined surface projects an edge on the plane to which it is perpendicular. It appears foreshortened on planes to which it is inclined. An inclined surface is shown in Figure 4.38. The degree of foreshortening is proportional to the inclination. While the surface may not appear true size in any view, it will have the same characteristic shape and the same number of edges in the views in which you see its shape.

4.9 OBLIQUE SURFACES

An **oblique surface** is tipped to all principal planes of projection. Since it is not perpendicular to any projection plane, it cannot appear on edge in any standard view. Since it is not parallel to any projection plane, it cannot appear true size in any standard view. An oblique surface always appears as a foreshortened surface in all three standard views. Figure 4.39 and Figure 4.40 show oblique surfaces.

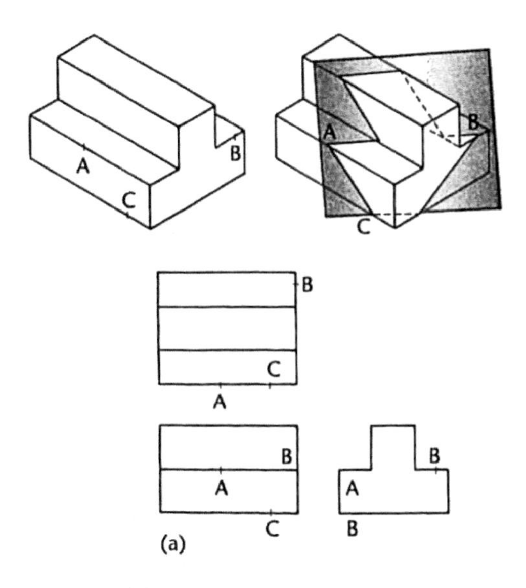

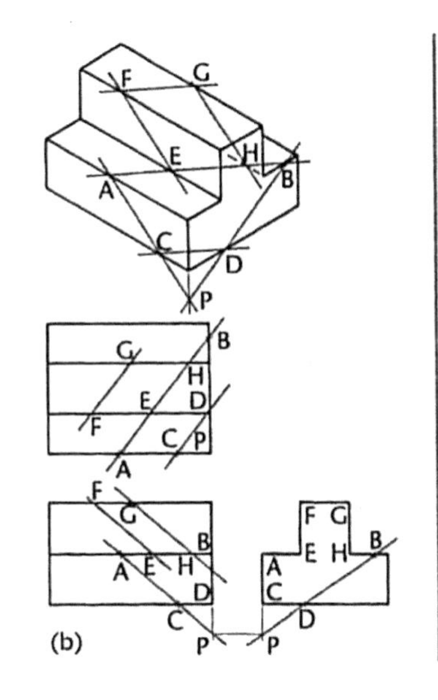

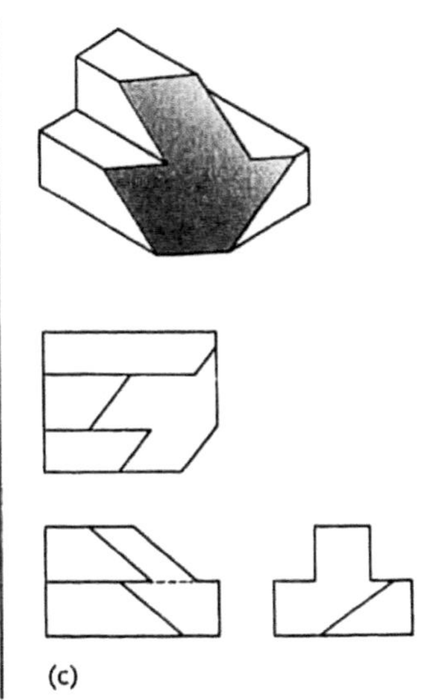

4.40 Oblique Surfaces

4.10 EDGES

The intersection of two plane surfaces of an object produces an **edge,** which shows as a straight line in the drawing. An edge is common to two surfaces, forming a boundary for each. If an edge is perpendicular to a plane of projection, it appears as a point; otherwise it appears as a line. If it is parallel to the plane of projection, it shows true length. If it is not parallel, it appears foreshortened. A straight line always projects as a straight line or as a point. The terms normal, inclined, and oblique describe the relationship of an edge to a plane of projection.

4.11 NORMAL EDGES

A **normal edge** is a line perpendicular to a plane of projection. It appears as a point on that plane of projection and as a true-length line on adjacent planes of projection (Figure 4.41).

4.12 INCLINED EDGES

An **inclined edge** is parallel to one plane of projection but inclined to adjacent planes. It appears as a true-length line on the plane to which it is parallel and as a foreshortened line on adjacent planes. The true-length view of an inclined line always appears as an angled line, but the foreshortened views appear as either vertical or horizontal lines (Figure 4.42).

4.13 OBLIQUE EDGES

An **oblique edge** is tipped to all planes of projection. Since it is not perpendicular to any projection plane, it cannot appear as a point in any standard view. Since it is not parallel to any projection plane, it cannot appear true length in any standard view. An oblique edge appears foreshortened and as an angled line in every view (Figure 4.43).

4.14 PARALLEL EDGES

When edges are parallel to one another on the object, they will appear as parallel lines in every view, unless they align one behind the other. This information can be useful when you are laying out a drawing, especially if it has a complex inclined or oblique surface that has parallel edges. Figure 4.44 shows an example of parallel lines in drawing views.

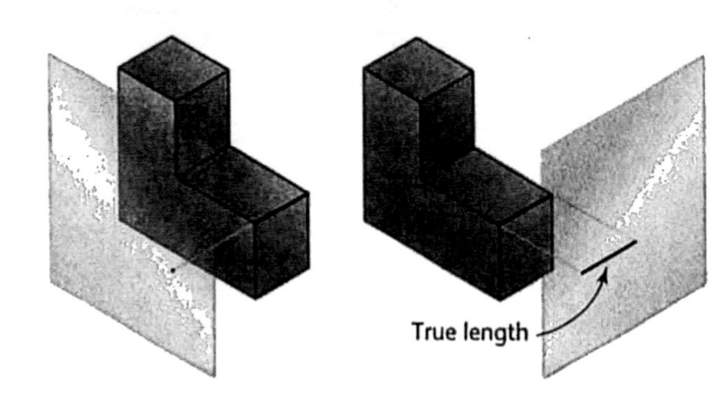

4.41 Projections of a Normal Edge

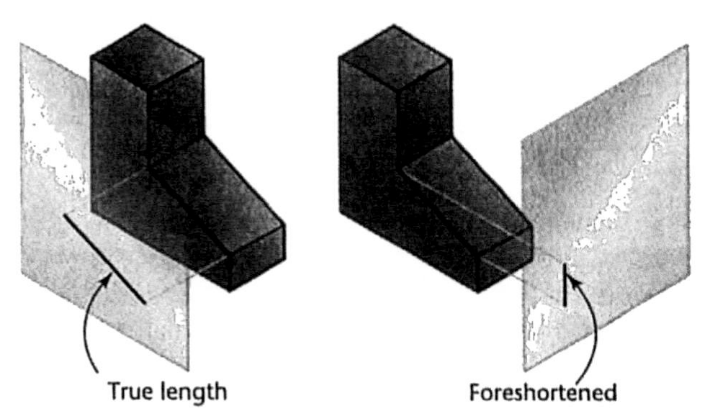

4.42 Projections of an Inclined Edge

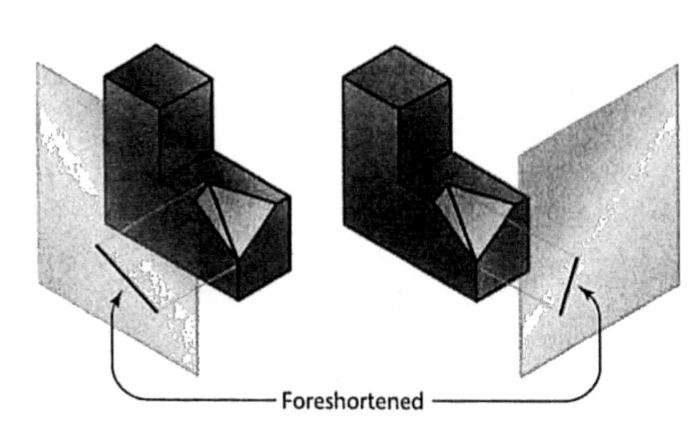

4.43 Projections of an Oblique Edge

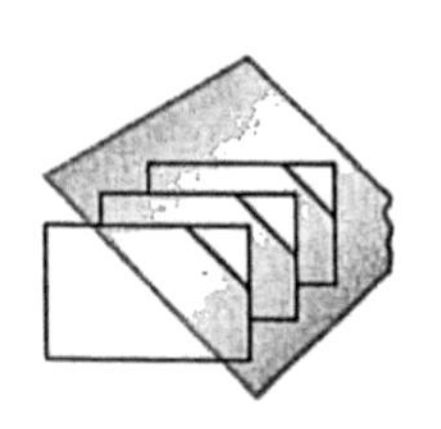

(a) Parallel planes intersected by another plane

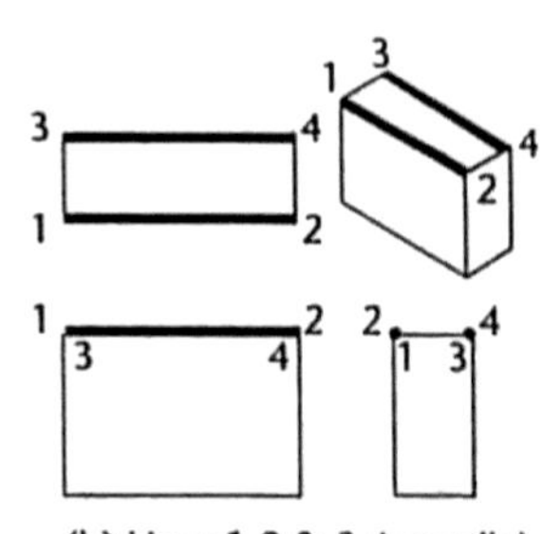

(b) Lines 1,2 & 3,4 parallel, and parallel to horizontal plane

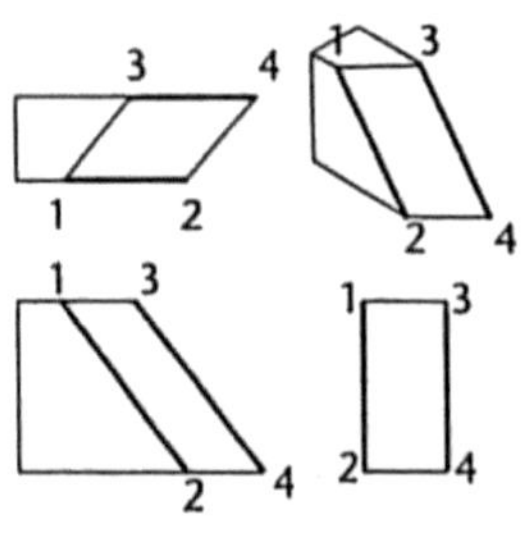

(c) Lines 1,2 & 3,4 parallel, & parallel to frontal plane

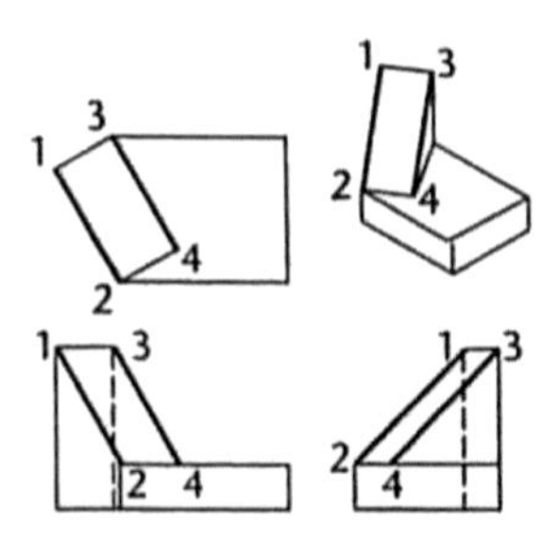

(d) Lines 1,2 & 3,4 parallel, and oblique to all planes

4.44 Parallel Lines

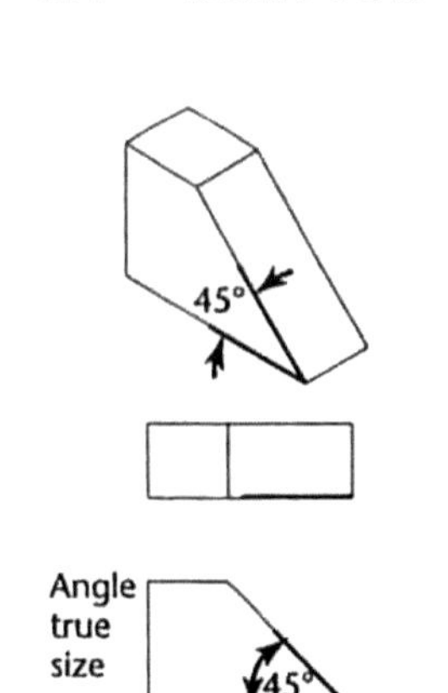

(a) Angle in normal plane

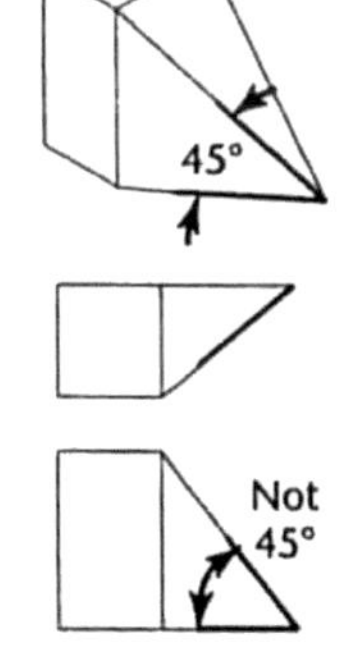

(b) Angle in inclined plane

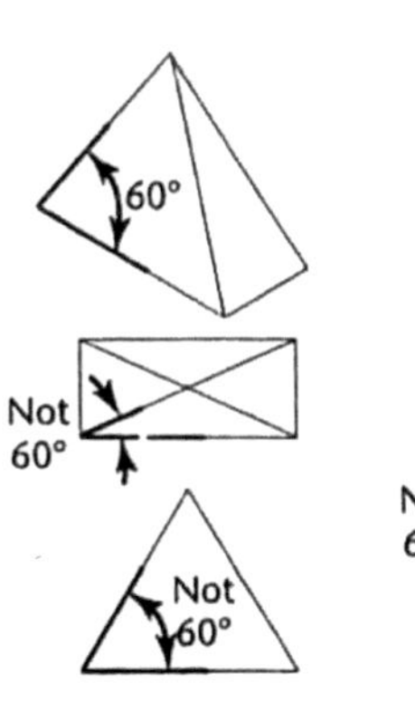

(c) Angle in inclined plane

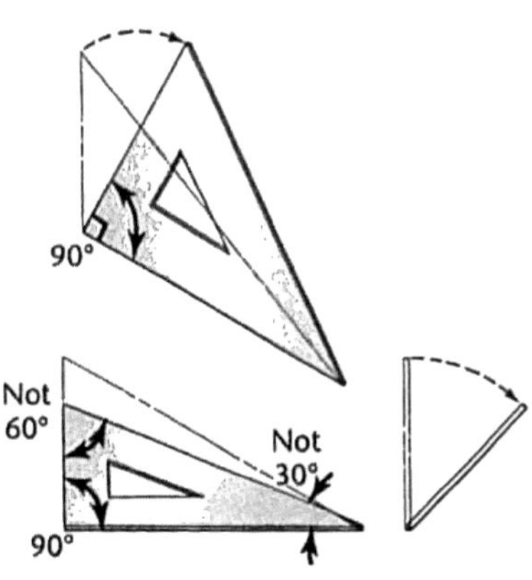

(d) Projections of 90° angles appear true size whenever one leg of the triangle appears true length

4.45 Angles

4.15 ANGLES

If an angle is in a normal plane (a plane parallel to a plane of projection) it will show as true size on the plane of projection to which it is parallel (Figure 4.45). If an angle is in an inclined plane, it may be projected either larger or smaller than the true angle, depending on its position. The 45° angle is shown oversize in the front view in Figure 4.45b, and the 60° angle is shown undersize in both views in Figure 4.45c.

A 90° angle will project as true size, even if it is in an inclined plane, provided that one leg of it is a normal line.

In Figure 4.45d the 60° angle is projected oversize and the 30° angle is projected undersize. Try this on your own using a 30° or 60° triangle as a model, or even the 90° corner of a sheet of paper. Tilt the triangle or paper to look at an oblique view.

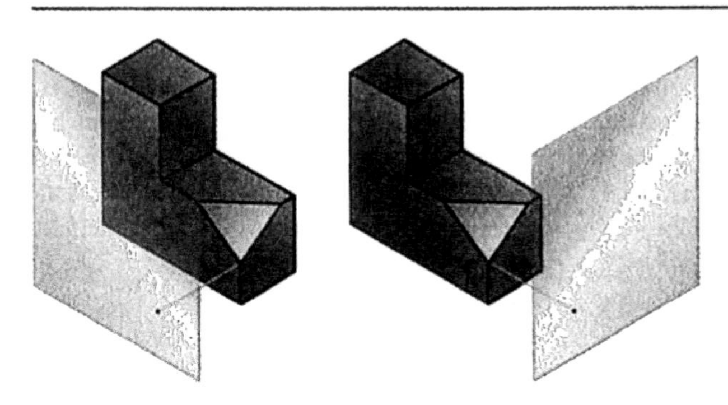

4.46 Views of a Point

4.16 VERTICES

A corner, or **point**, is the common intersection of three or more surfaces. A point appears as a point in every view. An example of a point on an object is shown in Figure 4.46.

4.17 INTERPRETING POINTS

A point located in a sketch can represent two things:

- A vertex
- The point view of an edge (two vertices lined up one directly behind the other)

4.18 INTERPRETING LINES

A straight visible or hidden line in a drawing or sketch has three possible meanings, as shown in Figure 4.47:

- An edge (intersection) between two surfaces
- The edge view of a surface
- The limiting element of a curved surface

Since no shading is used on orthographic views, you must examine all the views to determine the meaning of the lines. If you were to look at only the front and top views in Figure 4.46, you might believe line AB is the edge view of a flat surface. From the right-side view, you can see that there is a curved surface on top of the object.

If you look at only the front and side views, you might believe the vertical line CD is the edge view of a plane surface. The top view reveals that the line actually represents the intersection of an inclined surface.

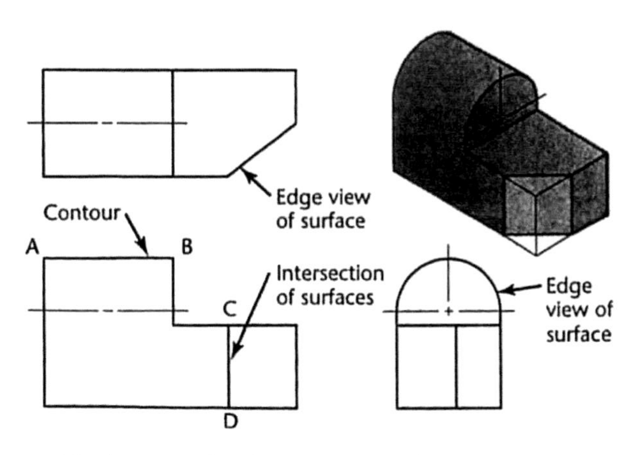

4.47 Interpreting Lines

4.19 SIMILAR SHAPES OF SURFACES

If a flat surface is viewed from several different positions, each view will show the same number of sides and a similar shape. This consistency of shapes is useful in analyzing views. For example, the L-shaped surface shown in Figure 4.48 appears L-shaped in every view in which it does not appear as a line. A surface will have the same number of sides and vertices and the same characteristic shape whenever it appears as a surface. Note how the U-shaped, hexagonal, and T-shaped surfaces in Figure 4.49 are recognizable in different views.

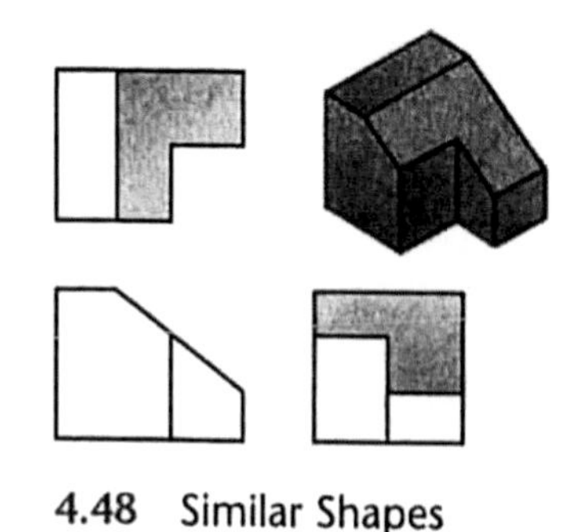

4.48 Similar Shapes

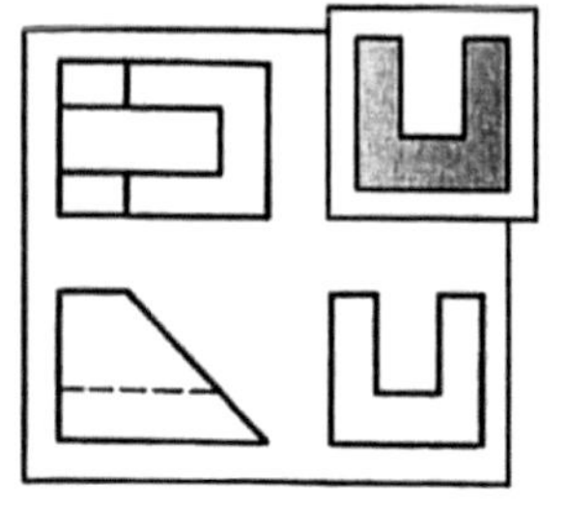

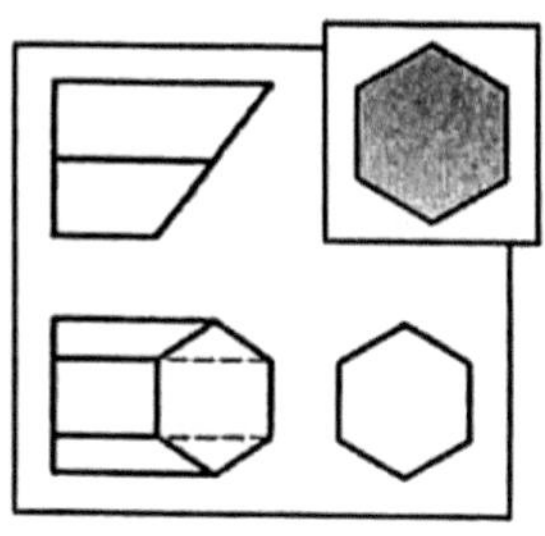

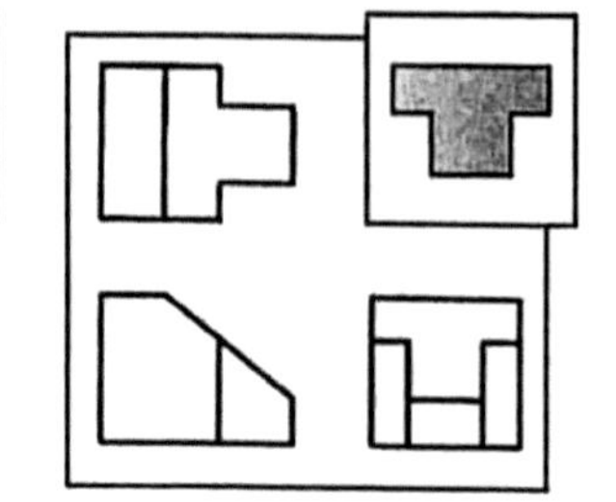

4.49 Similar Shapes

PRACTICE VISUALIZING

Look at the top view (a) and then examine some of the various objects it could represent. As you practice interpreting views, you will get better at visualizing three dimensional objects from projected views.

Notice that the top view alone does not provide all the information, but it does tell you that surfaces a, b and c are not in the same plane. There are many possibilities beyond those shown.

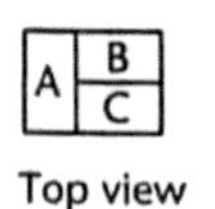

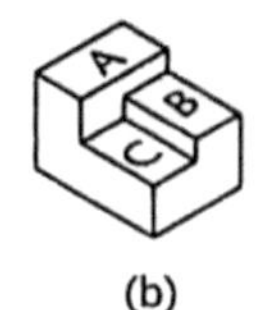

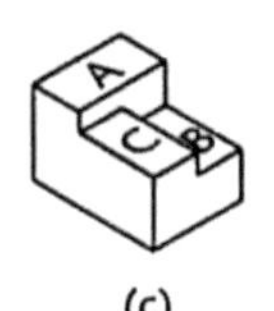

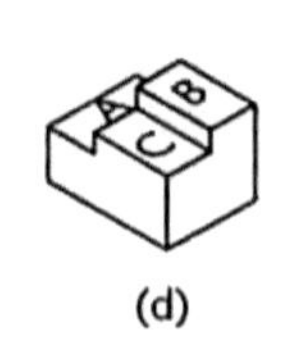

4.20 INTERPRETING VIEWS

One method of interpreting sketches is to reverse the mental process used in projecting them. The views of an angle bracket are shown in Figure 4.50a.

The front view (Figure 4.50b) shows the object's L-shape, its height and width, and the thickness of its members. The meanings of the hidden lines and centerlines are not yet clear, and you do not know the object's depth.

The top view (Figure 4.50c) shows the depth and width of the object. It also makes it clear that the horizontal feature is rounded at the right end and has a round hole. A hidden line at the left end indicates some kind of slot.

The right-side view (Figure 4.50d) shows the height and depth of the object. It reveals that the left end of the object has rounded corners at the top and clarifies that the hidden line in the front view represents an open-end slot in a vertical position.

Each view provides certain definite information about the shape of the object, and all are necessary to visualize it completely.

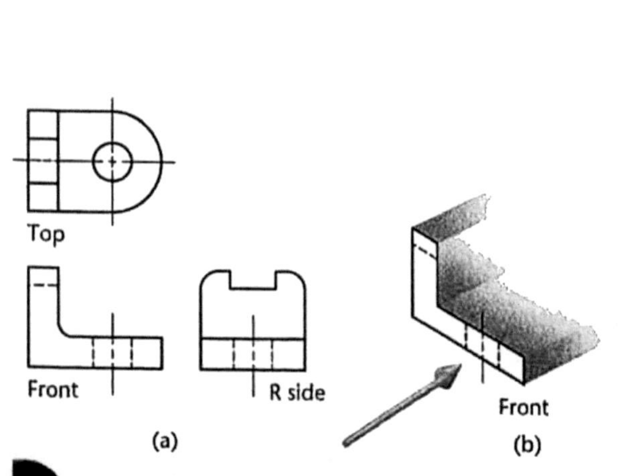

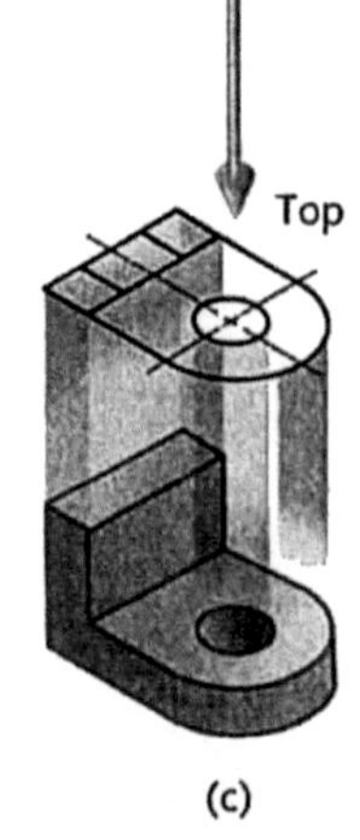

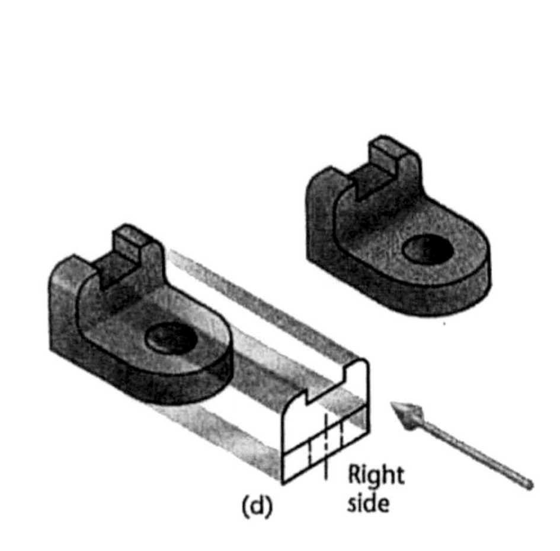

4.50 Visualizing from Given Views

READING A DRAWING

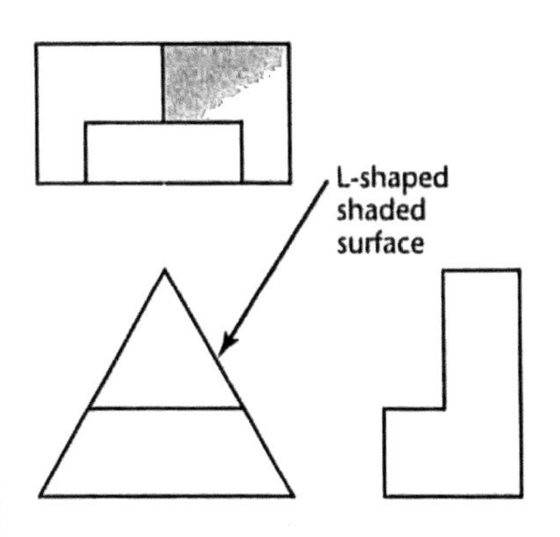

1 Visualize the object shown by the three views at left. Since no lines are curved, we know that the object is made up of plane surfaces.

The shaded surface in the top view is a six-sided L-shape. Since you do not see its shape in the front view—and every surface either appears as its shape or as a line—it must be showing on edge as a line in the front view. The indicated line in the front view also projects to line up with the vertices of the L-shaped surface.

Because we see its shape in the top view and because it is an angled line in the front view, it must be an inclined surface on the object. This means it will show its foreshortened shape in the side view as well, appearing L-shaped and six-sided. The L-shaped surface in the right-side view must be the same surface that was shaded in the top view.

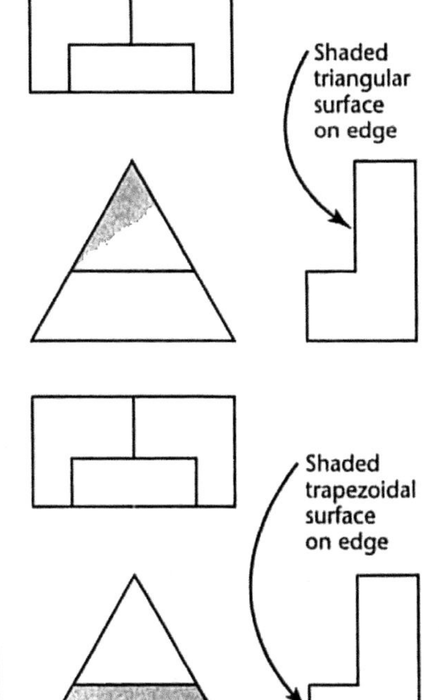

2 In the front view we see the top portion as a triangular-shaped surface, but no triangular shapes appear in either the top or the side view. The triangular surface must appear as a line in the top view and in the side view.

Sketch projection lines from the vertices of the surface where you see its shape. The same surface in the other views must line up along the projection lines. In the side view, it must be the line indicated. That can help you to identify it as the middle horizontal line in the top view.

3 The trapezoidal-shaped surface shaded in the front view is easy to identify, but there are no trapezoids in the top and side views. Again the surface must be on edge in the adjacent views.

4 On your own, identify the remaining surfaces using the same reasoning. Which surfaces are inclined, and which are normal? Are there any oblique surfaces?

If you are still having trouble visualizing the object, try picturing the views as describing those portions of a block that will be cut away, as illustrated below.

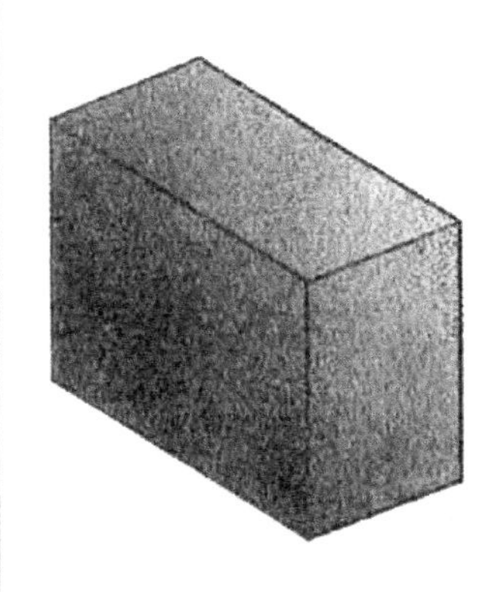

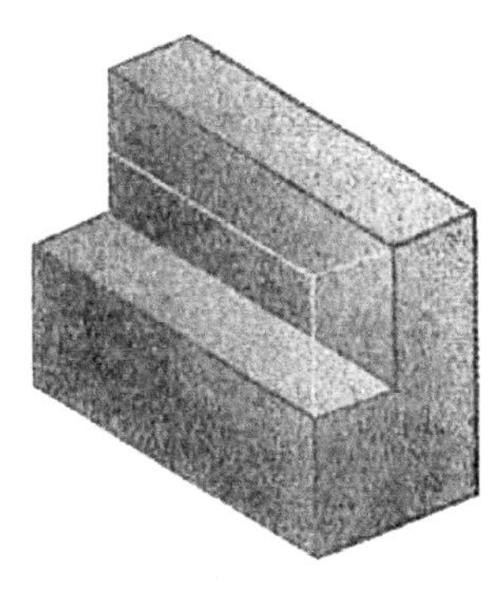

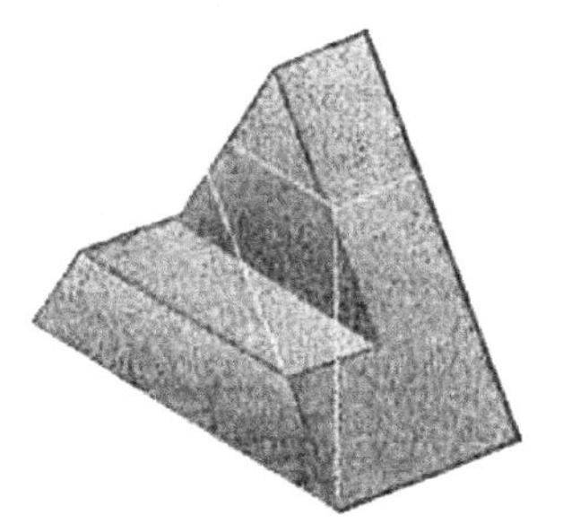

4.21 MODELS

One of the best aids to visualization is an actual model of the object. Models don't necessarily need to be made accurately or to scale. They may be made of any convenient material, such as modeling clay, soap, wood, wire, or Styrofoam, or any material that can easily be shaped, carved, or cut. Some examples of soap models are shown in Figure 4.51.

Rules for Visualizing from a Drawing: Putting It All Together

Reading a multiview drawing is like unraveling a puzzle. When you interpret a drawing, keep these things in mind:

- The closest surface to your view must have at least one edge showing as a visible line.
- A plane surface has a similar shape in any view or appears on edge as a straight line.
- Lines of the drawing represent either an intersection between two surfaces, a surface perpendicular to your view that appears "on edge," or the limiting element of a curved surface.
- No two adjacent areas divided by a visible line in an orthographic view can lie on the same plane in the actual object. Areas not adjacent in a view may lie in the same plane on the object.
- If a line appears hidden, a closer surface is hiding it.
- Your interpretation must account for all of the lines of the drawing. Every line has a meaning.

TIP

Making a Model

Try making a soap or clay model from projected views:

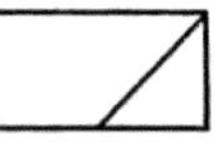

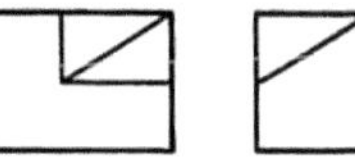

First, look at the three views of the object. Make your block of clay to the same principal dimensions (height, width, and depth) as shown in the views.

Score lines on the frontal surface of your clay block to correspond with those shown on the front view in the drawing. Then do the same for the top and right-side views.

Slice straight along each line scored on the clay block to get a 3D model that represents the projected views.

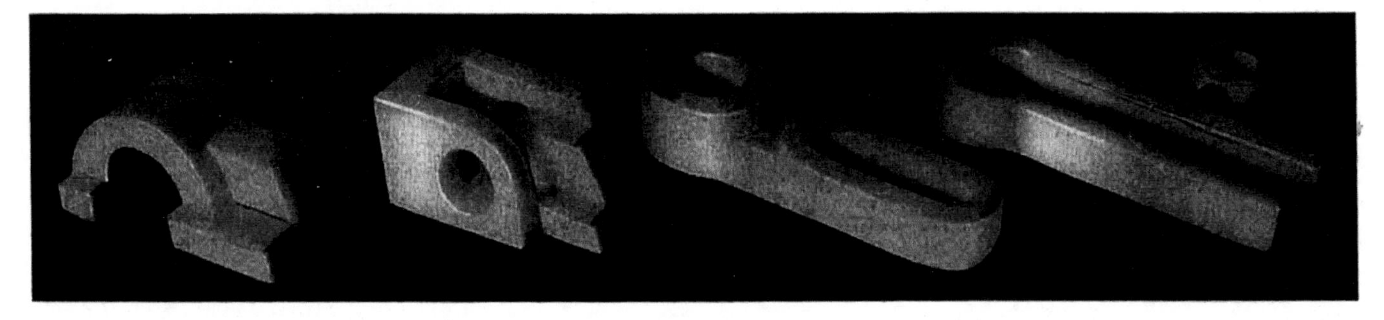

4.51 Soap Models

4.22 PROJECTING A THIRD VIEW

Ordinarily when you are designing a product or system, you have a good mental picture of what the object you are sketching will look like from different directions. However, skill in projecting a third view can be useful for two reasons. First, views must be shown in alignment in the drawing and projected correctly. Secondly, practice in projecting a third view from two given views is an excellent way to develop your visual abilities.

Numbering the vertices on the object makes projecting a third view easy. Points that you number on the drawing represent points on the object where three surfaces come together to form a vertex (and sometimes a point on a contour or the center of a curve).

Once you have located a point in two drawing views, its location in the third view is known. In other words, if a point is located in the front and top view, its location in the side view is a matter of projecting the height of the point in the glass box from the front view and the depth of the point in the glass box from the top view.

In order to number the points or vertices on the object and show those numbers in different views, you need to be able to identify surfaces on the object. Then project (or find) the points in each new view, surface by surface. You can use what you know about edges and surfaces to identify surfaces on the object when you draw views. This will help you to interpret drawings created by others as well as know how to project your own drawings correctly.

PROJECTING A THIRD VIEW

Follow the steps to project a third view.

The figure below is a pictorial drawing of an object to be shown in three views. It has numbers identifying each corner (vertex) and letters identifying some of the major surfaces. You are given the top and front view. You will use point numbers to project the side view.

1 To number points effectively, first identify surfaces and interpret the views that are given. Start by labeling visible surfaces whose shapes are easy to identify in one view. Then locate the same surface in the adjacent view. (The surfaces on the pictorial object have been labeled to make it easier.)

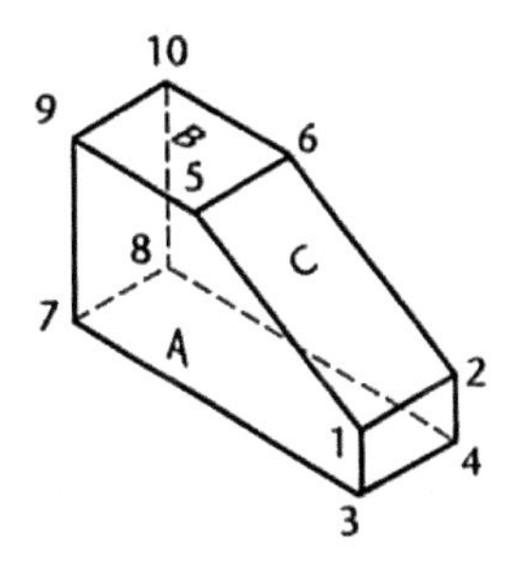

B C D A C A D

2 Surface A in the front view is a normal surface. It will appear as a horizontal line in the top view. The two rectangular surfaces B and C in the top view are a normal surface and an inclined surface. They will show as a horizontal line and an inclined line in the front view, respectively.

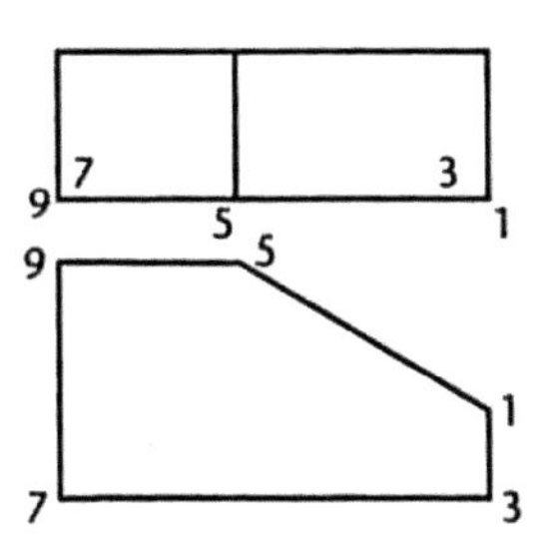

3 After identifying the surfaces, label the vertices of a surface that has an easily recognized shape, in this case, surface A.

Label its vertices with numbers at each corner as shown. If a point is directly visible in the view, place the number outside the corner.

If the point is not directly visible in that view, place the numeral inside the corner. Using the same numbers to identify the same points in different views will help you to project known points in two views to unknown positions in a third view.

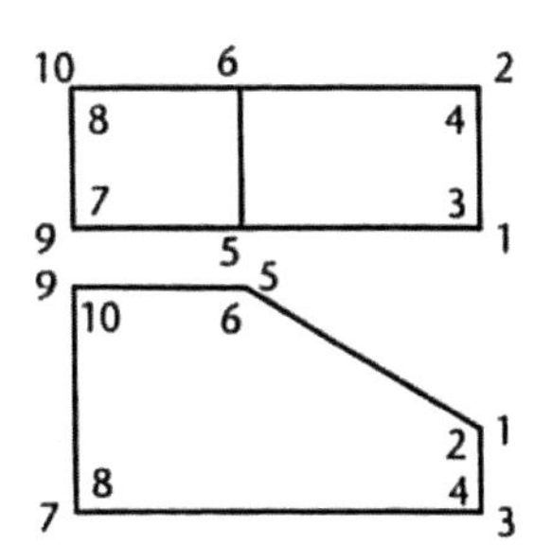

4 Continue on, surface by surface, until you have numbered all of the vertices in the given views as shown. Do not use two different numbers for the same vertex.

5 Try to visualize the right-side view you will create. Then construct the right-side view point by point, using very light lines. Locate point 1 in the side view by drawing a light horizontal projection line from point 1 in the front view. Use the edge view of surface A in the top view as a reference plane to transfer the depth location for point 1 to the side view as shown.

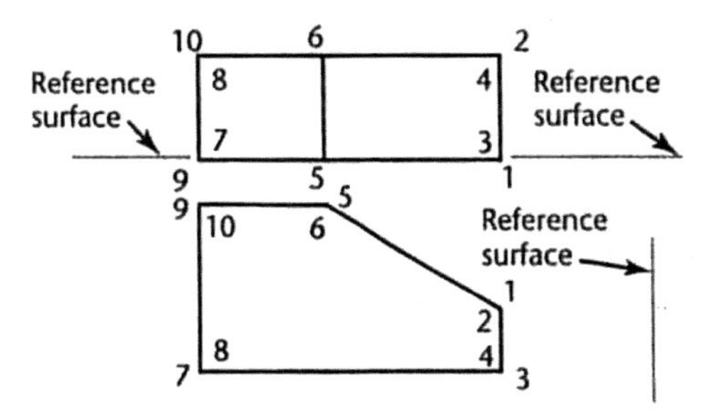

6 Project points 2, 3, and 4 in a similar way to complete the vertical end surface of the object.

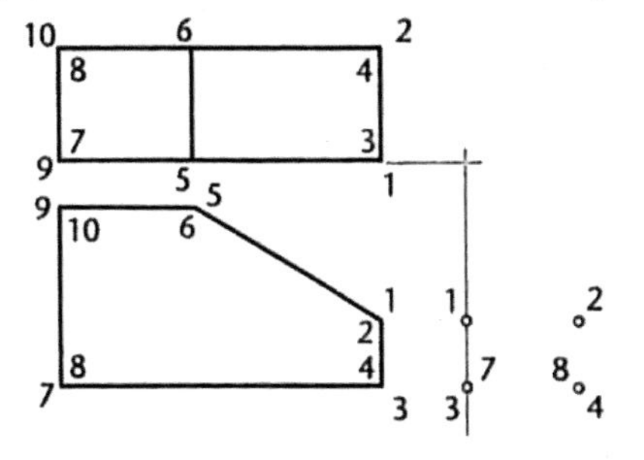

7 Project the remaining points using the same method, proceeding surface by surface.

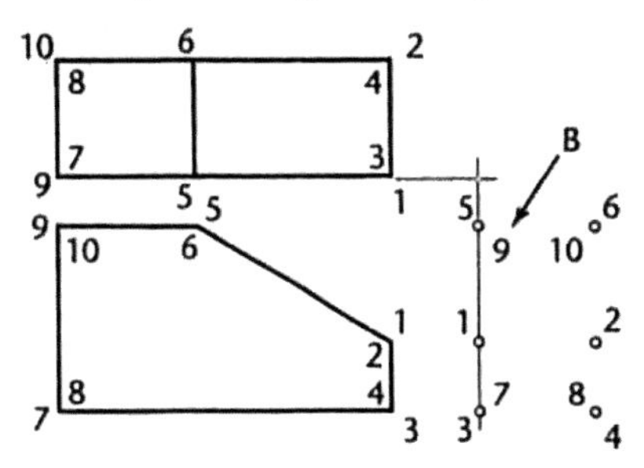

PROJECTING A THIRD VIEW

Continue the steps to project a third view.

8 Use the points that you have projected into the side view to draw the surfaces of the object as in this example.

If surface A extended between points 1-3-7-9-5 in the front view where you can see its shape clearly, it will extend between those same points in every other view.

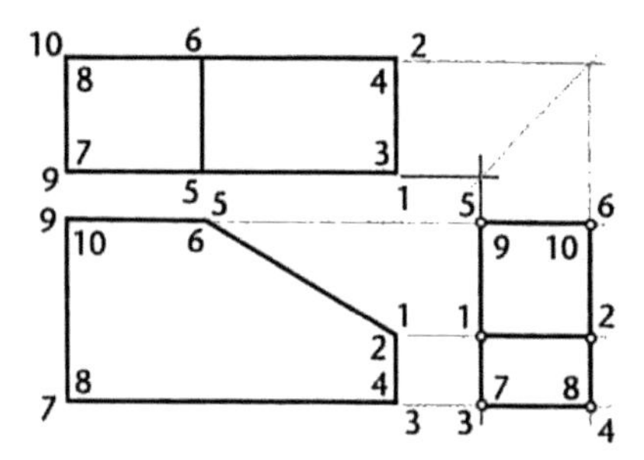

When you connect these points in the side view, they form a vertical line.

This makes sense, because A is a normal surface. As is the rule with normal surfaces, you will see its shape in one standard view (the front in this case) and it will appear as a horizontal or vertical line in the other views.

Continue connecting vertices to define the surfaces on the object, to complete the third view.

9 Inspect your drawing to see if all of the surfaces are shown and darken the final lines.

Consider the visibility of surfaces. Surfaces that are hidden behind other surfaces should be shown with hidden lines.

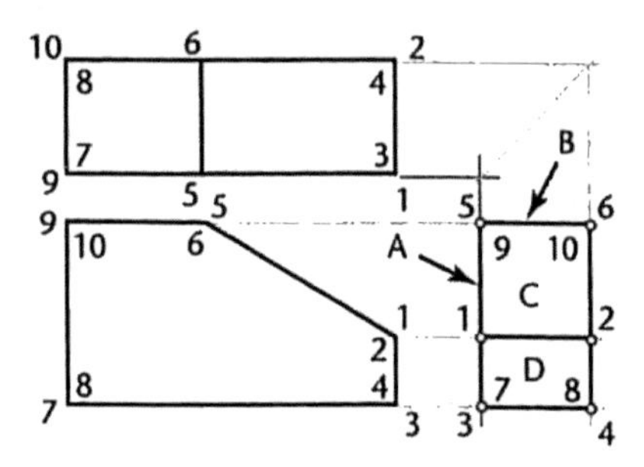

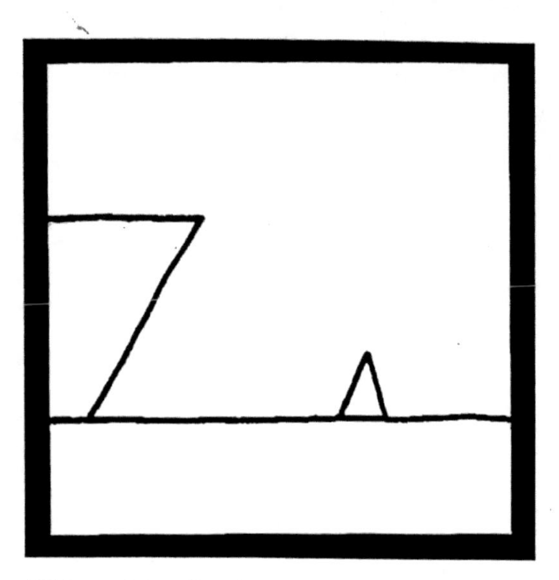

"Ship Arriving Too Late to Save Drowning Witch." This well-known drawing by artist Roger Price is an example of how a single orthographic view can be difficult to interpret. *Courtesy of "Droodles, The Classic Collection."*

4.23 BECOMING A 3D VISUALIZER

To the untrained person, orthographic projections might not convey the idea of a 3D shape, but with some practice you should now be able to look at projected front, top, and right-side views and envision that they represent the width, depth, and height of an object. Understanding how points, lines, and surfaces can be interpreted and how normal, inclined, or oblique surfaces appear from different views helps you interpret orthographic views to let you form a mental image of the 3D object they represent.

Having an understanding of how orthographic views represent an object gives you the power to start capturing your own 3D concepts on paper in a way that others can accurately interpret. Keep in mind the idea of an unfolded "glass box" to explain the arrangement of views. This clarifies how the views relate to one another and why you can transfer certain dimensions from adjacent views. Using standard practices to represent hidden lines and centerlines helps you further define surfaces, features, and paths of motion.

The better you understand the foundation concepts of projected views, the more fluent you will be in the language of 3D representation and the skill of spatial thinking, regardless of whether you sketch by hand or use CAD.

USING A MITER LINE

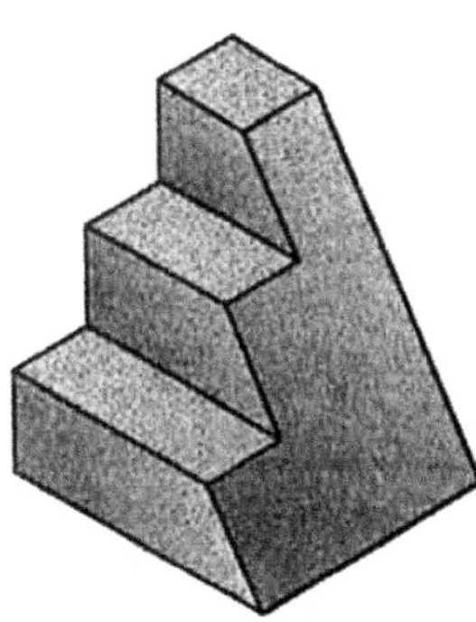

Given two completed views you can use a miter line to transfer the depths and draw the side view of the object shown at left.

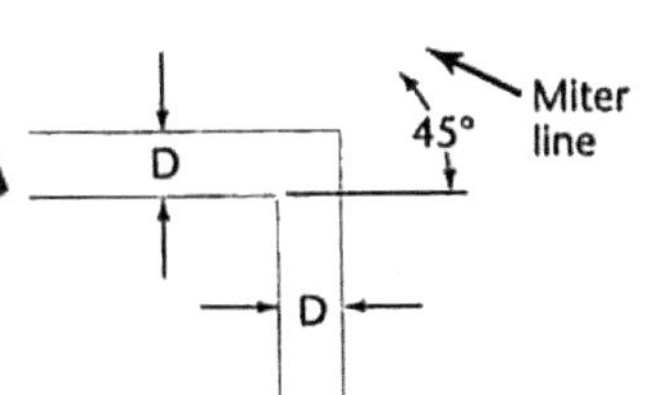

1 Locate the miter line a convenient distance away from the object to produce the desired spacing between views.

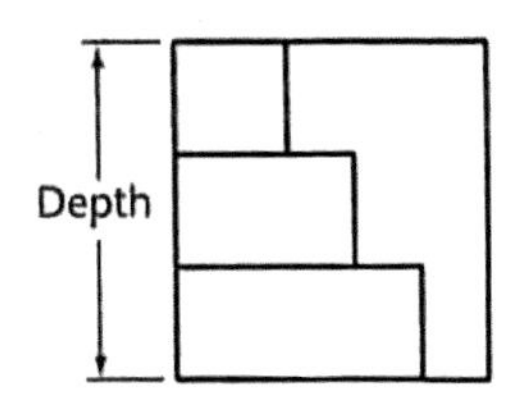

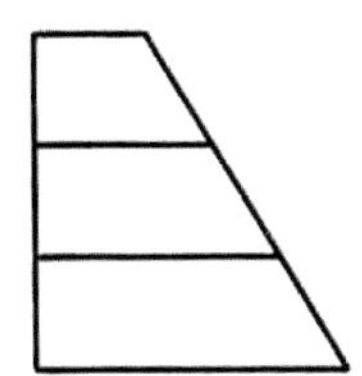

2 Sketch light lines projecting depth locations for points to the miter line and then down into side view as shown.

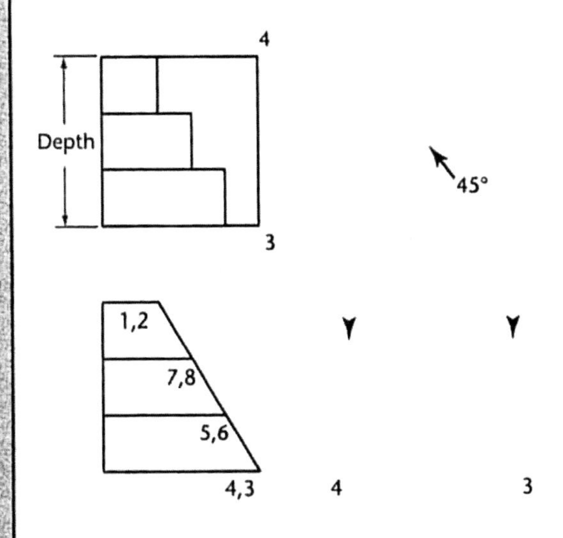

3 Project the remaining points.

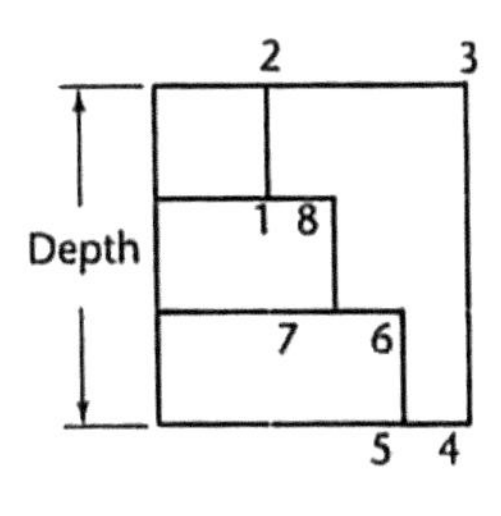

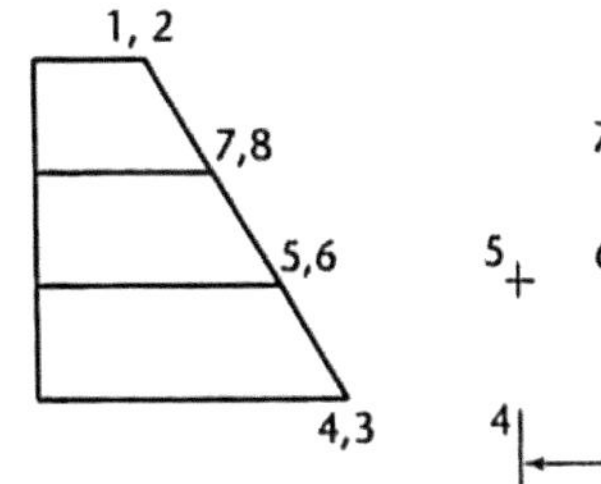

4 Draw view locating each vertex of surface on the projection line and the miter line. To move the right-side view to the right or left, move the top view upward or downward by moving the miter line closer to or farther from the view. You don't need to draw continuous lines between the top and side views via the miter line. Instead, make short dashes across the miter line and project from these. The 45° miter-line method is also convenient for transferring a large number of points, as when plotting a curve.

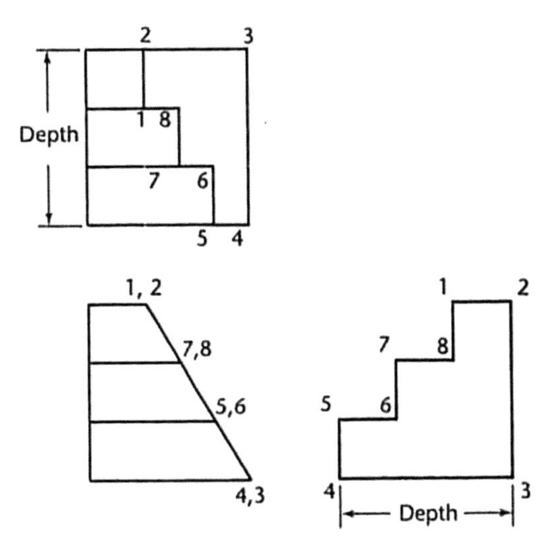

PLACING VIEWS FROM A 3D MODEL

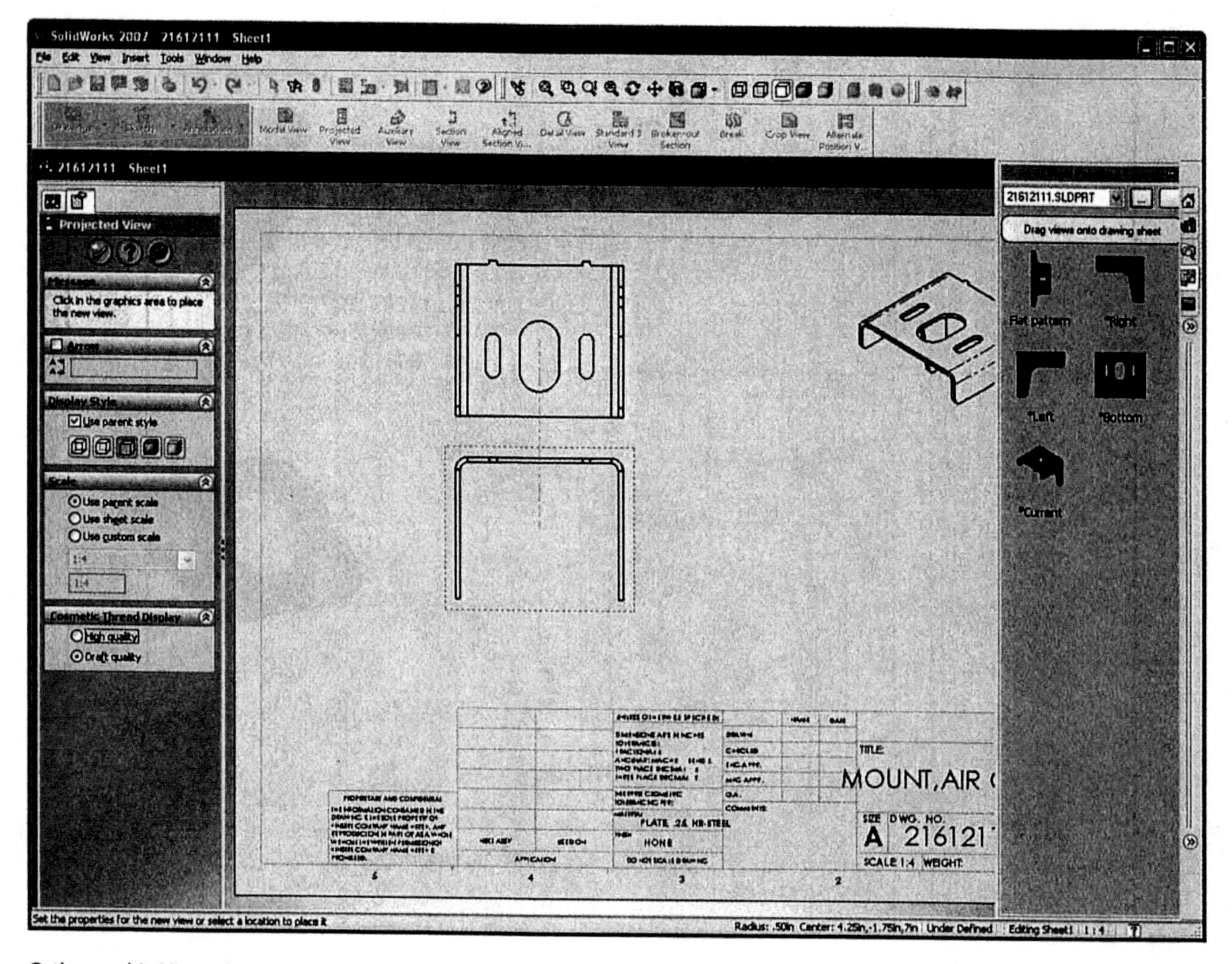

Orthographic Views Generated from a CAD Model. *Courtesy of Solidworks Corporation.*

Once a 3D model is created, most CAD packages allow you to place orthographic views generated from the model. To place a projected view is as easy as selecting the base view and then choosing where to place the projected view. You can also usually turn off hidden lines in each individual view based on whether or not they add useful information. Using CAD to place the 3D views also makes it easy to show views in alignment.

Most 3D CAD software allows you to configure it to show the views in either third-angle or first-angle projection.

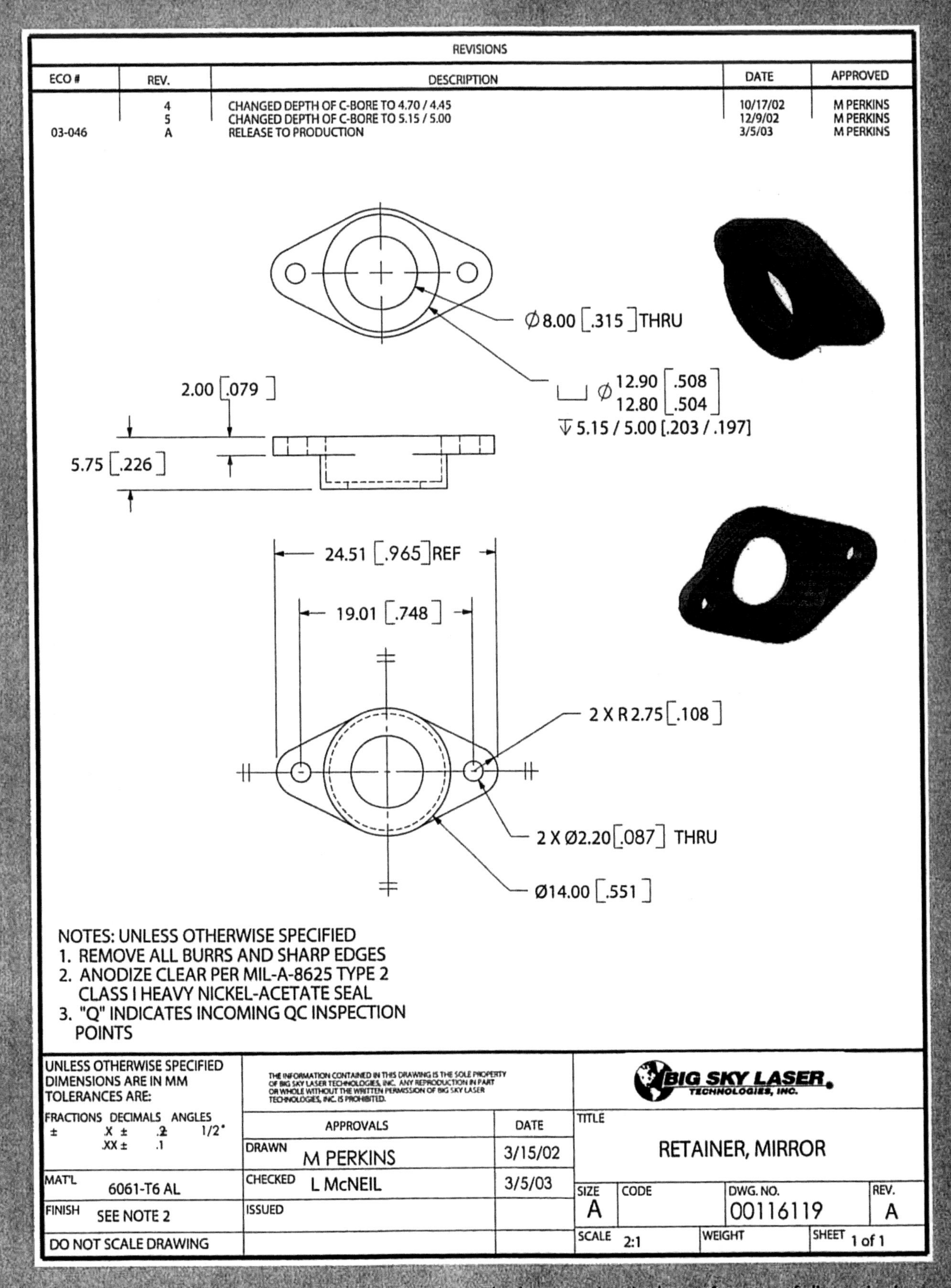

Top, Front and Bottom Views of a Mirror Retainer. The bottom view is shown for ease of dimensioning. *Courtesy of Big Sky Laser.*

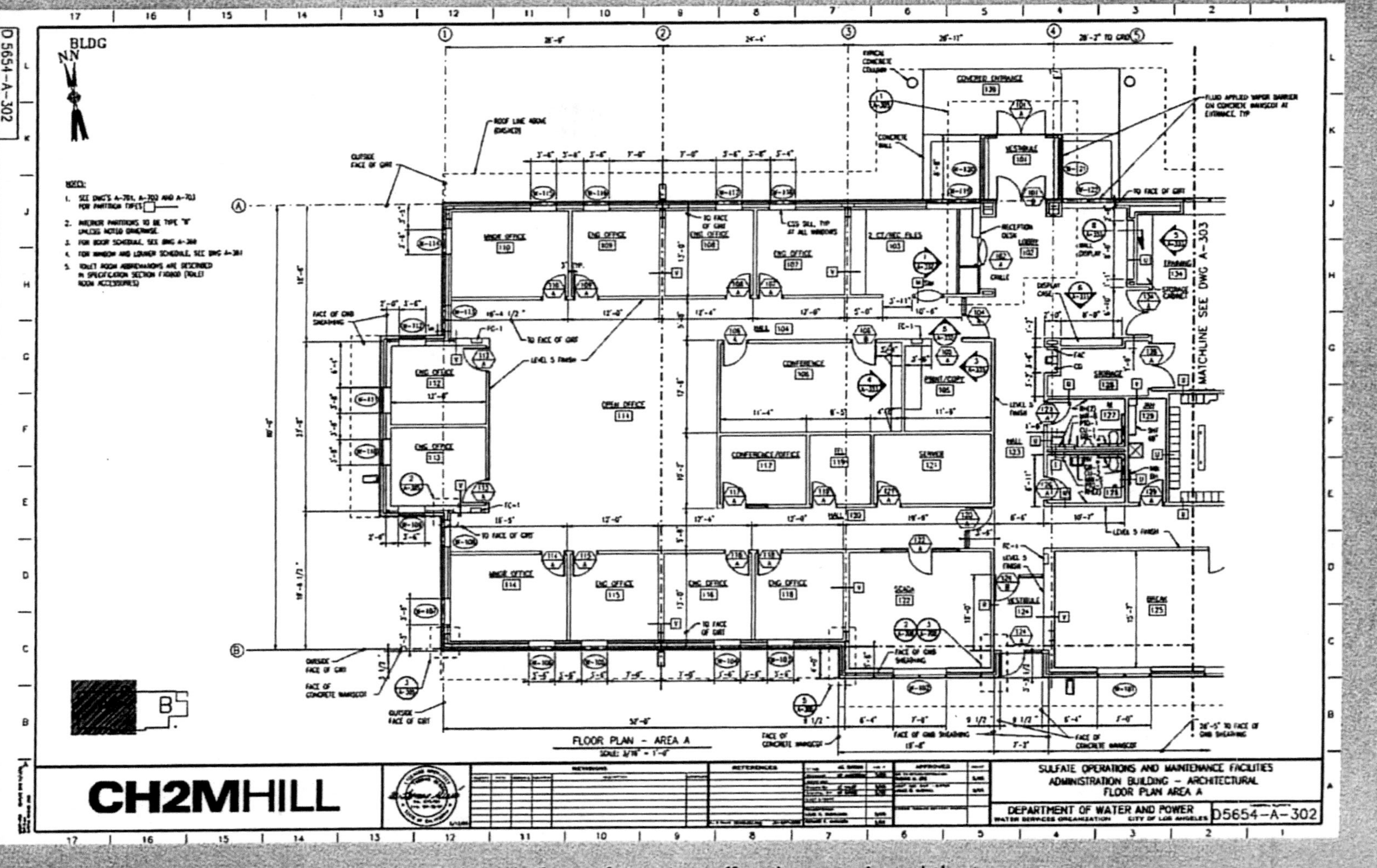

Architectural floor plans show the building as though the roof were cut off and you projected the top orthographic view. *Courtesy of CH2M HILL.*

KEY WORDS

- Multiview Projection
- Principal Views
- Width
- Height
- Depth
- Plane of Projection
- Orthographic
- Frontal Plane
- Horizontal Plane
- Profile Plane
- Glass Box
- Folding Lines
- Three Regular Views
- Necessary Views
- Third-Angle Projection
- First-Angle Projection
- Projection Symbols
- Surfaces
- Plane
- Normal Surface
- Inclined Surface
- Oblique Surface
- Edge
- Normal Edge
- Inclined Edge
- Oblique Edge
- Point

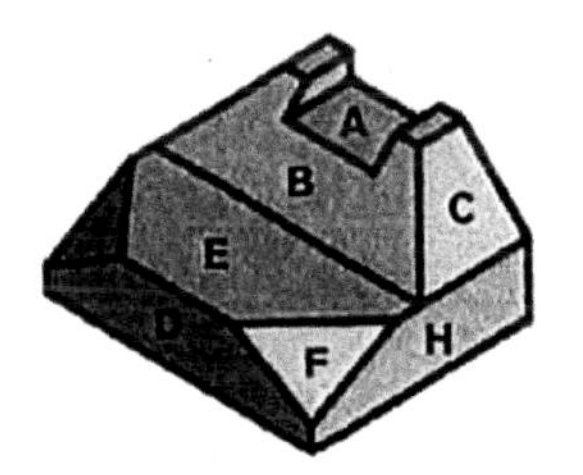

Key to Figure 4.35
Normal Surfaces: A, D, E, H
Inclined Surfaces: B, C
Oblique Surface: F

CHAPTER SUMMARY

- Orthographic drawings are the result of projecting the image of a 3D object onto one of six standard planes of projection. The six standard views are often thought of as an unfolded glass box. The arrangement of the views in relation to one another is important. Views must project to line up with adjacent views, so that any point in one view projects to line up with that same point in the adjacent view. The standard arrangement of views shows the top, front, and right side of the object.
- Visualization is an important skill for engineers. You can build your visual abilities through practice and through understanding terms describing objects. For example, surfaces can be normal, inclined, or oblique. Normal surfaces appear true size in one principal view and as an edge in the other two principal views. Inclined surfaces appear as an edge view in one of the three principal views. Oblique surfaces do not appear as an edge view in any of the principal views.
- Choice of scale is important for representing objects clearly on the drawing sheet.
- Hidden lines are used to show the intersections of surfaces, surfaces that appear on edge, and the limits of curved surfaces that are hidden from the viewing direction.
- Centerlines are used to show the axis of symmetry for features and paths of motion, and to indicate the arrangement for circular patterns.
- Creating CAD drawings involves applying the same concepts as paper drawing. The main difference is that drawing geometry is stored more accurately using a computer than in any hand drawing. CAD drawing geometry can be reused in many ways and plotted to any scale as necessary.

REVIEW QUESTIONS

1. Sketch the symbol for third-angle projection.
2. List the six principal views of projection.
3. Sketch the top, front, and right-side views of an object of your design having normal, inclined, and oblique surfaces.
4. In a drawing that shows the top, front, and right-side view, which two views show depth? Which view shows depth vertically on the sheet? Which view shows depth horizontally on the drawing sheet?
5. What is the definition of a normal surface? An inclined surface? An oblique surface?
6. What are three similarities between using a CAD program to create 2D drawing geometry and sketching on a sheet of paper? What are three differences?
7. What dimensions are the same between the top and front view: width, height, or depth? Between the front and right-side view? Between the top and right-side view?
8. List two ways of transferring depth between the top and right-side views.
9. If surface A contained corners 1, 2, 3, 4, and surface B contained corners 3, 4, 5, 6, what is the name of the line where surfaces A and B intersect?

MULTIVIEW PROJECTION EXERCISES

The following projects are intended to be sketched freehand on graph paper or plain paper. Sheet layouts such as A-1, found in the back of this book, are suggested, but your instructor may prefer a different sheet size or arrangement. Use metric or decimal inch as assigned. The marks shown on some exercises indicate rough units of either 1/2" and 1/4" (or 10 mm and 5 mm). All holes are through holes. If dimensions are required, study Chapter 9. Use metric or decimal inch dimensions if assigned by the instructor.

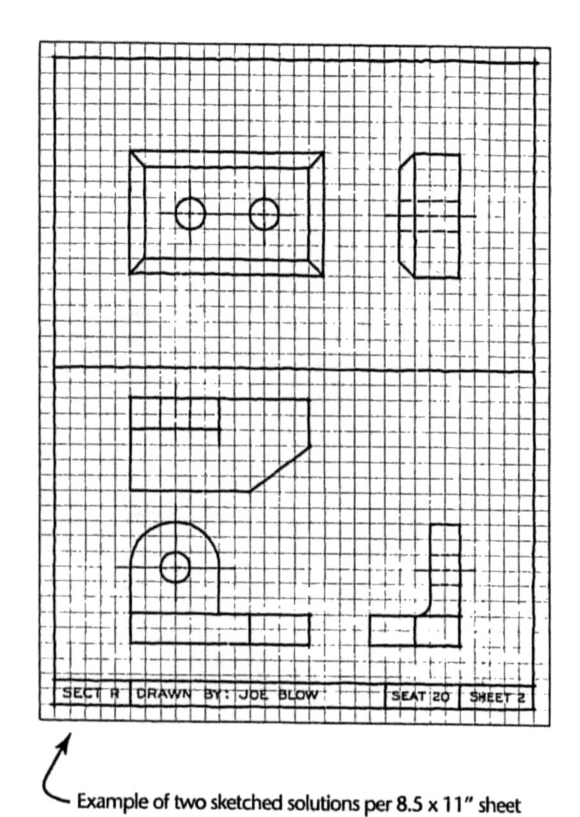

Example of two sketched solutions per 8.5 x 11" sheet

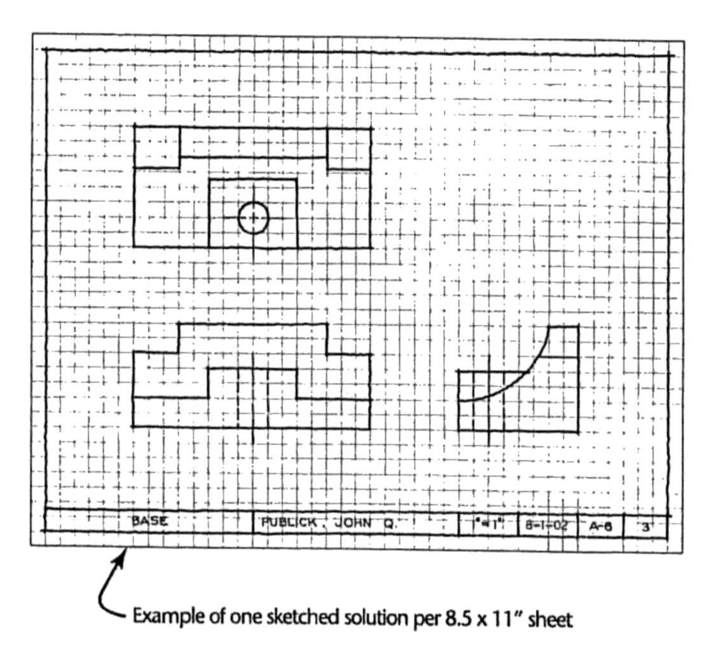

Example of one sketched solution per 8.5 x 11" sheet

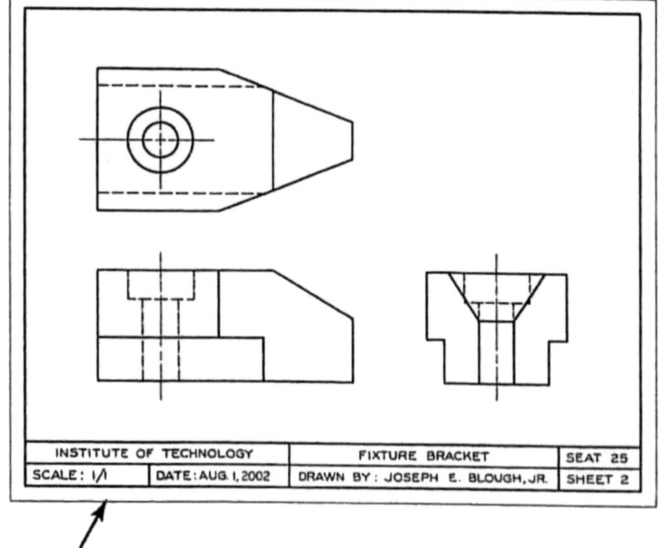

Example of one CAD/instrument solution per 8.5 x 11" sheet

Example Exercise

EXERCISES

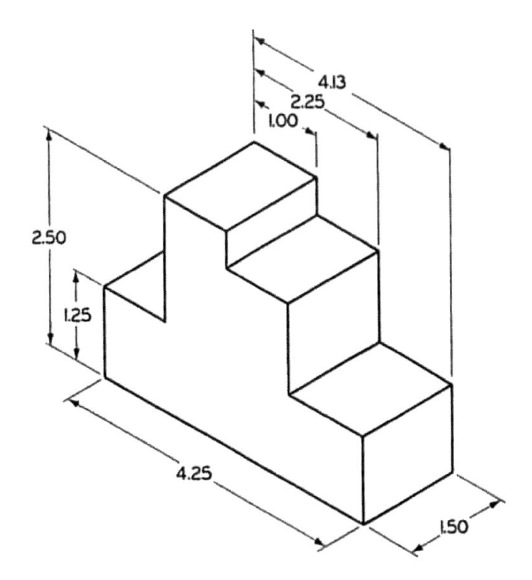

Exercise 4.1 Spacer. Draw and sketch all necessary views.

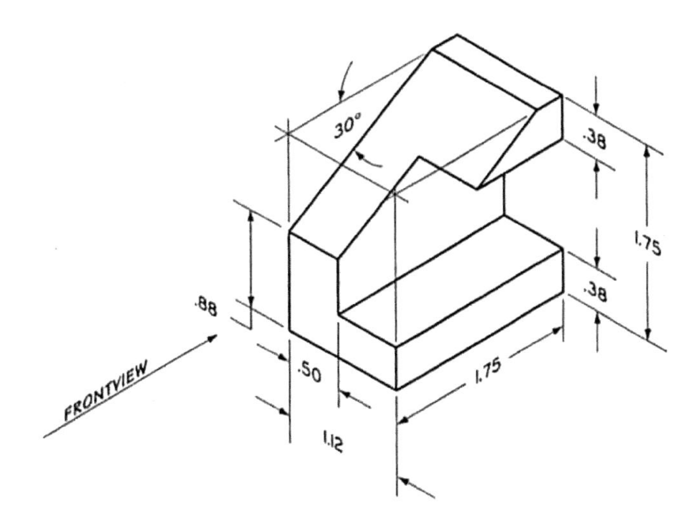

Exercise 4.2 Slide. Draw and sketch all necessary views.

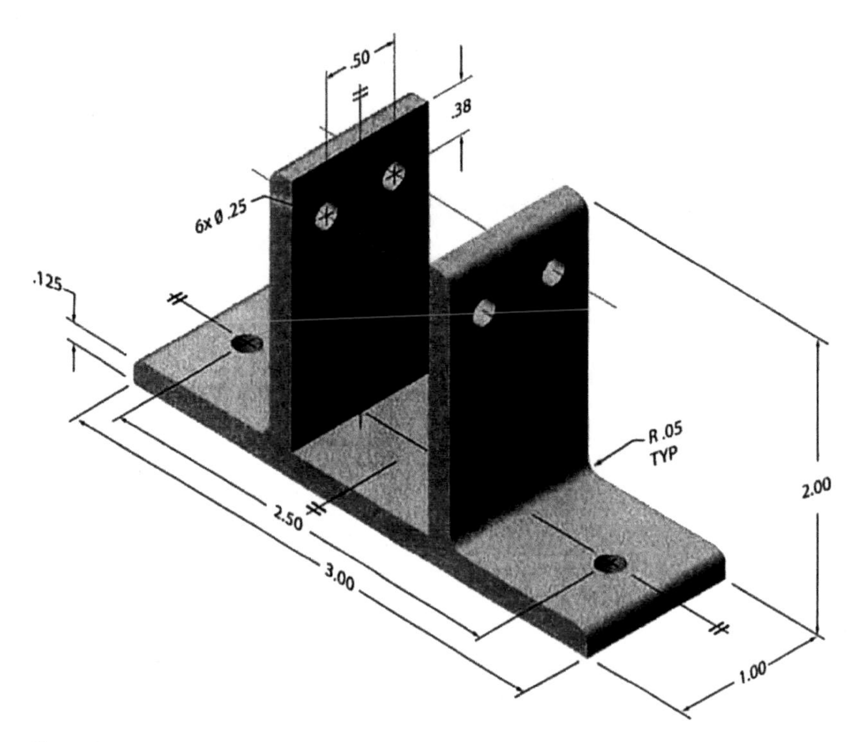

Exercise 4.3 Wall bracket
Create a drawing with the necessary orthographic views for the wall bracket.

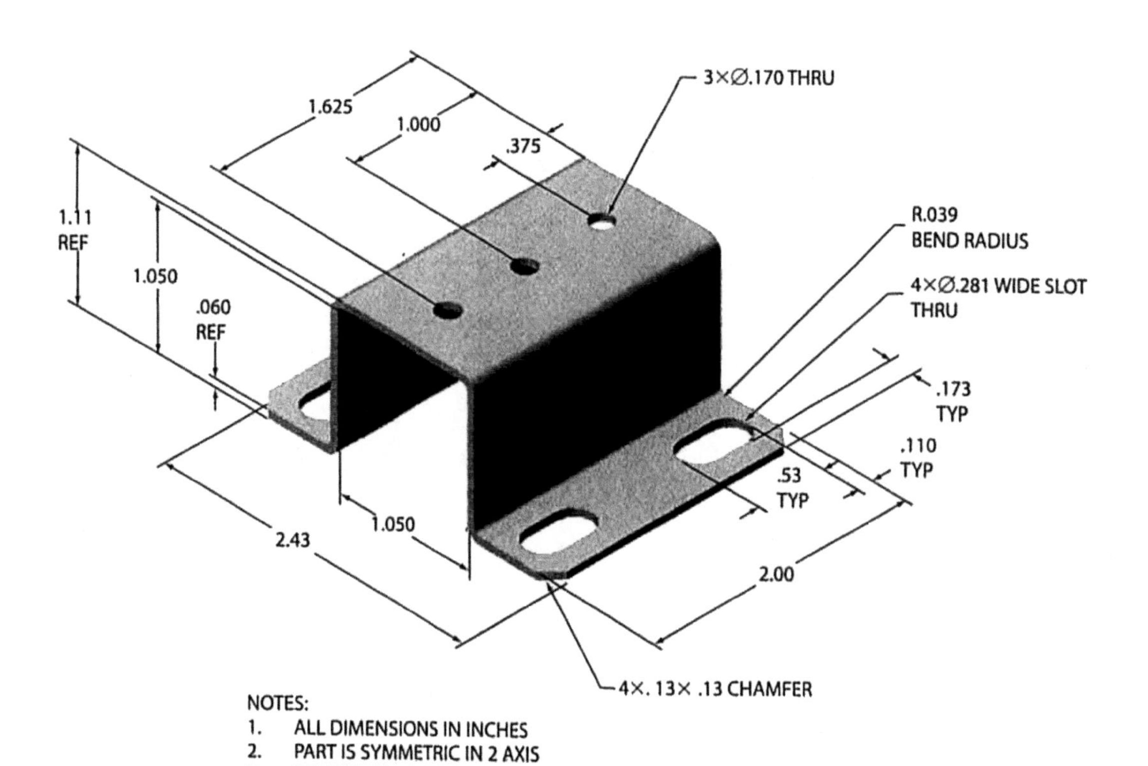

Exercise 4.4 Sheet metal bracket
Create a drawing of the necessary orthographic views for the sheet metal bracket.

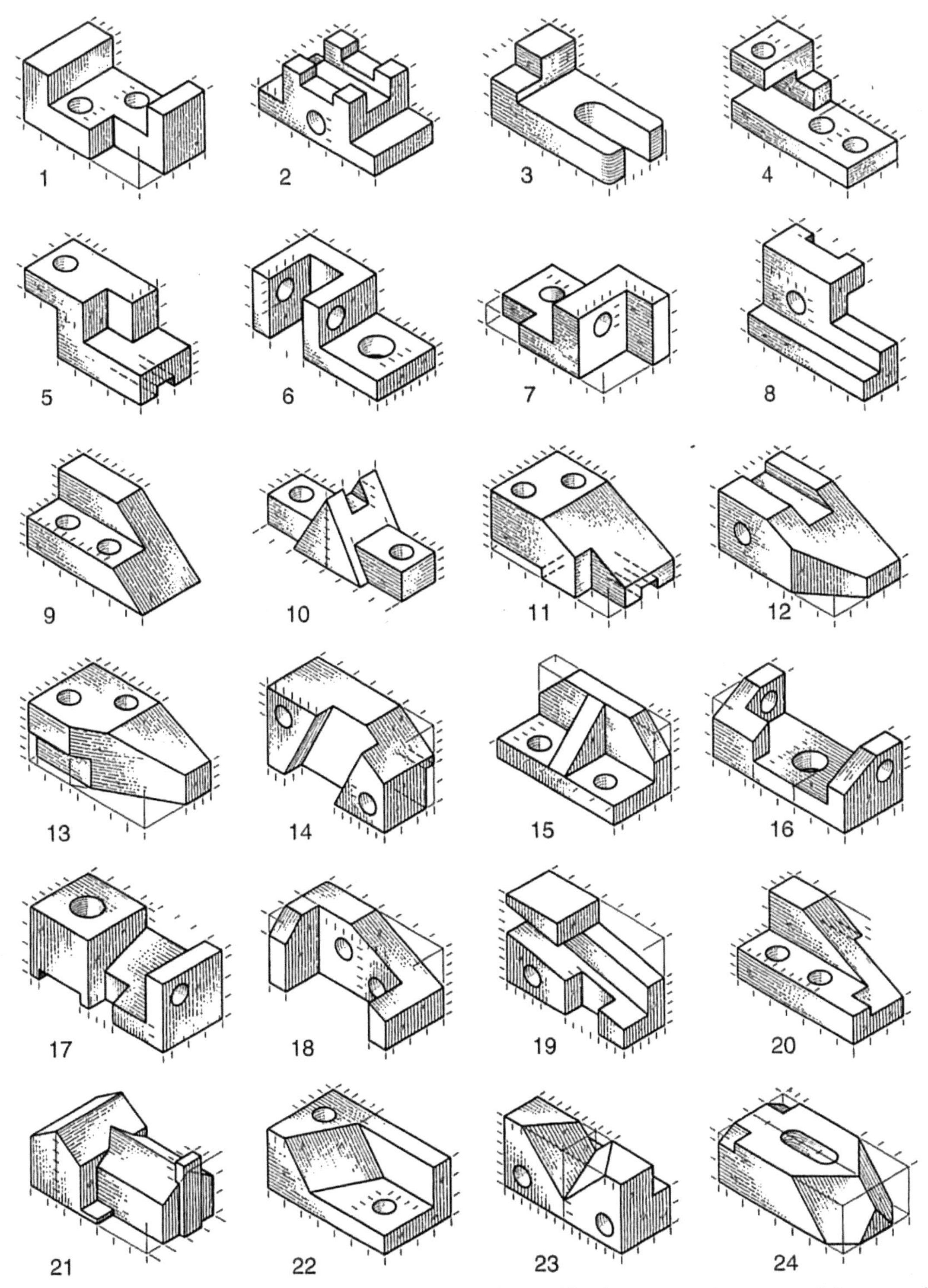

Exercise 4.5 Multiview Sketching Problems. Sketch necessary orthographic views on graph paper or plain paper, showing either one or two problems per sheet as assigned by your instructor. These exercises are designed to fit on 8½ × 11" size A, or metric A4 paper. The units shown may be either .500" and .250" or 10 mm and 5 mm. All holes are through holes.

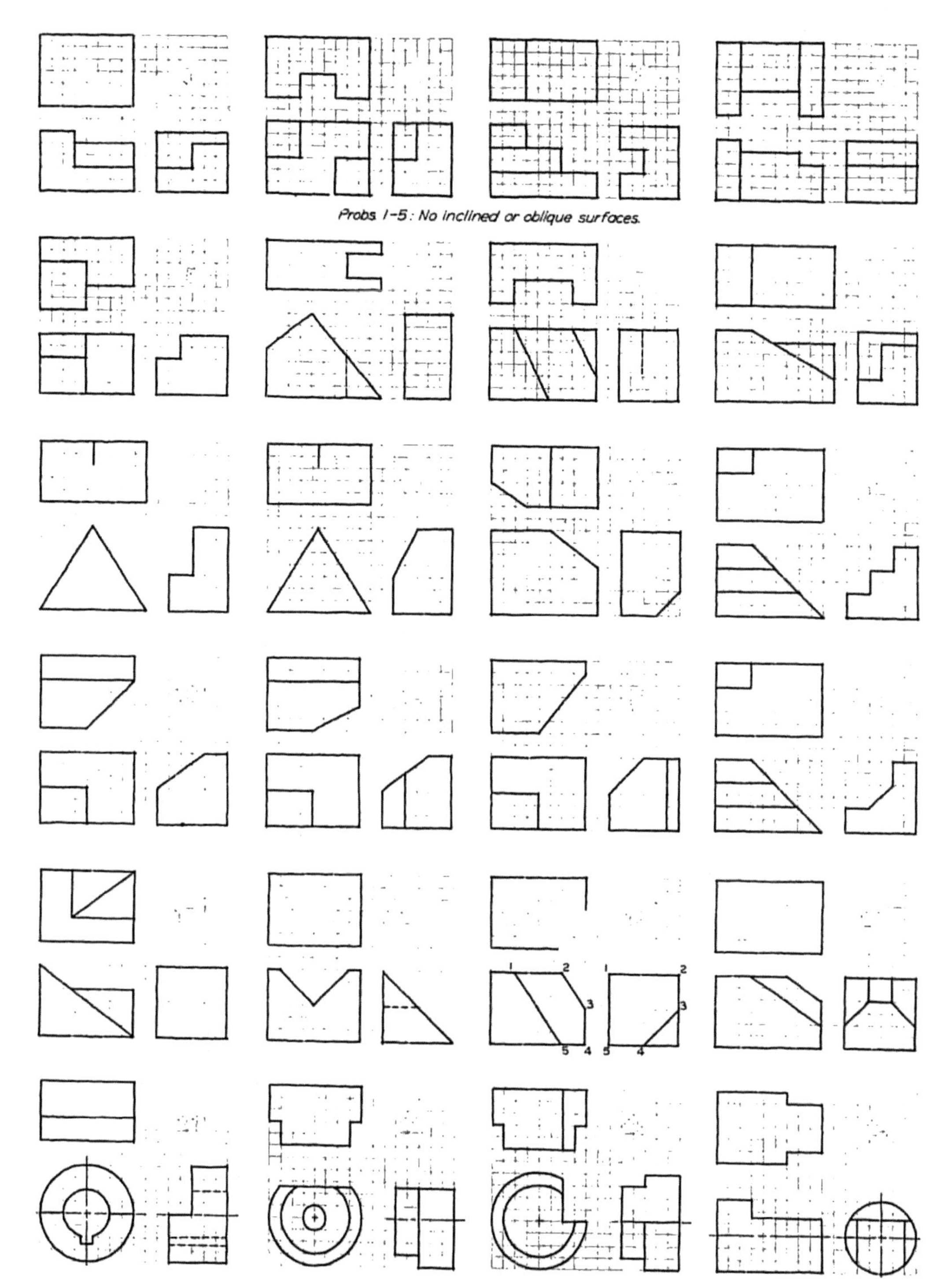

Exercise 4.6 Missing-Line Sketching Problems. (1) Sketch given views on graph paper or plain paper showing either one or two problems per sheet as assigned by your instructor. These exercises are designed to fit on 8½ × 11" size A or metric A4 paper. Add missing lines. The squares may be either .250" or 5 mm. See instructions on page 150. (2) Sketch in isometric on isometric paper or in oblique on cross-section paper, if assigned.

Exercise 4.7 Third-View Sketching Problems. Sketch the given views and add the missing views as indicated on graph paper or plain paper. These exercises are designed to fit on 8½ × 11" size A or metric A4 paper. The squares may be either .25" or 5 mm. The given views are either front and right-side views or front and top views. Hidden holes with centerlines are drilled holes.

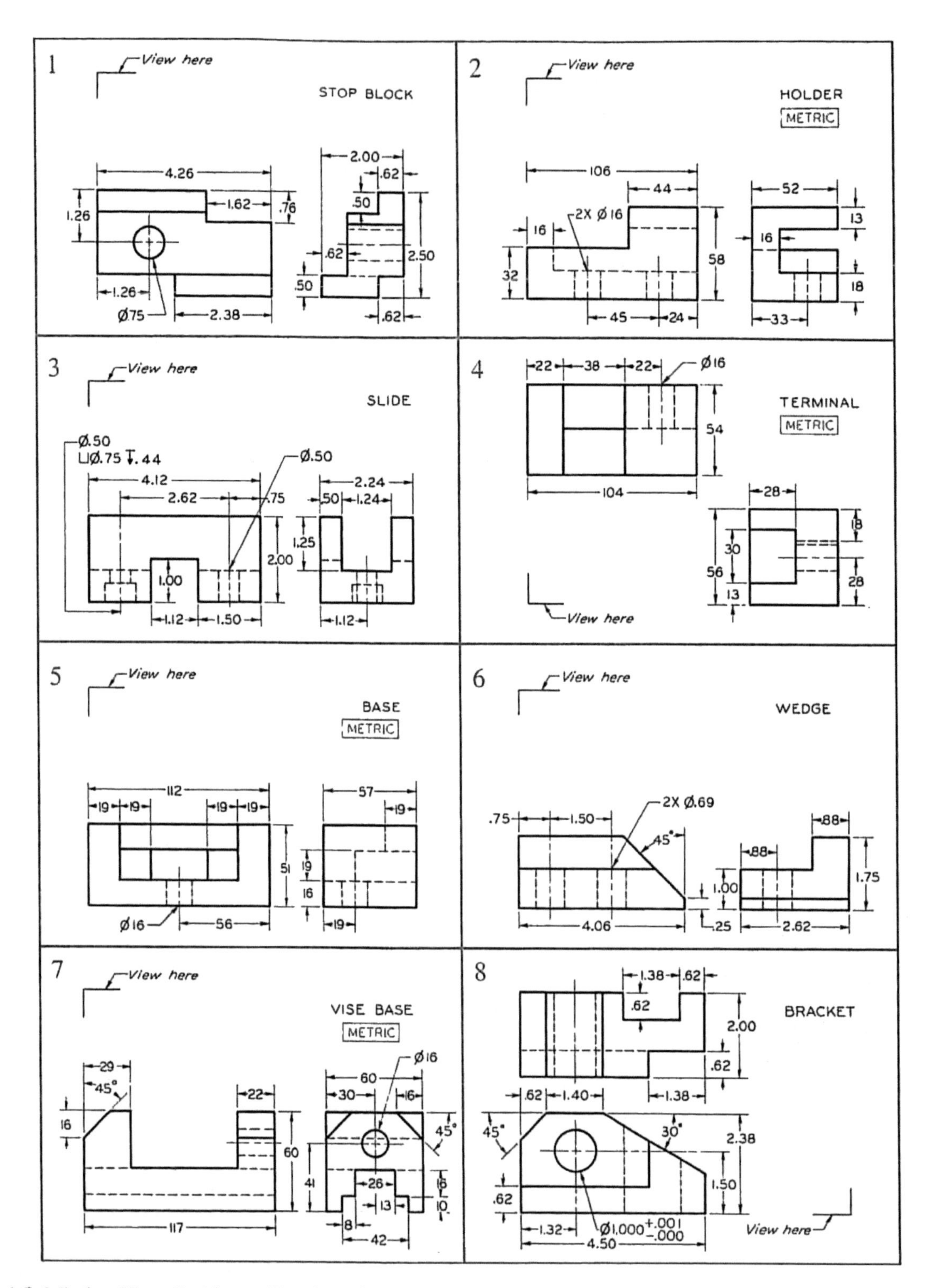

Exercise 4.8 Missing-View Problems. Sketch or draw the given views, and add the missing view. If dimensions are required, study Chapter 9. These exercises are designed to fit on 8½ × 11" size A or metric A4 paper. Use metric or decimal inch dimensions as assigned by the instructor. Move dimensions to better locations where possible. In Exercises 1–5, all surfaces are normal surfaces.

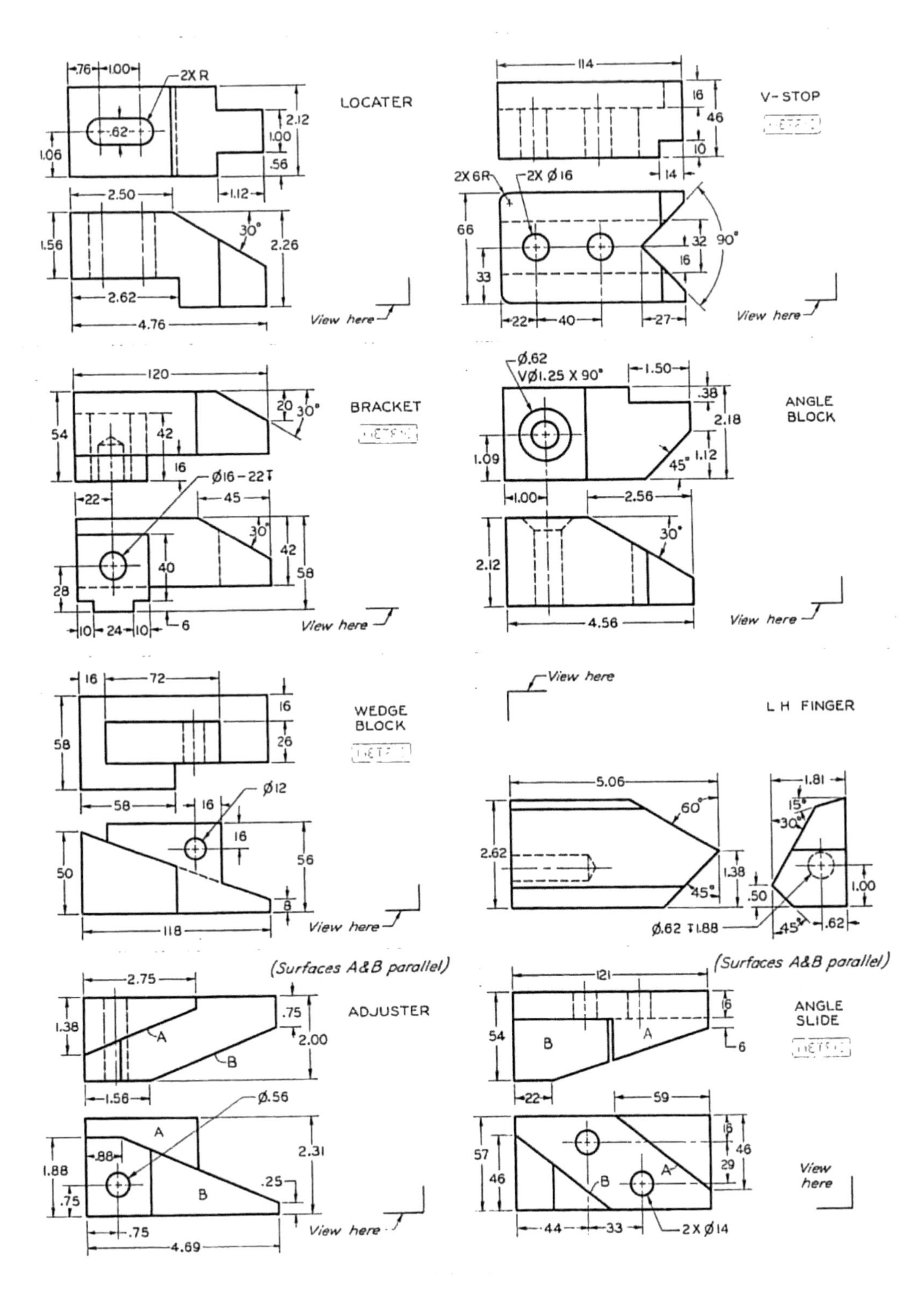

Exercise 4.9 Missing-View Problems. Sketch or draw the given views, and add the missing view. These exercises are designed to fit on 8½ × 11" size A or metric A4 paper. If dimensions are required, study Chapter 9. Use metric or decimal inch dimensions as assigned by the instructor. Move dimensions to better locations where possible.

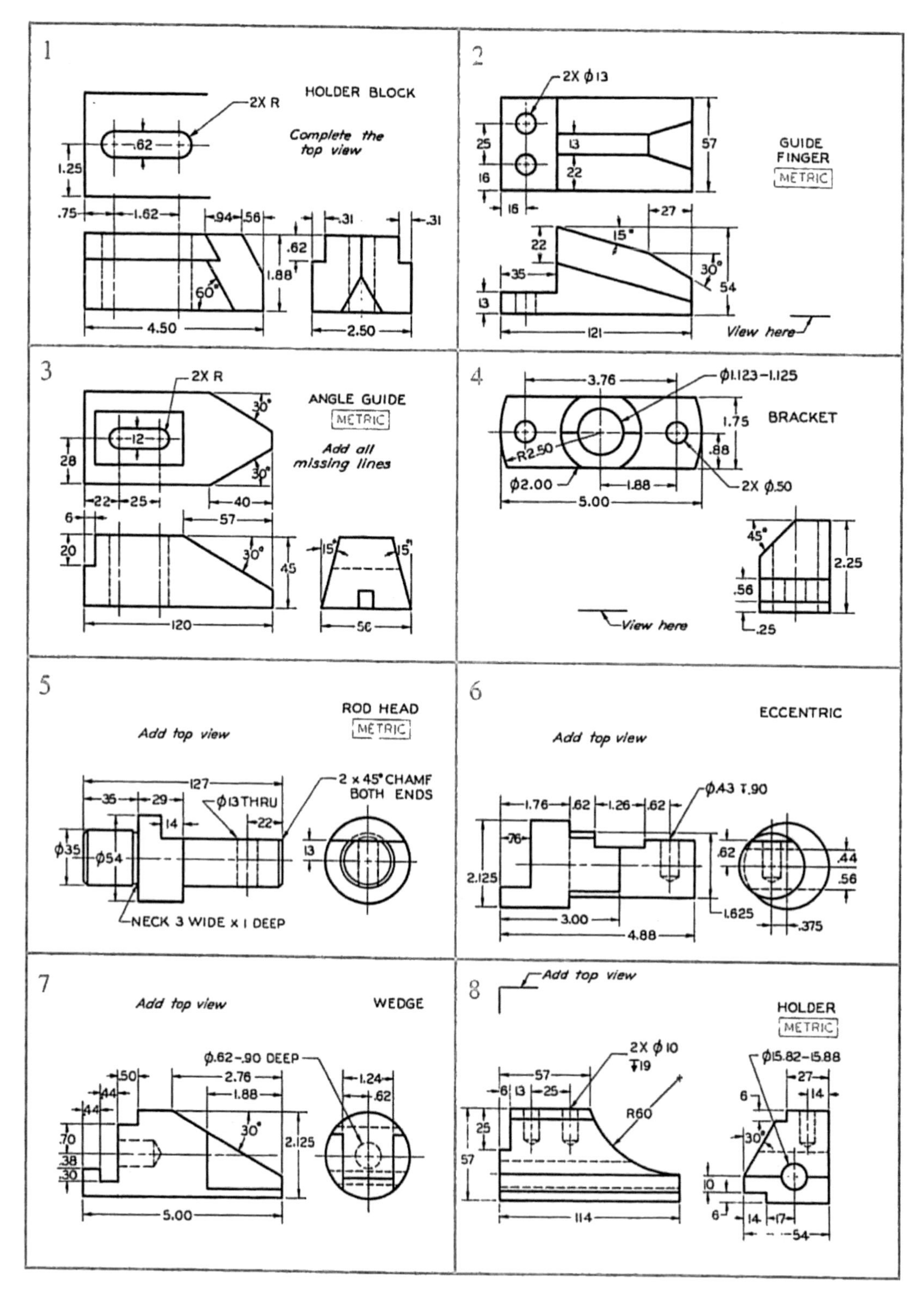

Exercise 4.10 Missing-View Problems. Sketch or draw the given views, and add the missing view. These exercises are designed to fit on 8½ × 11" size A or metric A4 paper. If dimensions are required, study Chapter 9. Use metric or decimal inch dimensions as assigned by the instructor. Move dimensions to better locations where possible.

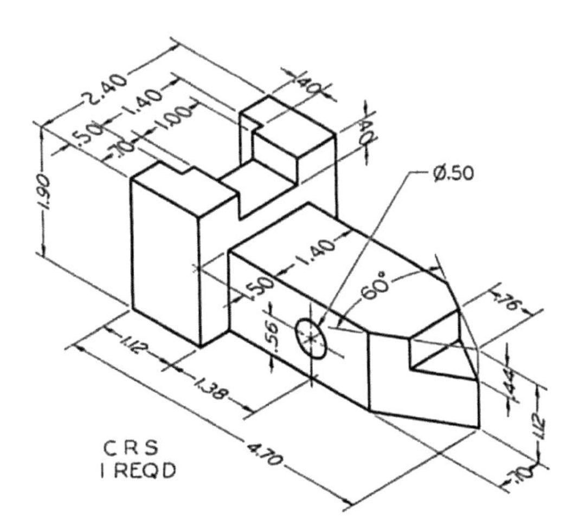

Exercise 4.11 Safety Key. Draw the necessary orthographic views on 8½ × 11" size A or metric A4 paper. Use a title block or title strip as assigned by your instructor.

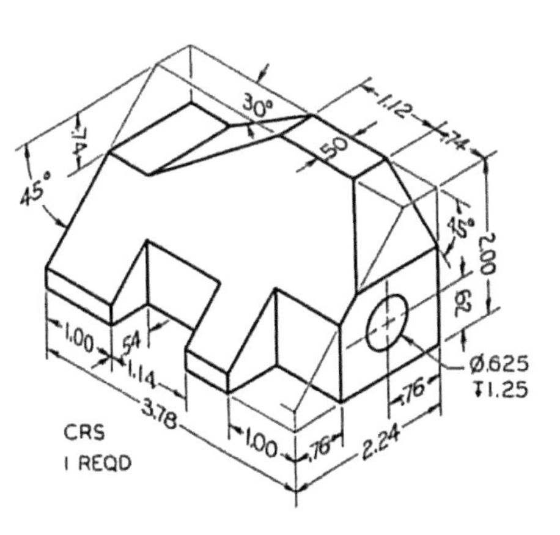

Exercise 4.12 Tool Holder. Draw the necessary orthographic views on 8½ × 11" size A or metric A4 paper. Use a title block or title strip as assigned by your instructor.

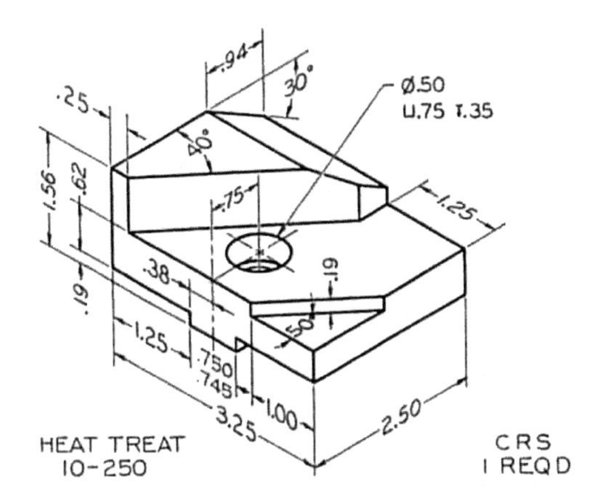

Exercise 4.13 Index Feed. Draw the necessary orthographic views on 8½ × 11" size A or metric A4 paper. Use a title block or title strip as assigned by your instructor.

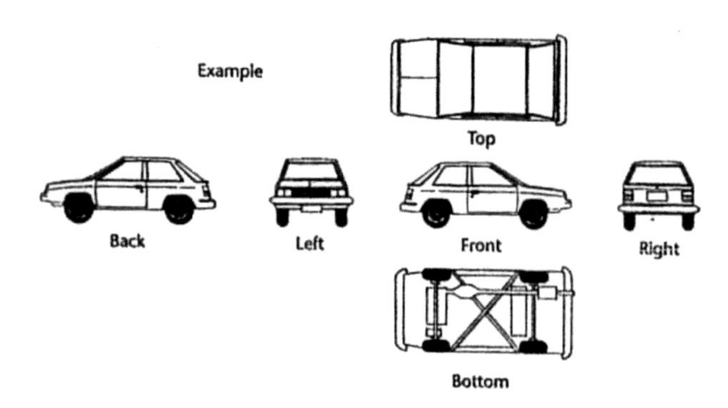

SELECTING THE BEST FRONT VIEW

1. Shows shape clearly.
2. Usual, stable, or operating position.
3. Orient long shapes horizontally.
4. The right and top views are generally preferred to the bottom and left views.

For the drawings below, six views of the part are shown. In each problem, circle the best choice for the front view. Given your choice of front view, list the letter for the following views.

Tape Dispenser
Top ________
Right ________
Left ________
Bottom________
Rear ________

Base block
Top ________
Right ________
Left ________
Bottom________
Rear ________

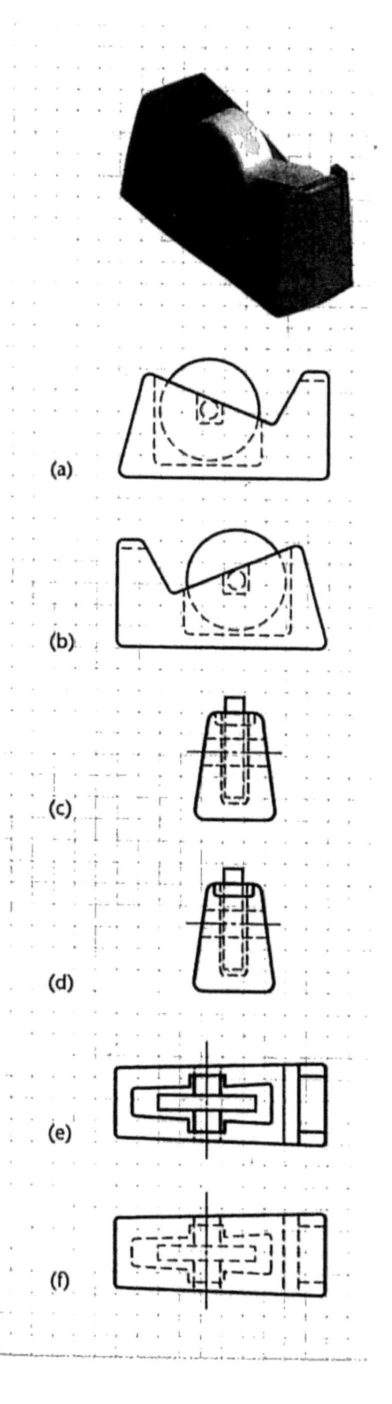

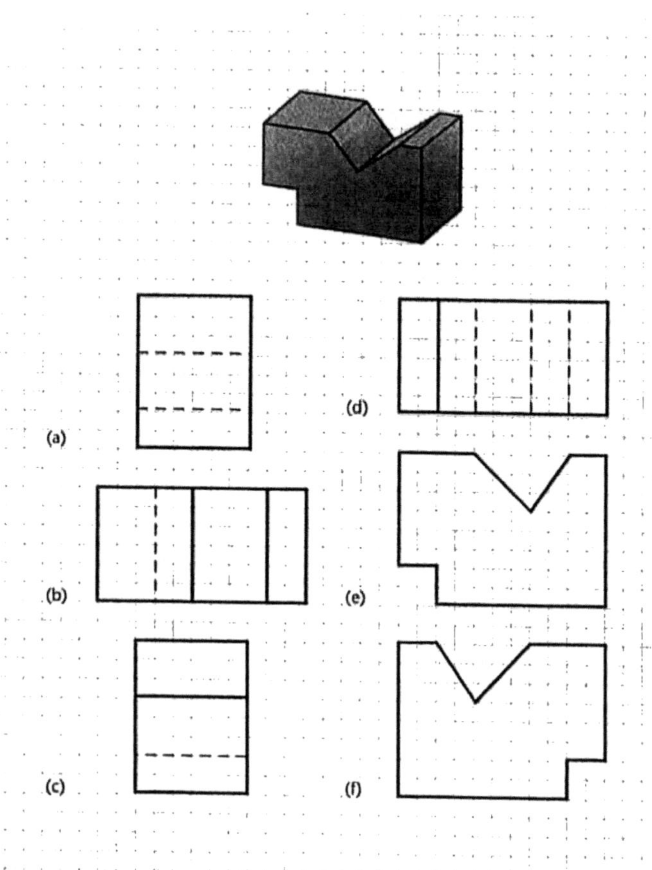

ADJACENT VIEWS

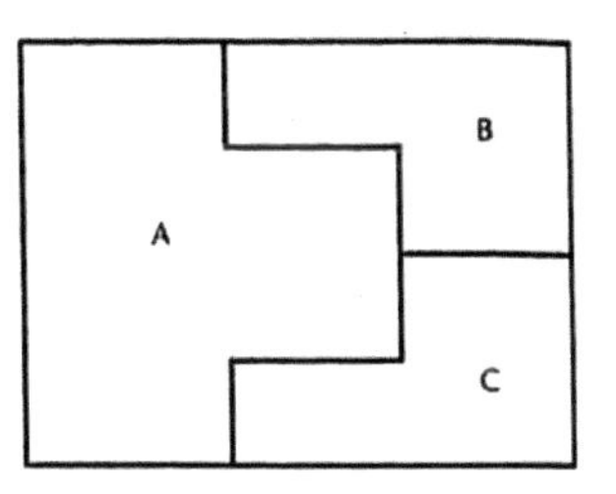

Top view

In the top view shown here, lines divide the view into three adjacent areas. No two adjacent areas lie in the same plane because each line represents an edge (or intersection) between surfaces. While each area represents a surface at a different level, you can't tell whether *A*, *B*, or *C* is the highest surface or what shape the surfaces may be until you see the other necessary views of the object.

The same reasoning applies to the adjacent areas in any given view. Since an area or surface in a view can be interpreted in different ways, other views are necessary to determine which interpretation is correct.

Below is one shape that the top view above might represent. Make a rough sketch of the front view for each description. Sketch two more possible interpretations for this top view and write their descriptions.

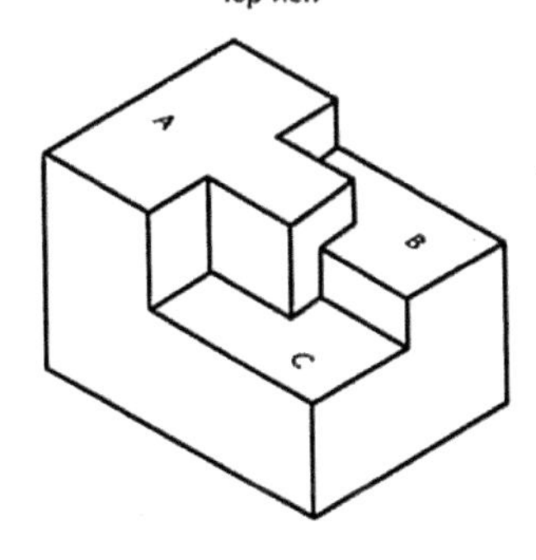

- Surface *B* is highest, and *C* and *A* are both lower.
- One or more surfaces are inclined.
- Surface *B* is highest, and surfaces *A* and *C* are lower.
- Surface *C* is highest, and *B* is lower than *A*.

NORMAL AND OBLIQUE SURFACES

Oblique surface *C* appears in the top view and front view with its vertices labeled 1-2-3-4.

- Locate the same vertices and number them in the side view.
- Shade oblique surface *C* in the side view. (Note that any surface appearing as a line in any view cannot be an oblique surface.)
- How many inclined surfaces are there on the part shown?______________
- How many normal surfaces?______________

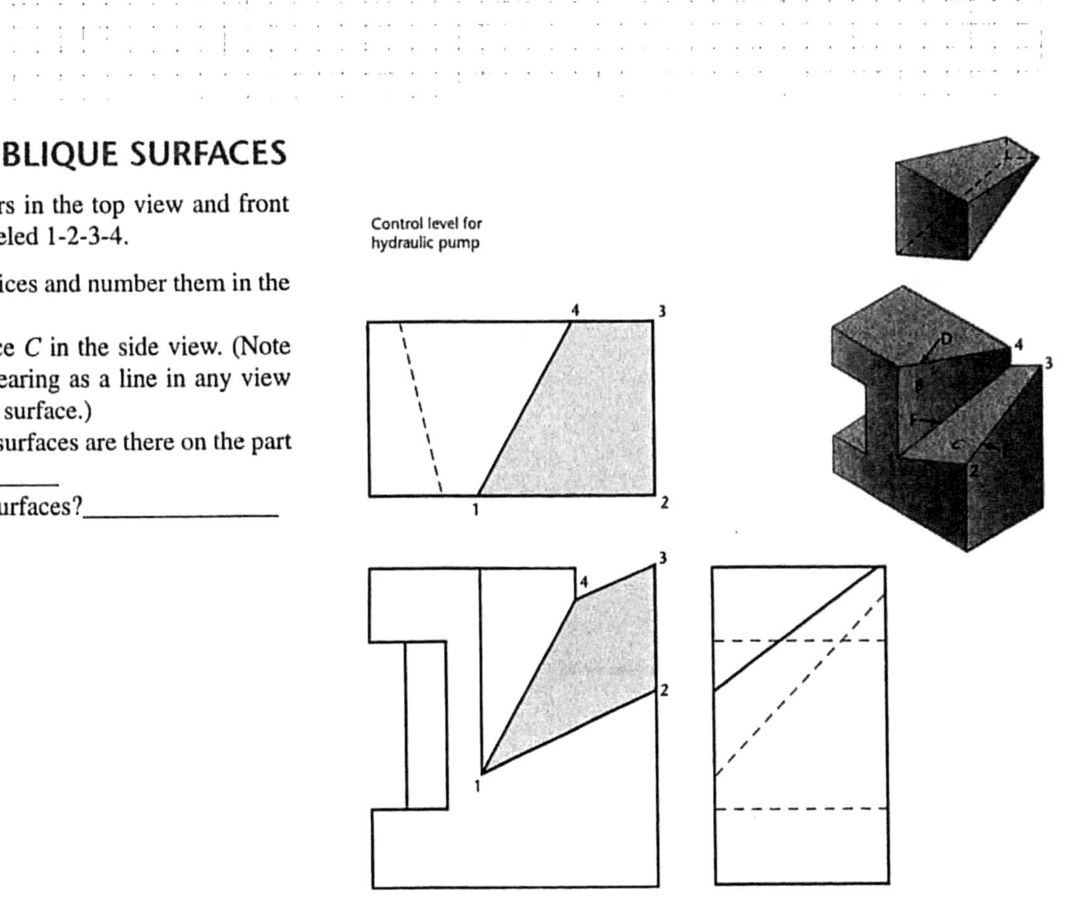

SKETCHING THREE VIEWS

A pictorial sketch of a lever bracket that requires three views is shown. Follow the steps to sketch the three views:

1. Block in the enclosing rectangles for the three views. You can either use overall proportions by eye or if you know the dimensions you can use your scale to sketch accurately sized views. Spacing your views equally from the edge of the rectangle and from each other, sketch horizontal lines to establish the height of the front view and the depth of the top view. Sketch vertical lines to establish the width of the top and front views and the depth of the side view. Make sure that this is in correct proportion to the height, and remember to maintain a uniform space between views. Remember that the space between the front and right-side view is not necessarily equal. Transfer the depth dimension from the top view to the side view; use the edge of a strip of paper or a pencil as a measuring stick. The depth in the top and side views should always be equal.
2. Block in all details lightly.
3. Sketch all arcs and circles lightly.
4. Darken all final lines.

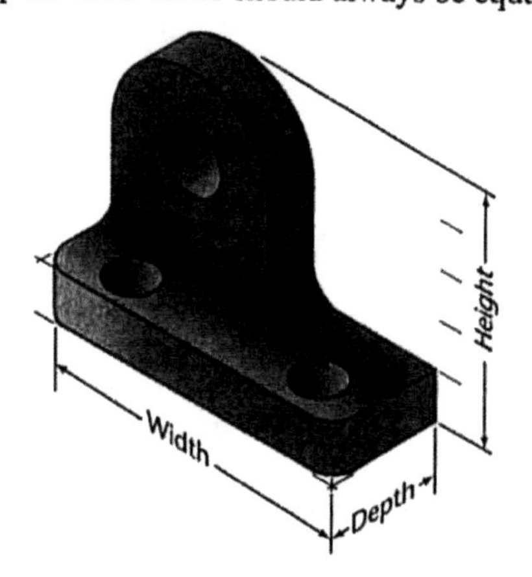

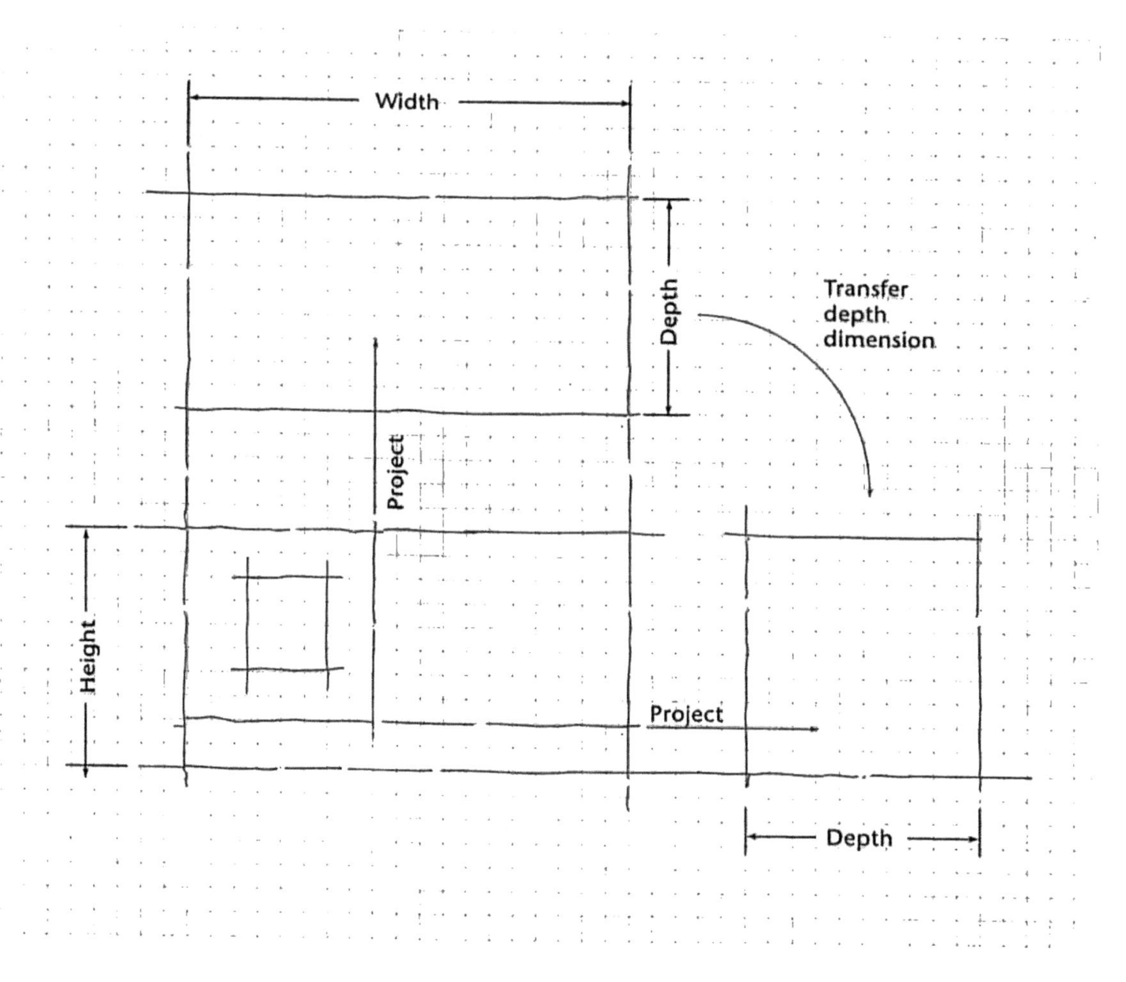

PRACTICE SKETCHING HIDDEN AND CENTERLINES

Practice sketching centerlines on the features shown.

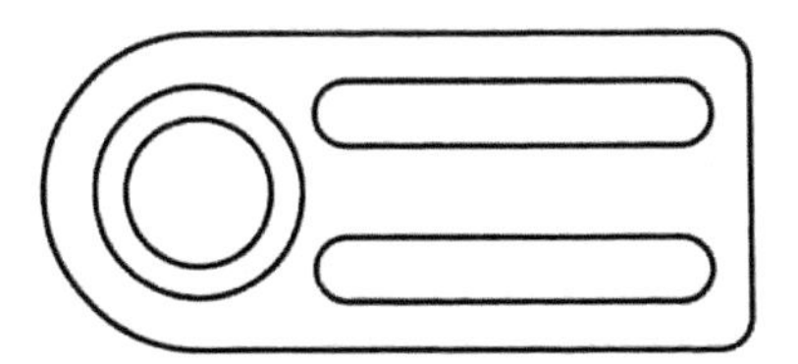

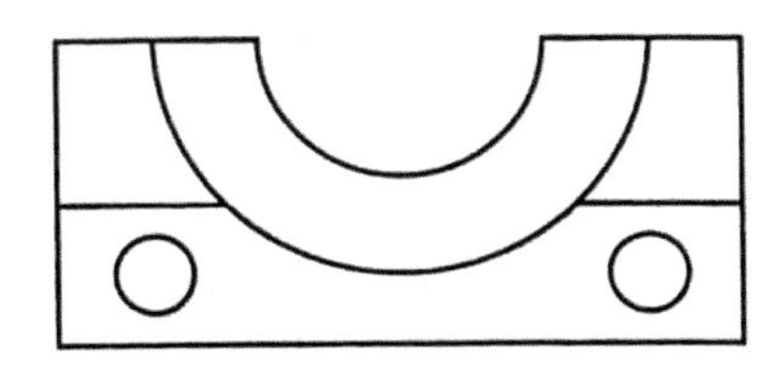

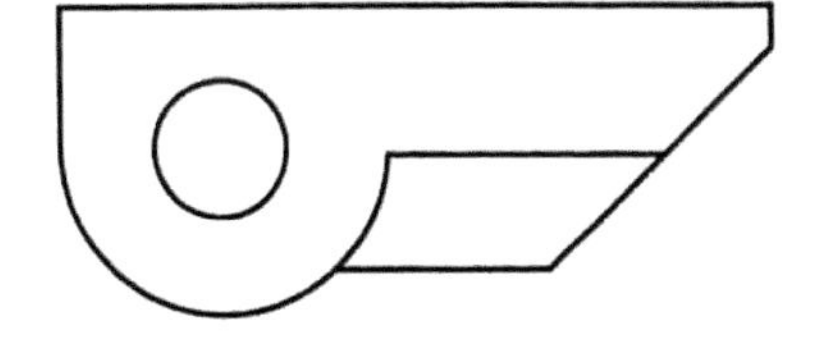

Draw the hidden and centerlines for the parts shown.

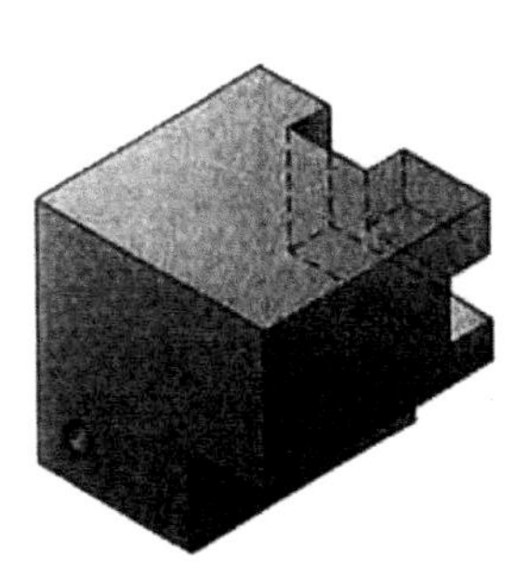

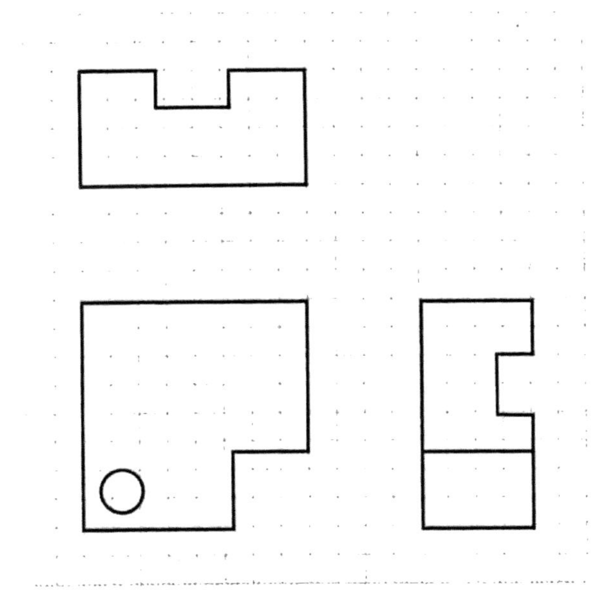

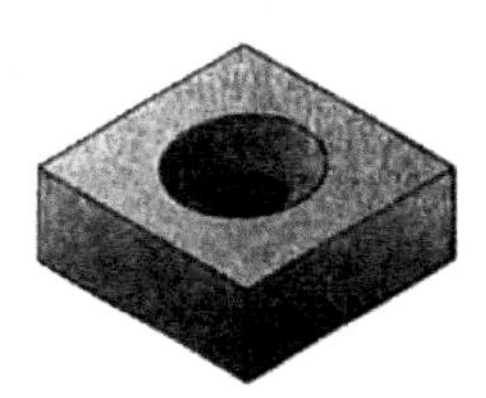

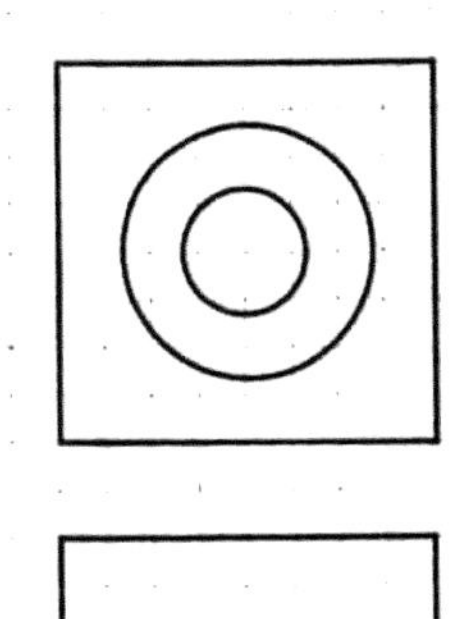

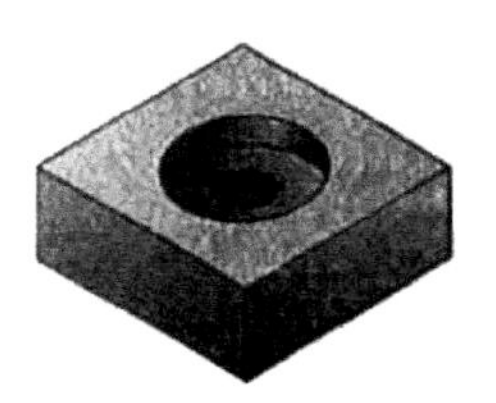

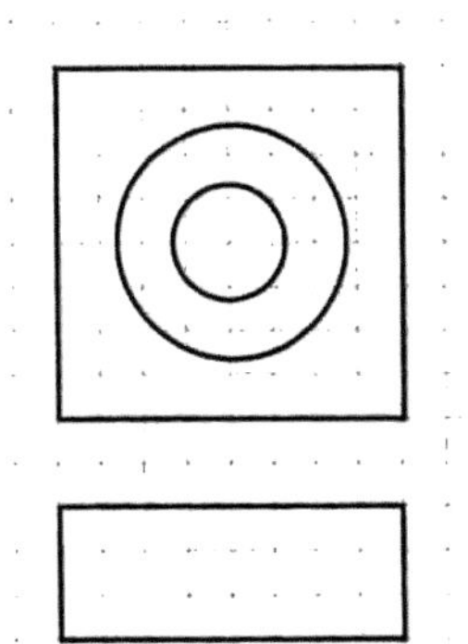

Dudley, 1st Edition – Chapter 1

Textbook: Handbook of Practical Gear Design
Chapter Title: Gear Design Trends

Gear-Design Trends

Gears are used in most types of machinery. Like nuts and bolts, they are a common machine element which will be needed from time to time by almost all machine designers. Gears have been in use for over three thousand years,* and they are an important element in all manner of machinery used in current times.

Gear design is a highly complicated art. The constant pressure to build less expensive, quieter running, lighter weight, and more powerful machinery has resulted in a steady change in gear designs. At present much is known about gear load-carrying capacity, and many complicated processes for making gears are available.

The industrialized nations are all doing gear research work in university laboratories and in manufacturing companies. Even less-developed countries are doing a certain amount of research work in the mathematics of gears and in gear applications of particular interest. At the 1981 International Sympo-

*The book *Evolution of the Gear Art*, 1969, by D. W. Dudley gives a brief review of the history of gears through the ages. Consult the reference section at the end of the book for complete data on this book and on other important references.

TABLE 1.1 Glossary of Gear Nomenclature, Chap. 1

Gear A geometric shape that has teeth uniformly spaced around the circumference. In general, a gear is made to mesh its teeth with another gear. (A *sprocket* looks like a gear but is intended to drive a chain instead of another gear.)
Pinion When two gears mesh together, the *smaller* of the two is called the *pinion.* The larger is called the *gear.*
Ratio *Ratio* is an abbreviation for gear-tooth ratio, which is the number of teeth on the gear divided by the number of teeth on its mating pinion.
Module A measure of tooth size in the metric system. In units, it is millimeters of pitch diameter *per tooth.* As the tooth size increases, the module also increases. Modules usually range from 1 to 25.
Diametral pitch A measure of tooth size in the English system. In units, it is the *number of teeth* per inch of pitch diameter. As the tooth size increases, the diametral pitch *decreases.* Diametral pitches usually range from 25 to 1.
Circular pitch The circular distance from a point on one gear tooth to a like point on the next tooth, taken along the *pitch circle.* Two gears must have the same circular pitch to mesh with each other. As they mesh, their pitch circles will be tangent to one another.
Pitch diameter The diameter of the pitch circle of a gear.
Addendum The radial height of a gear tooth above the pitch circle.
Dedendum The radial depth of a gear tooth below the pitch circle.
Whole depth The total radial height of a gear tooth. (Whole depth = addendum + dedendum.)
Pressure angle The slope of the gear tooth at the pitch-circle position. (If the pressure angle were 0°, the tooth flank would be radial.)
Helix angle The inclination of the tooth in a lengthwise direction. (If the helix angle is 0°, the tooth is parallel to the axis of the gear—and is really a spur-gear tooth.)
Spur gears Gears with teeth straight and parallel to the axis of rotation.
Helical gears Gears with teeth that spiral around the body of the gear.
External gears Gears with teeth on the outside of a cylinder.
Internal gears Gears with teeth on the inside of a hollow cylinder. (The mating gear for an internal gear must be an external gear.)
Bevel gears Gears with teeth on the outside of a conical-shaped body (normally used on 90° axes).
Worm gears Gearsets in which one member of the pair has teeth wrapped around a cylindrical body like screw threads. (Normally this gear, called the *worm,* has its axis at 90° to the worm-gear axis.)
Face gears Gears with teeth on the *end* of the cylinder.
Spiroid gears A family of gears in which the tooth design is in an intermediate zone between bevel-, worm-, and face-gear design. The Spiroid design is patented by the Spiroid Division of Illinois Tool Works, Chicago, Illinois.
Transverse section A section through a gear *perpendicular to the axis* of the gear.
Axial section A section through a gear in a lengthwise direction that *contains the axis* of the gear.
Normal section A section through the gear that is *perpendicular to the tooth* at the pitch circle. (For spur gears, a normal section is also a transverse section.)

Notes: 1. For terms relating to gear materials, see Chap. 4.
2. For terms relating to gear manufacture, see Chap. 5.
3. For terms relating to the specifics of gear design and rating, see Chaps. 2 and 3.
4. For a simple introduction to gears, see Chap. 9, Sec. 9.1.

TABLE 1.2 Gear Terms, Symbols, and Units, Chap. 1

Term	Metric		English		First reference or definition
	Symbol	Units*	Symbol	Units*	
Module	m	mm	—	—	Eq. (1.1)
Pressure angle	α	deg	ϕ	deg	Figs. 1.18, 1.33
Number of teeth or threads	z	—	N	—	
Number of teeth, pinion	z_1	—	N_P or n	—	Eq. (1.7)
Number of teeth, gear	z_2	—	N_G or N	—	Eq. (1.7)
Ratio (gear or tooth ratio)	u	—	m_G	—	$u = z_2/z_1$ ($m_G = N_G/N_P$)
Diametral pitch	—	—	P_d or P	in.$^{-1}$	Eq. (1.1)
Pi	π	—	π	—	$\pi \cong 3.1415927$
Pitch diameter, pinion	d_{p1}	mm	d	in.	Fig. 1.18, Eq. (1.3)
Pitch diameter, gear	d_{p2}	mm	D	in.	Fig. 1.18, Eq. (1.3)
Base (circle) diameter, pinion	d_{b1}	mm	d_b	in.	Fig. 1.18
Base (circle) diameter, gear	d_{b2}	mm	D_b	in.	Fig. 1.18
Outside diameter, pinion	d_{a1}	mm	d_o	in.	Fig. 1.18 (abbrev. O.D.)
Outside diameter, gear	d_{a2}	mm	D_o	in.	Fig. 1.18
Form diameter	d'_f	mm	d_f	in.	Fig. 1.18
Circular pitch	p	mm	p	in.	Fig. 1.18, Eq. (1.2)
Addendum	h_a	mm	a	in.	Fig. 1.18
Dedendum	h_f	mm	b	in.	Fig. 1.18
Face width	b	mm	F	in.	Fig. 1.18
Whole depth	h	mm	h_t	in.	Fig. 1.18
Working depth	h'	mm	h_k	in.	Fig. 1.18
Clearance	c	mm	c	in.	Fig. 1.18
Chordal thickness	$\bar{s}$	mm	t_c	in.	Fig. 1.18
Chordal addendum	$\bar{h}_a$	mm	a_c	in.	Fig. 1.18
Tooth thickness	s	mm	t	in.	Fig. 1.18
Center distance	a	mm	C	in.	Eq. (1.4)
Circular pitch, normal	p_n	mm	p_n	in.	Fig. 1.19, Eq. (1.5)
Circular pitch, transverse	p_t	mm	p_t	in.	Fig. 1.19, Eq. (1.5)
Lead angle	γ	deg	λ	deg	Fig. 1.19, Eq. (1.33)
Helix angle	β	deg	ψ	deg	Fig. 1.19, Eq. (1.8)
Pitch, base or normal	p_b or p_{bn}	mm	p_b or p_N	in.	Fig. 1.19
Diametral pitch, normal	—	—	P_n	in.$^{-1}$	Eq. (1.10)
Axial pitch	p_x	mm	p_x	in.	Eq. (1.11)
Inside (internal) diameter	d_i	mm	D_i	in.	Fig. 1.20, (abbrev. I.D.)
Root diameter, pinion	d_{f1}	mm	d_R	in.	Fig. 1.29
Root diameter, gear	d_{f2}	mm	D_R	in.	Fig. 1.20
Pitch angle, pinion	δ'_1	deg	γ	deg	Fig. 1.21, Eq. (1.16)
Pitch angle, gear	δ'_2	deg	Γ	deg	Fig. 1.21, Eq. (1.16)
Root angle, pinion	δ_{f1}	deg	γ_R	deg	Fig. 1.21
Root angle, gear	δ_{f2}	deg	Γ_R	deg	Fig. 1.21
Face angle, pinion	δ_{a1}	deg	γ_o	deg	Fig. 1.21
Face angle, gear	δ_{a2}	deg	Γ_o	deg	Fig. 1.21
Dedendum angle	θ_f	deg	δ	deg	Fig. 1.21
Shaft angle	Σ	deg	Σ	deg	Fig. 1.21, Eq. (1.18)
Cone distance	R	mm	A	in.	Fig. 1.21
Circular (tooth) thickness	s	mm	t	in.	Fig. 1.21
Backlash	j	mm	B	in.	Fig. 1.21
Throat diameter of worm	d_{t1}	mm	d_t	in.	Fig. 1.29
Throat diameter of worm gear	d_{t2}	mm	D_t	in.	Fig. 1.29
Lead of worm	p_{zW}	mm	L_W	in.	Eq. (1.32)

*Abbreviations for units: mm = millimeters, in. = inches, deg = degrees.

sium on Gearing and Power Transmissions in Tokyo, Japan, 162 papers were presented by delegates from 24 nations. After excellent world gear conferences in Paris, France, in 1977 and Dubrovnik, Yugoslavia, in 1978, the even larger Tokyo gear conference in 1981 was a high-water mark in the gear art.

Most machine designers do not have the time to keep up with all the developments in the field of gear design. This makes it hard for them to quickly design gears which will be competitive with the best that are being used in their field. There is a great need for *practical* gear-design information. Even though there is a wealth of published information on gears, gear designers often find it hard to locate the information they need quickly. This book is written to help gear designers get the vital information they need as easily as possible.

Those who are just starting to learn about gears need to start by learning some basic words that have special meanings in the gear field. The glossary in Table 1.1 is intended to give simple definitions of these terms as they are understood by gear people. Table 1.2 shows the metric and English gear symbols for the terms which are used in Chap. 1. The Chap. 5 glossary (Table 5.1) defines gear-manufacturing terms. See AGMA 112.05 (1976)* and AGMA 600.01 (1979) for English and metric gear nomenclature.

In the last chapter—an appendix chapter—Sec. 9.1 gives a simple explanation of some basics of gearing. Beginners will probably find it helpful to read Sec. 9.1 and the tables mentioned above before they continue with the Chap. 1 text.

MANUFACTURING TRENDS

Before plunging into the formulas for calculating gear dimensions, it is desirable to make a brief survey of how gears are presently being made and used in different applications.

The methods used to manufacture gears depend on design requirements, machine tools available, quantity required, cost of materials, and *tradition*. In each particular field of gear work, certain methods have become established as the *standard* way of making the gears. These methods tend to change from time to time, but the tradition of the industry tends to act as a brake to restrain any abrupt changes that result from technological developments. The gear designer, studying gears as a whole, can get a good perspective of gear work by reviewing the methods of manufacture in each field.

*AGMA stands for the American Gear Manufacturers Association, and AGMA 112.05 refers to a specific published standard of AGMA. Complete titles of AGMA standards mentioned in the text are given in the references at the end of the book.

1.1 Small, Low-Cost Gears for Toys, Gadgets, and Mechanisms

There is a large field of gear work in which tooth stresses are of almost no consequence. Speeds are slow, and life requirements do not amount to much. Almost any type of "cog" wheel which could transmit rotary motion might be used. In this sort of situation, the main thing the designer must look for is low cost and high volume of production.

The simplest type of gear drive—such as those used in toys—frequently uses punched gears. Pinions with small numbers of teeth may be die-cast or extruded. If loads are light enough and quietness of operation is desired, injection-molded gears and pinions may be used. Molded-plastic gears from a toy train are shown in Fig. 1.1. Things like film projectors, oscillating fans, cameras, cash registers, and calculators frequently need quiet-running gears to transmit insignificant amounts of power. Molded-plastic gears are widely used in such instances. It should be said, though, that the devices just mentioned often need precision-cut gears where loads and speeds become appreciable.

Die-cast gears of zinc alloy, brass, or aluminum are often used to make

FIG. 1.1 Plastic gears in a toy.

small, low-cost gears. This process is particularly favored where the gear wheel is integrally attached to some other element, such as a sheave, cam, or clutch member. The gear teeth and the special contours of whatever is attached to the gear may all be finished to close accuracy merely by die-casting the part in a precise metal mold. Many low-cost gadgets on the market today would be very much more expensive if it were necessary to machine all the complicated gear elements that are in them.

Metal forming is more and more widely used as a means of making small gear parts. Pinions and gears with small numbers of teeth may be cut from rod stock with cold-drawn or extruded teeth already formed in the rod. Figure 1.2 shows some rods with cold-drawn teeth. Small worms may have cold-rolled threads. The forming operations tend to produce parts with very smooth, work-hardened surfaces. This feature is important in many devices

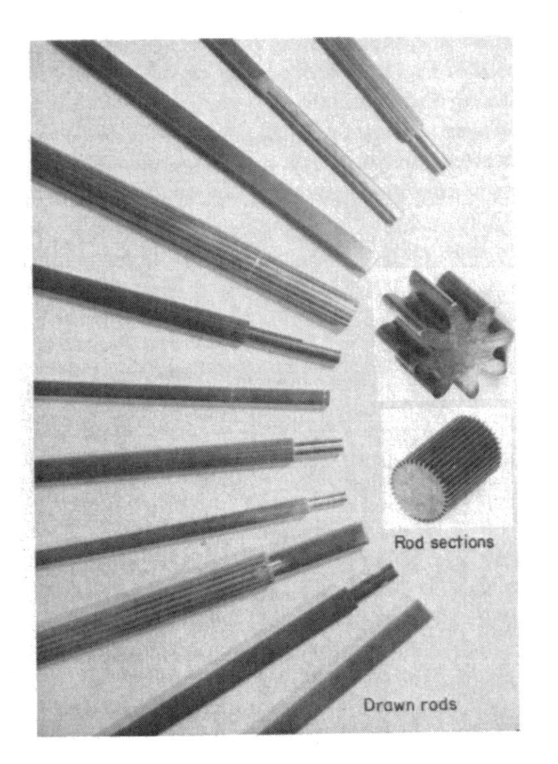

FIG. 1.2 Samples of cold-drawn rod stock for making gears and pinions. (*Courtesy of Rathbone Corp., Palmer, MA, U.S.A.*)

FIG. 1.3 Assortment of gears used in small mechanisms. (*Courtesy of Winzeler Manufacturing and Tool Co., Chicago, IL, U.S.A.*)

where the friction losses in the gearing tend to be the main factor in the power consumption of the device.

Figure 1.3 shows an assortment of stamped and molded gears used in small mechanisms.

1.2 Appliance Gears

Home appliances like washing machines, food mixers, and fans use large numbers of small gears. Because of competition, these gears must be made for only a relatively few cents apiece. Yet they must be quiet enough to suit a discriminating homemaker and must be able to endure for many years with little or no more lubrication than that given to them at the factory.

Medium-carbon-steel gears finished by conventional cutting used to be the standard in this field. Cut gears are still in widespread use, but the cutting is often done by high-speed automatic machinery. The worker on the cutting machine does little more than bring up trays of blanks and take away trays of finished parts.

Modern appliances are making more and more use of gears other than cut steel gears. Figure 1.4 shows some sintered-iron gears from an automatic washer. Sintered-iron gears are very inexpensive (in large quantities), run quietly, and frequently wear less than comparable cut gears. The sintered metal is porous and may be impregnated with a lubricant. It may also be

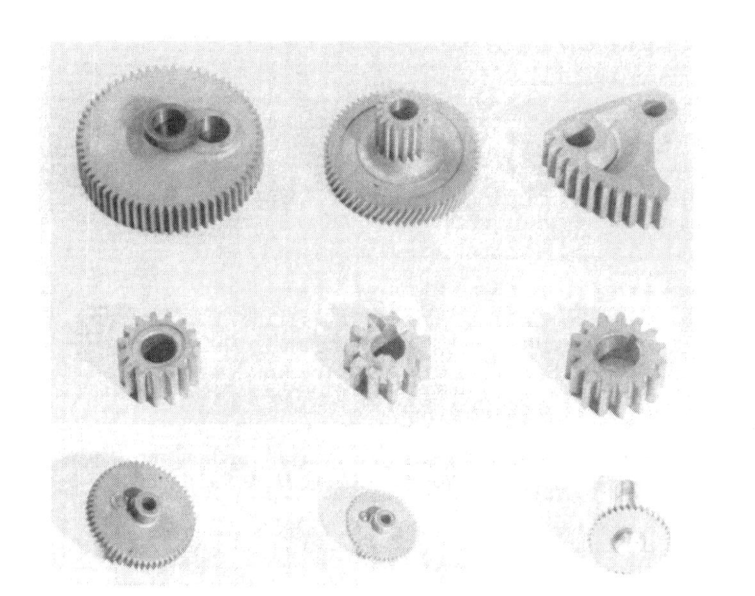

FIG. 1.4 Sintered-iron gears used in an automatic washing machine.

impregnated with copper to improve its strength. Gear teeth and complicated gear-blank shapes may all be completely finished in the sintering process. The tools needed to make a sintered gear may cost as much as $50,000, but this does not amount to much if 100,000 or more gears are to be made on semiautomatic machines.

Laminated gears using phenolic resins and cloth or paper have proved very good where noise reduction is a problem. The laminates in general have much more load-carrying capacity than molded-plastic gears. Nonmetallic gears with cut teeth do not suffer nearly so much from tooth inaccuracy as do metal gears. Under the same load, a laminated phenolic-resin gear tooth will bend about thirty times as much as a steel gear tooth! It has often been possible to take a set of steel gears which were wearing excessively because of tooth-error effects, replace one member with a nonmetallic gear, and have the set stand up satisfactorily.

Nylon gear parts have worked very well in situations in which wear resulting from high sliding velocity is a problem. The nylon material seems to have some of the characteristics of a solid lubricant. Nylon gearing has been used in some processing equipment where the use of a regular lubricant would pollute the material being processed.

1.3 Machine Tools

Accuracy and power-transmitting capacity are quite important in machine-tool gearing. Metal gears are usually used. The teeth are finished by some precise metal-cutting process.

The machine tool is often literally full of gears. Speed-change gears of the spur or helical type are used to control feed rate and work rotation. Index drives to work or table may be worm or bevel. Sometimes they are spur or helical. Many bevel gears are used at the right-angle intersections between shafts in bases and shafts in columns. Worm gears and spiral gears are also commonly used at these places.

Machine-tool gearing is often finish-machined in a medium-hard condition (250 to 300 HV or 25 to 30 HRC). Mild-alloy steels are frequently used because of their better machinability and physical properties. Cast iron is often favored for change gears because of the ease in casting the gear blank to shape, its excellent machinability, and its ability to get along with scanty lubrication.

The higher cutting speeds involved in cutting with tungsten carbide tools have forced many machine-tool builders to put in harder and more accurate gears. Shaving and grinding are commonly used to finish machine-tool gears to high tooth accuracy. In a few cases, machine-tool gears are being made with such top-quality features as full hardness (700 HV or 60 HRC), profile modification, and surface finish of about 0.5 μm or 20 μin.

The machine-tool designer has a hard time calculating gear sizes. Loads vary widely depending on feeds, speeds, size of work, and material being cut. It is anybody's guess what the user will do with the machine. Because machine tools are quite competitive in price, the overdesigned machine may be too expensive to sell. In general, the designer is faced with the necessity of putting in gears which have more capacity than the average load, knowing that the machine tool is apt to be neglected or overloaded on occasions.

Figure 1.5 shows an example of machine-tool gears.

1.4 Control Gears

The guns on ships, helicopters, and tanks are controlled by gear trains with the backlash held to the lowest possible limits. The primary job of these gears is to transmit motion. What power they may transmit is secondary to their job of precise control of angular motion. In power gearing, a worn-out gear is one with broken teeth or bad tooth-surface wear. In the control-gear field, a worn-out gear may be one whose thickness has been reduced by as small an amount as 0.01 mm (0.0004 in.)!

FIG. 1.5 Example of change gears on a machine tool. (*Courtesy of Warner & Swasey Co., Cleveland, OH, U.S.A.*)

Some of the most spectacular control gears are those used to drive radio telescopes and satellite tracking antennas. These gears are so large that the only practical way to make them is to cut rack sections and then bend each section into an arc of a circle. Figure 1.6 shows an example of a very large antenna drive made this way. The teeth on the rack sections are cut so that they will have the correct tooth dimensions when they are bent to form part of the circular gear.

The radar units on an aircraft carrier use medium-pitch gears of fairly large size. Radar-unit gearing is generally critical on backlash, must handle rather high momentary loads, and must last for many years with somewhat marginal lubrication. (Radar gears are often somewhat in the open, and therefore can only use grease lubrication.)

Control gears are usually spur, bevel, or worm. Helical gears are used to a limited extent. Control gears are often in the fine-pitch range—1.25 module (20 diametral pitch) or finer. Figure 1.7 shows a control device with many small gears.

In a few cases, control gears become quite large. The gears which train a main battery must be very rugged. The reaction on the gears when a main battery is fired can be terrific.

Control gears are usually made of medium-alloy, medium-carbon steels. In many cases, they are hardened to a medium hardness before final machining. In other cases, they are hardened to a moderately high hardness after final machining. These gears need hardness mainly to limit wear. Any hardening done after final finishing of the teeth must be done in such a way as to give only negligible dimensional change or distortion. To eliminate backlash, it is necessary to size the gear teeth almost perfectly (or use special "antibacklash" gear arrangements).

Shaving and/or grinding are used to control tooth thickness to the very close limits needed in control gearing. The inspection of control gears is usually based on checking machines which measure the variation in center distance when a master pinion or rack is rolled through mesh with the gear being checked. A spring constantly holds the master and the gear being checked in tight mesh. The chart obtained from such a checking machine gives a very clear picture of both the tooth thickness of the gear being checked and the variations in backlash as the gear is rolled through mesh. If

FIG. 1.6 Gears are used to position large radio telescopes. (*Courtesy of Harris Corporation, Melbourne, FL, U.S.A.*)

FIG. 1.7 Control device with many small gears.

the backlash variation can be held to acceptable limits in control-gear sets, there is usually no need to know the exact involute and spacing accuracy. In some types of radar and rocket-tracking equipment, control-gear teeth must be spaced very accurately with regard to accumulated error. For instance, it may be necessary to have every tooth all the way around a gear wheel within its true position within about 10 seconds of arc. On a 400-mm (15.7-in.) wheel, this would mean that every tooth had to be correctly spaced with respect to every other tooth within 0.01 mm (0.0004 in.). This kind of accuracy can be achieved only by special gear-cutting techniques. Inspection of such gears generally requires the use of gear-checking machines that are equipped to measure accumulated spacing error. The equipment commonly used will measure any angle to within 1 second of arc or better.

1.5 Vehicle Gears

The automobile normally uses spur and helical gears in the transmission and bevel gears in the rear end. If the car is a front-end drive, bevel gears may still be used or helical gears may be used. Automatic transmissions are now widely used. This does not eliminate gears, however. Most automatic transmissions have more gears than manual transmissions.

Automotive gears are usually cut from low-alloy-steel forgings. At the time of tooth cutting, the material is not very hard. After tooth cutting, the gears are case-carburized and quenched. Quenching dies are frequently used to minimize distortion. The composition of heats (batches) of steel is

watched carefully, and all steps in manufacture are closely controlled with the aim of having each gear in a lot behave in the same way when it is carburized and quenched. Even if there is some distortion, it can be compensated for in machining, provided that each gear in a lot distorts uniformly and by the same amount. Since most automotive-gear teeth are not ground or machined after final hardening, it is essential that the teeth be quite accurate in the as-quenched condition. The only work that is done after hardening is the grinding of journal surfaces and sometimes a small amount of lapping. Finished automotive gears usually have a surface hardness of about 700 HV or 60 HRC and a core hardness of 300 HV or 30 HRC.

A variety of machines are used to cut automotive-gear teeth. In the past, shapers, hobbers, and bevel-gear generators were the conventional machines. New types of these machines are presently favored. For instance, multistation shapers, hobbers, and shaving machines are used. A blank is loaded on the machine at one station. While the worker is loading other stations, the piece is finished. This means that the worker spends no time waiting for work to be finished.

There is a wide variety of special design machines to hob, shape, shave, or grind automotive gears. The high production of only one (or two or three) gear design makes it possible to simplify a general-purpose machine tool and then build a kind of *processing center* where these functions are performed:

Incoming blank is checked for correct size.

Incoming blank is automatically loaded into the machine.

Teeth are cut (or finished) very rapidly.

Finished parts are checked for accuracy and sorted into categories of accept, rework, or reject.

Outgoing parts are loaded onto conveyor belts.

Sometimes the processing center may be developed to the point where cutting, heat treating, and finishing are all done in one processing center.

The gears for the smaller trucks and tractors are made somewhat like automotive gears, but the volume is not quite so great and sizes are larger. Examples of large-vehicle gears are shown in Figs. 1.8 and 1.9. Large tractors, large trucks, and "off-the-road" earth-moving vehicles use much larger gears and have much lower volume than automotive gears. More conventional machine tools are used. Special-purpose processing centers are generally not used.

Vehicle gears are heavily loaded for their size. Fortunately, their heaviest loads are of short duration. This makes it possible to design the gears for limited life at maximum motor torque and still have a gear that will last many years under average driving torque.

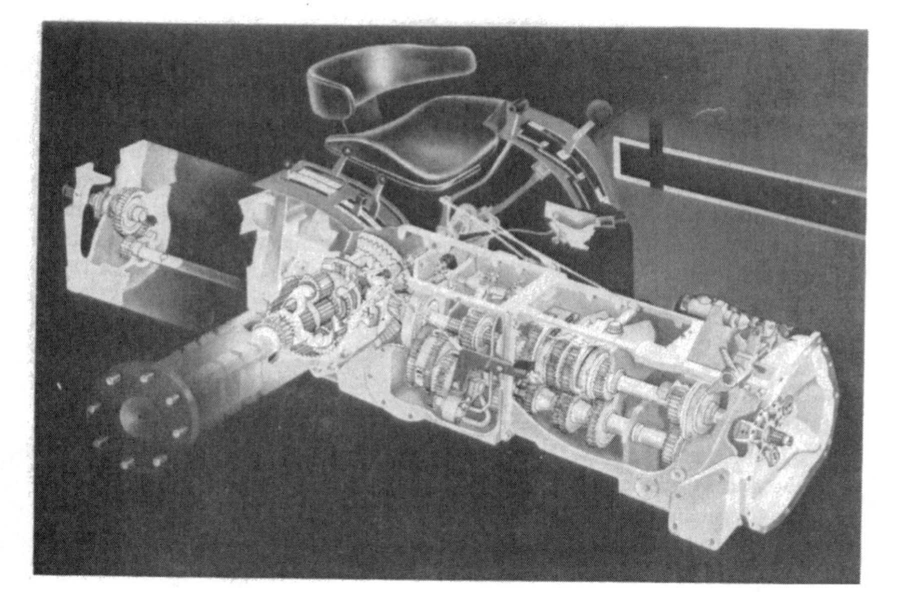

FIG. 1.8 Tractor power train. (*Courtesy of International Harvester Co., Chicago, IL, U.S.A.*)

Although carburizing has been widely used as a means of hardening automotive gears, other heat treatments are being used on an increased scale. Combinations of carburizing and nitriding are used. Processes of this type produce a shallower case for the same length of furnace time, but tend to make the case harder and the distortion less. Induction hardening is being used on some flywheel-starter gears as well as some other gears. Flame hardening is also used to a limited extent.

1.6 Transportation Gears

Buses, subways, mine cars, and railroad diesels all use large quantities of spur and helical gears. The gears range up to 0.75 meter (30 in.) or more in diameter. See Fig. 1.10. Teeth are sometimes as coarse as 20 module ($1\frac{1}{4}$ diametral pitch). Plain carbon and low-alloy steels are usually used. Much of the gearing is case-carburized and ground. Through-hardening is also used extensively. A limited use is being made of induction-hardened gears.

Transportation gears are heavily loaded. Frequently their heaviest loads last for a long period of time. Diesels that pull trains over high mountain ranges have long periods of operation at maximum torque. In some appli-

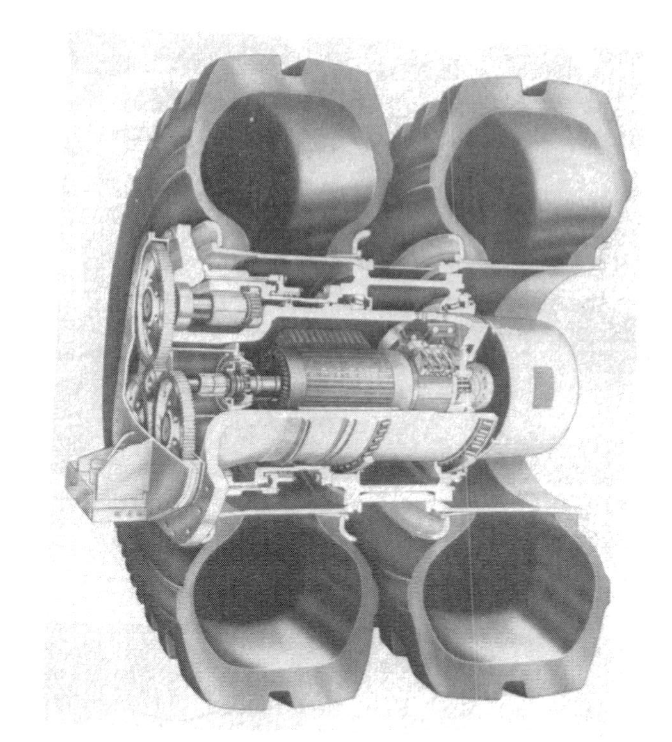

FIG. 1.9 Motorized wheels drive large earth-moving trucks. (*Courtesy of General Electric Co., Erie, PA, U.S.A.*)

FIG. 1.10 Railroad wheel drive. (*Courtesy of Electro-Motive Div., General Motors Corp., La Grange, IL, U.S.A.*)

cations, severe but infrequent shock loads are encountered. Shallow-hardening, medium-carbon steels seem to resist shock better than gears with a fully hardened carburized case. Both furnace and induction-hardening techniques are used to produce shallow-hardened teeth with high shock resistance.

Gear cutting is done mostly by conventional hobbing or shaping machines. Some gears are shaved and then heat-treated, while others are heat-treated after cutting and then ground. In this field of work, the volume of production is much lower and the size of parts is much larger than in the vehicle-gear field. Both these conditions make it harder to keep heat-treat distortion so well under control that the teeth may be finished before hardening.

The machine tools used to make transportation gears are quite conventional in design. The volume of production in this field is not large enough to warrant the use of the faster and more elaborate types of machine tools used in the vehicle-gear field.

1.7 Marine Gears

Powerful, high-speed, large-sized gears power merchant marine and navy fighting ships. Propeller drives on cargo ships use bull gears up to 5 meters (200 in.) in diameter. First reduction pitch-line speeds on some ships go up

to 100 m/s (meters per second) or 20,000 fpm (feet per minute). Single propeller drives in a navy capital ship go up to 40,000 or more kW (kilowatts) of power. Some of the new cargo ships now in service have 30,000 kW [40,000 hp (horsepower)] per screw. Figure 1.11 shows a typical marine gear unit.

Marine gears are almost all made by finishing the teeth after hardening. Extreme accuracy in tooth spacing is required to enable the gearing to run satisfactorily at high speed. As many as 6000 pair of teeth may go through one gear mesh in a second's time!

There is an increasing use of carburized and ground gears in the marine field. The fully hard gear is smaller and lighter. This lowers pitch-line velocity and helps to keep the engine room reasonably small. Hard gears resist pitting and wear better than medium-hard gears (through-hardened gears).

Single helical gears have long been used for electric-power-generating equipment. Double helical gears have generally been used for the main propulsion drive of large ships. Single helical gears—with special thrust runners on the gears themselves—are coming into fairly common use even on larger ships.

FIG. 1.11 Partially assembled double-reduction marine gear unit. (*Courtesy of Transamerica Delaval Inc., Trenton, NJ, U.S.A.*)

Spur and bevel gear drives are often used on small ships, but they are seldom used on large turbine-drive ships. (Some slower-speed diesel-drive ships use spur gearing.)

Marine gears generally use a medium-alloy carbon steel. It is difficult or impossible to heat-treat plain carbon and low-alloy carbon steels in large sizes and get hardnesses over about 250 HV (250 HB) and a satisfactory metallurgical structure. Special welding techniques for large gear wheels make it possible to use medium-alloy steels and get good through-hardened gears up to 5 meters (200 in.) with gear-tooth hardness at 320 HV (300 HB) or higher. The smaller pinions are not welded and can usually be made up to 375 HV (350 HB) or higher in sizes up to 0.75 meter (30 in.).

There is growing use of carburized and ground gears for ship propulsion. Some of these gears are now being made as large as 2 meters (80 in.) in size. In a very large marine drive, it is quite common to have fully hard, carburized gears in the first reduction and medium-hard, through-hardened gears in the second reduction. Figure 1.12 shows a large marine drive unit being lowered into position.

The designer of marine gearing has to worry about both noise and load-

FIG. 1.12 Lowering a very powerful marine gear unit into the engine space of a supertanker. (*Courtesy of Transamerica Delaval Inc., Trenton, NJ, U.S.A.*)

carrying capacity. Although the tooth loads are not high compared with those on aircraft or transportation gearing, the capacity of the medium-hardness gearing to carry load is not high either. Considering that during its lifetime a high-speed pinion on a cargo ship may make 10 to 11 billion cycles of operation at full-rated torque, it can be seen that load-carrying capacity is very important.

Gear noise is a more or less critical problem on all ship gearing. The auxiliary gears that drive generators are frequently located quite close to passenger quarters. The peculiar high-pitched whine of a high-speed gear has a damaging effect on either an engine-room operator or a passenger who may be quartered near the engine room.

On fighting ships, there is the added problem of keeping enemy submarines from picking up water-borne noises and of operating your own ship quietly enough to be able to hear water-borne noises from an enemy.

Quietness is achieved on marine gears by making the gears with large numbers of teeth and cutting the teeth with extreme accuracy. A typical marine pinion may have around 60 teeth and a tooth-to-tooth spacing accuracy of 5 μm (0.0002 in.) maximum. A pinion of the same diameter used in a railroad-gear application would have about 15 teeth and a tooth-spacing accuracy of 12 μm (0.0005 in.). In the comparison just made, the marine-gear teeth would be only one-quarter the size of the railroad-gear teeth.

1.8 Aerospace Gears

Gears are used in a wide variety of applications on aircraft. Propellers are usually driven by single- or double-reduction gear trains. Accessories such as generators, pumps, hydraulic regulators, and tachometers are gear-driven. Many gears are required to drive these kinds of accessories even on *jet* engines—which have no propellers. Additional gears are used to raise landing wheels, open bomb-bay doors, control guns, operate computers for gun- or bomb-sighting devices, and control the pitch of propellers. Helicopters have a considerable amount of gearing to drive main rotors and tail rotors. (See Fig. 1.13.) Space vehicles often use power gears between the turbine and the booster fuel pumps.

The most distinctive types of aircraft gears are the power gears for propellers, accessories, and helicopter rotors. The control and actuating types of gears are not too different from what would be used for ground applications of a similar nature (except that they are often highly loaded and made of extra hard, high-quality steel).

Aircraft power gears are usually housed in aluminum or magnesium castings. The gears have thin webs and light cross sections in the rim or hub. Accessory gears are frequently made integral with an internally splined hub.

FIG. 1.13 Cutaway view of a helicopter gear unit. The last two stages are epicyclic. The first stage is spiral bevel. (*Courtesy of Bell Helicopter Textron, Fort Worth, TX, U.S.A.*)

Spur or bevel gears are usually used for accessory drives. Envelope clearances required to mount the accessory driven by the gear often make it necessary to use large center distance but narrow face width. This fact plus the thrust problem tends to rule out helical accessory gears. Propeller-drive gears have wider face widths. Spur, bevel, and helical gear drives are all currently in use. To get maximum power capacity with small size and lightweight gearboxes, it is often desirable to use an epicyclic gear train. In this kind of arrangement, there is only one output gear, but several pinions drive against it.

Aerospace power gears are usually made of high-alloy steel and fully hardened (on the tooth surface) by either case carburizing or nitriding. In some designs it is feasible to cut, shave, case-carburize, and grind only the journals. Many designs have such thin, nonsymmetrical webs as to require grinding after hardening. In general, piston-engine gearing runs more slowly

than gas-turbine gearing. This makes some difference in the required accuracy. So far, many more piston-engine gears have been successfully finished before hardening than have gas-turbine gears.

The tooth loads and speeds are both very high on modern aircraft gears. The designer must achieve high tooth strength and high wear resistance. In addition, the thin oils used for low-temperature starting of military aircraft make the scoring type of lubrication failure a critical problem. Several special things are done to meet the demands of aircraft-gear service. Pinions are often made *long* addendum and gears *short* addendum to adjust the tip sliding velocities and to strengthen the pinion. Pinion tooth thicknesses are often increased at the expense of the gear to strengthen the pinion. High pressure angles, such as $22\frac{1}{2}°$, $25°$, and $27\frac{1}{2}°$, are often used to reduce the contact stress on the tooth surface and increase the width of the tooth at the base. Involute-profile modifications are generally used to compensate for bending and to keep the tips from cutting the mating part.

The most highly developed aerospace gears are those used in rocket engines. The American projects *Vanguard*, *Mercury*, *Gemini*, and *Apollo* succeeded in boosting heavy payloads into orbit and eventually putting men on the moon. Unusual aerospace-gear capability was developed to meet the special requirements of power gears and control gears in space vehicles.

Materials and dimensional tolerances must be held under close control. A gear failure can frequently result in the loss of human life. The gear designer and builder both have a grave responsibility to furnish gears that are always sure to work satisfactorily. Extensive ground and flight testing are required to prove new designs.

The machine tools used to make aircraft gears are conventional hobbers, shapers, bevel-gear generators, shavers, and gear grinders. For the close tolerances, the machinery must be in the very best condition and precision tooling must be used. A complete line of checking equipment is needed so that involute profile, tooth spacing, helix angle, concentricity, and surface finish can be precisely measured.

1.9 Industrial Gearing

A wide range of types and sizes of gears that are used in homes, factories, and offices come under the "industrial" category (see examples in Figs. 1.14 and 1.15). In general, these gears involve electric power from a motor used to drive something. The driven device may be a pump, conveyor, or liquid stirring unit. It may also be a garage door opener, an air compressor for office refrigeration, a hoist, a winch, or a drive to mix concrete on a truck hauling the concrete to the job.

Industrial gearing is relatively low speed and low horsepower. Typical

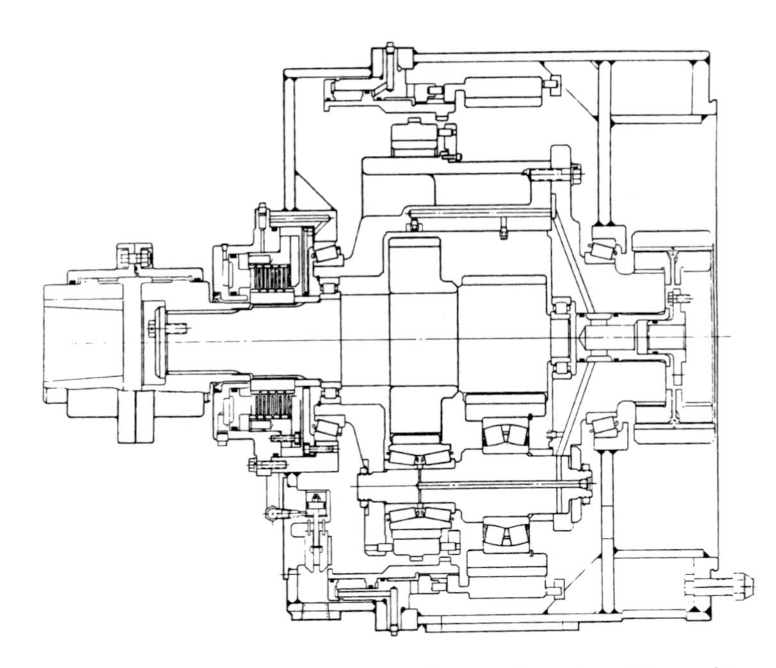

FIG. 1.14 Truck-mounted, two-speed gear unit used to drive a piston pump for oil-well fracturing. (*Courtesy of Sier-Bath Gear Co., Inc., North Bergen, NJ, U.S.A.*)

pitch-line speeds range from about 0.5 m/s to somewhat over 20 m/s (100 fpm to 4000 fpm). The types of gears may be spur, helical, bevel, worm, or Spiroid.* The power may range from less than 1 kW up to a few hundred kW. Typical input speeds are those of the electric motor, like 1800, 1500, 1200 and 1000 rpm (revolutions per minute).

The industrial field also includes drives with hydraulic motors. The field is characterized more by relatively low pitch-line speeds and power inputs than by the means of making or using the power.

Much of the gearing used in industrial work is made with through-hardened steel used *as cut*. There is, though, growing use of fully hardened gears where the size of the gearing or the life of the gearing is critical.

In the past, industrial gearing has not generally required long life or high reliability. The trend now—in the more important factory installations—is to obtain gears for moderately long life and reliability. For instance, a pump

*Spiroid is a registered trademark of the Spiroid Division of the Illinois Tool Works, Chicago, IL, U.S.A.

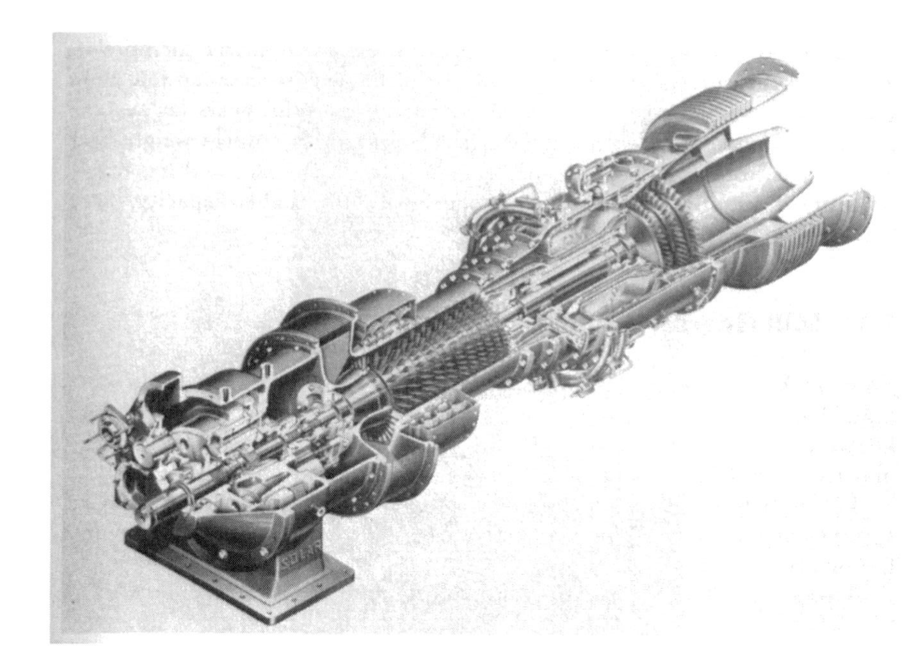

FIG. 1.15 Two-stage epicyclic gear, close-coupled to a high-speed gas turbine. This unit is used to drive a generator. (*Courtesy of Solar Turbines Incorporated, San Diego, CA, U.S.A.*)

drive with an 80 percent probability of running OK for 1000 hours might have been quite acceptable in the 1960s. The pump buyer in the 1980s may be more concerned with the cost of *downtime* and parts replacement, and may want to get gears good enough to have a 95 percent probability of lasting for 10,000 hours at rated load.

1.10 Gears in the Oil and Gas Industry

The production of petroleum products for the energy needs of the world requires a considerable amount of high-power, high-speed gearing. Gear units are used on oil platforms, pumping stations, drilling sites, refineries, and power stations. Usually the drive is a turbine, but it may be a large diesel engine. The power range goes from about 750 kW to over 50,000 kW. Pitch-line speeds range from 20 to 200 m/s (4000 to 40,000 fpm).

Bevel gears are used to a limited extent. Sometimes a stage of bevel gearing is needed to make a 90° turn in a power drive. (As an example, a horizontal-axis turbine may drive a vertical-axis compressor.)

Hardened and ground gears are widely used. With better facilities to grind and measure large gears and better equipment to case-harden large gears has come a strong tendency to design the powerful gears for turbine and diesel applications with fully hardened teeth. This reduces weight and size considerably. Pitch-line speeds become lower. Less space and less frame structure is required in a power package with the higher-capacity, fully hardened gears.

1.11 Mill Gears

Large mills make cement, grind iron ore, make rubber, roll steel, or do some other basic function. See Fig. 1.16. It is common to have a few thousand kilowatts of power going through two or more gear stages to drive some massive processing drum or rolling device.

The mill is usually powered by electric motors, but diesel engines or turbines may be used. The characteristic of mill drives is high power (and frequently unusually high torques).

A process mill will often run continuously for months at a time. Down-

FIG. 1.16 Large mill gear drive. Note twin power paths to bull gear. (*Courtesy of the Falk Corporation, Milwaukee, WI, U.S.A.*)

FIG. 1.17 Jumbo-size drive gear for a large mill application. (*Courtesy of David Brown Gear Industries, Huddersfield, England*)

time is critical because the output ceases completely when a mill unit is shut down.

Spur and helical gears are generally used. Pitch-line speeds are usually quite low. A first stage may be going around 20 m/s (4000 fpm), but the final stage in a mill may run as slowly as 0.1 to 1.0 m/s (20 to 200 fpm). Some bevel gears are also used—where axes must be at 90°.

The larger mill gears are generally made medium hard, but they may even be of a low hardness. The very large sizes involved often make it impractical to use the harder gears. There is an increasing use, though, of fully hardened mill gears up to gear pitch diameters of about 2 meters.

Mill gears are commonly made in sizes up to 11 meters. Such giant gears have to be made in two or more segments. Figure 1.17 shows a large two-segment mill gear. The segments are bolted together for cutting and then unbolted for shipping. (It is impractical to ship round pieces of metal over about 5 meters in diameter.)

SELECTION OF THE RIGHT KIND OF GEAR

The preceding sections gave some general information on how gears are made and used in different fields. In this part of the chapter, we shall concentrate on the problem of selecting the *right* kind of gear. The first step in designing a set of gears is to pick the right kind.

In many cases, the geometric arrangement of the apparatus which needs a gear drive will considerably affect the selection. If the gears must be on parallel axes, then spur or helical gears are the ones to use. Bevel and worm gears can be used if the axes are at right angles, but they are not feasible with parallel axes. If the axes are nonintersecting and nonparallel, then crossed-helical gears, hypoid gears, worm gears, or Spiroid gears may be used. Worm gears, though, are seldom used if the axes are not at right angles to each other. Table 1.3 shows in more detail the principal kinds of gears and how they are mounted.

There are no dogmatic rules that tell the designer just which gear to use. Frequently the choice is made after weighing the advantages and disadvantages of two or three kinds of gears. Some generalizations, though, can be made about gear selection.

In general, external helical gears are used when both high speeds and high horsepowers are involved. External helical gears have been built to carry as much as 45,000 kW of power on a single pinion and gear. And this is not the limit for designing helical gears—bigger ones could be built if anyone needed them. It is doubtful if any other kind of gear could be built and used successfully to carry this much power on a single mesh.

Bevel gears are ordinarily used on right-angle drives when high efficiency is needed. These gears can usually be designed to operate with 98 percent or better efficiency. Worm gears often have a hard time getting above 90 percent efficiency. Hypoid gears do not have as good efficiency as bevel gears, but they make up for this by being able to carry more power in the same space—provided the speeds are not too high.

TABLE 1.3 Kinds of Gears in Common Use

Parallel axes	Intersecting axes	Nonintersecting nonparallel axes
Spur external	Straight bevel	Crossed-helical
Spur internal	Zerol bevel	Single-enveloping worm
Helical external	Spiral bevel	Double-enveloping worm
Helical internal	Face gear	Hypoid
		Spiroid

Worm gears are ordinarily used on right-angle drives when very high ratios are needed. They are also widely used in low to medium ratios as packaged speed reducers. Single-thread worms and worm gears are used to provide the indexing accuracy on many machine tools. The critical job of indexing hobbing machines and gear shapers is nearly always done by a worm-gear drive.

Spur gears are relatively simple in design and in the machinery used to manufacture and check them. Most designers prefer to use them wherever design requirements permit.

Spur gears are ordinarily thought of as slow-speed gears, while helical gears are thought of as high-speed gears. If noise is not a serious design problem, spur gears can be used at almost any speeds which can be handled by other types of gears. Aircraft gas-turbine spur gears sometimes run at pitch-line speeds above 50 m/s (10,000 fpm). In general, though, spur gears are not used much above 20 m/s (4000 fpm).

1.12 External Spur Gears

Spur gears are used to transmit power between parallel shafts. They impose only radial loads on their bearings. The tooth profiles are ordinarily curved in the shape of an involute. Variations in center distance do not affect the trueness of the gear action unless the change is so great as to either jam the teeth into the root fillets of the mating member or withdraw the teeth almost out of action.

Spur-gear teeth may be hobbed, shaped, milled, stamped, drawn, sintered, cast, or shear-cut. They may be given a finishing operation such as grinding, shaving, lapping, rolling, or burnishing. Speaking generally, there are more kinds of machine tools and processes available to make spur gears than to make any other gear type. This favorable situation often makes spur gears the choice where cost of manufacture is a major factor in the gear design.

The standard measure of spur-gear tooth size in the metric system is the *module*. In the English system, the standard measure of tooth size is *diametral pitch*. The meanings are:

Module is millimeters of pitch diameter per tooth.

Diametral pitch is number of teeth per inch of pitch diameter (a reciprocal function).

Mathematically,

$$\text{Module} = \frac{25.400}{\text{diametral pitch}} \tag{1.1}$$

or

$$\text{Diametral pitch} = \frac{25.400}{\text{module}}$$

Curiously, module and diametral pitch are size dimensions which cannot be directly measured on a gear. They are really reference values used to calculate other size dimensions which are measurable.

Gears can be made to any desired module or diametral pitch, provided that cutting tools are available for that tooth size. To avoid purchasing cutting tools for too many different tooth sizes, it is desirable to pick a progression of modules and design to these except where design requirements force the use of special sizes. The following commonly used modules are recommended as a start for a design series: 25, 20, 15, 12, 10, 8, 6, 5, 4, 3, 2.5, 2.0, 1.5, 1, 0.8, 0.5. Many shops are equipped with English-system diametral pitches in this series: 1, $1\frac{1}{4}$, $1\frac{1}{2}$, $1\frac{3}{4}$, 2, $2\frac{1}{2}$, 3, 4, 5, 6, 8, 10, 12, 16, 20, 24, 32, 48, 64, 128. Considering the trend toward international trade, it is desirable to purchase new gear tools in standard metric sizes, so that they will be handy for gear work going to any part of the world. *(The standard measuring system of the world is the metric system.)*

Most designers prefer a 20° pressure angle for spur gears. In the past the $14\frac{1}{2}$° pressure angle was widely used. It is not popular today because it gets into trouble with undercutting much more quickly than the 20° tooth when small numbers of pinion teeth are needed. Also, it does not have the load-carrying capacity of the 20° tooth. A pressure angle of $22\frac{1}{2}$° or 25° is often used. Pressure angles above 20° give higher load capacity but may not run quite as smoothly or quietly.

Figure 1.18 shows the terminology used with a spur gear or a spur rack (a rack is a section of a spur gear with an *infinitely* large pitch diameter).

The following formulas apply to spur gears in all cases:

Circular pitch	= pi × module	Metric	(1.2*a*)
	= pi ÷ diametral pitch	English	(1.2*b*)
Pitch diameter	= no. of teeth × module	Metric	(1.3*a*)
	= no. of teeth ÷ diametral pitch	English	(1.3*b*)

The *nominal* center distance is equal to the sum of the pitch diameter of the pinion and the pitch diameter of the gear divided by 2:

$$\text{Center distance} = \frac{\text{pinion pitch dia.} + \text{gear pitch dia.}}{2} \tag{1.4}$$

Since the center distance is a machined dimension, it may not come out to be exactly what the design calls for. In addition, it is common practice to use a slightly larger center distance to increase the operating pressure angle. For instance, if the actual center is made 1.7116 percent larger, gears cut

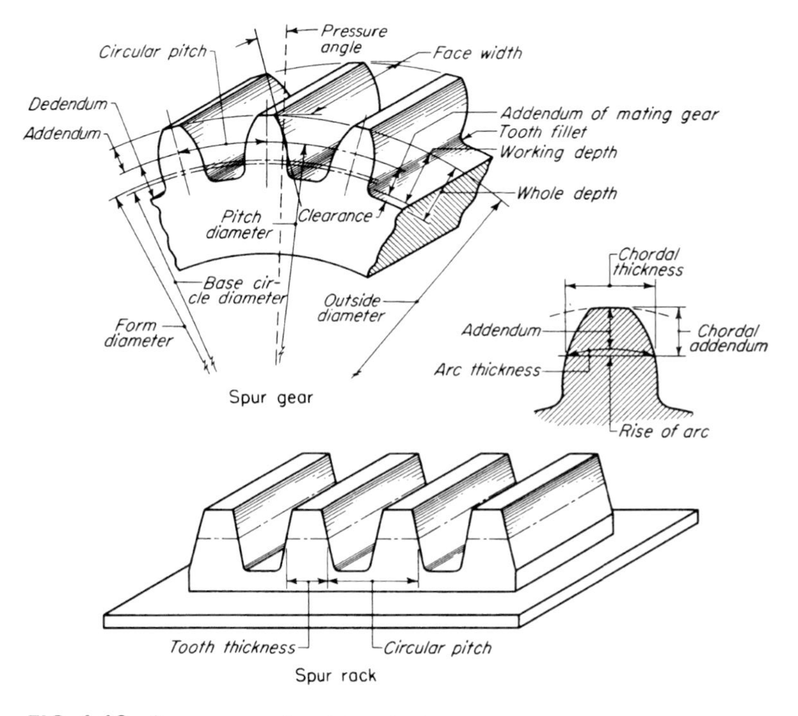

FIG. 1.18 Spur-gear and rack terminology.

with 20° hobs or shaper-cutters will run at $22\frac{1}{2}$° pressure angle. (See Sec. 8.1 for methods of design for special center distances.)

For the reasons just mentioned, it is possible to have two center distances, a *nominal* center distance and an *operating* center distance. Likewise, there are two pitch diameters. The pitch diameter for the tooth-cutting operation is the nominal pitch diameter and is given by Eq. (1.3). The operating pitch diameter is

$$\text{Pitch dia. (operating) of pinion} = \frac{2 \times \text{operating cent. dist.}}{\text{ratio} + 1} \tag{1.5}$$

$$\text{Pitch dia. (operating) of gear} = \text{ratio} \times \text{pitch dia. (operating) of pinion} \tag{1.6}$$

where

$$\text{Ratio} = \frac{\text{no. gear teeth}}{\text{no. pinion teeth}} \tag{1.7}$$

1.13 External Helical Gears

Helical gears are used to transmit power or motion between parallel shafts. The helix angle must be the same in degrees on each member, but the hand of the helix on the pinion is opposite to that on the gear. (A RH pinion meshes with a LH gear, and a LH pinion meshes with a RH gear.)

Single helical gears impose both thrust and radial load on their bearings. Double helical gears develop equal and opposite thrust reactions which have the effect of canceling out the thrust load. Usually double helical gears have a gap between helices to permit a runout clearance for the hob, grinding wheel, or other cutting tool. One kind of gear shaper has been developed that permits double helical teeth to be made *continuous* (no gap between helices).

Helical-gear teeth are usually made with an involute profile in the *transverse* section (the transverse section is a cross section perpendicular to the gear axis). Small changes in center distance do not affect the action of helical gears.

Helical-gear teeth may be made by hobbing, shaping, milling, or casting. Sintering has been used with limited success. Helical teeth may be finished by grinding, shaving, rolling, lapping, or burnishing.

The size of helical-gear teeth is specified by module for the metric system and by diametral pitch for the English system. The helical tooth will frequently have some of its dimensions given in the *normal section* and others given in the *transverse section.* Thus standard cutting tools could be specified for either section—but not for *both* sections. If the helical gear is small (less than 1 meter pitch diameter), most designers will use the same pressure angle and standard tooth size in the normal section of the helical gear as they would use for spur gears. This makes it possible to hob helical gears with standard spur gear hobs. (It is not possible, though, to cut helical gears with standard spur-gear shaper-cutters.)

Helical gears often use 20° as the standard pressure angle in the normal section. However, higher pressure angles, like $22\frac{1}{2}°$ or 25°, may be used to get extra load-carrying capacity.

Figure 1.19 shows the terminology of a helical gear and a helical rack. In the transverse plane, the elements of a helical gear are the same as those of a spur gear. Equations (1.1) to (1.7) apply just as well to the transverse plane of a helical gear as they do to a spur gear. Additional general formulas for helical gears are:

$$\text{Normal circ. pitch} = \text{circ. pitch} \times \text{cosine helix angle} \tag{1.8}$$

$$\text{Normal module} = \text{transverse module} \times \text{cosine helix angle} \tag{1.9}$$

$$\text{Normal diam. pitch} = \text{trans. diam. pitch} \div \text{cosine helix angle} \tag{1.10}$$

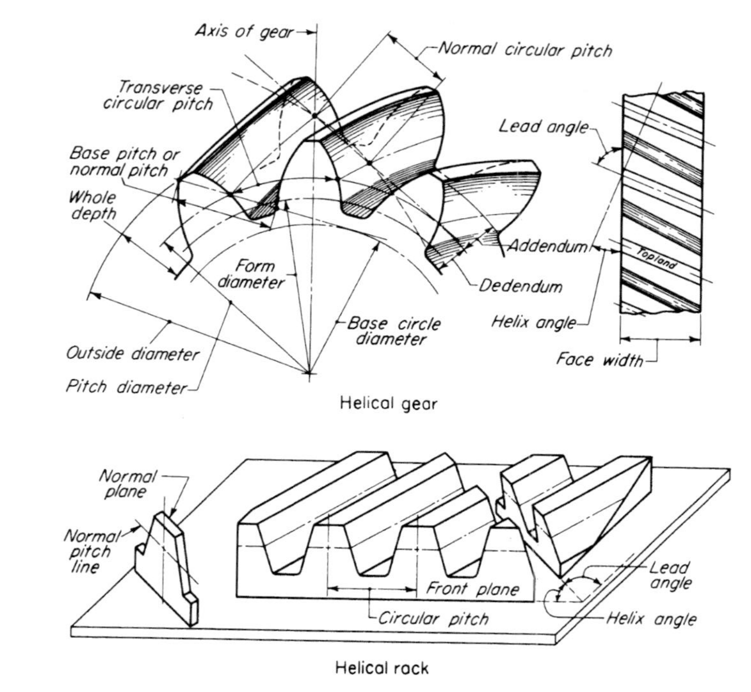

FIG. 1.19 Helical-gear and rack terminology.

$$\text{Axial pitch} = \text{circ. pitch} \div \text{tangent helix angle} \tag{1.11}$$

$$= \text{norm. circ. pitch} \div \text{sine helix angle} \tag{1.12}$$

1.14 Internal Gears

Two internal gears will not mesh with each other, but an external gear may be meshed with an internal gear. The external gear must not be larger than about two-thirds the pitch diameter of the internal gear when full-depth 20° pressure angle teeth are used. The axes on which the gears are mounted must be parallel.

Internal gears may be either spur or helical. Even double helical internal gears are used occasionally.

An internal gear is a necessity in an epicyclic type of gear arrangement. The short center distance of an internal gearset makes it desirable in some applications where space is very limited. The shape of an internal gear forms

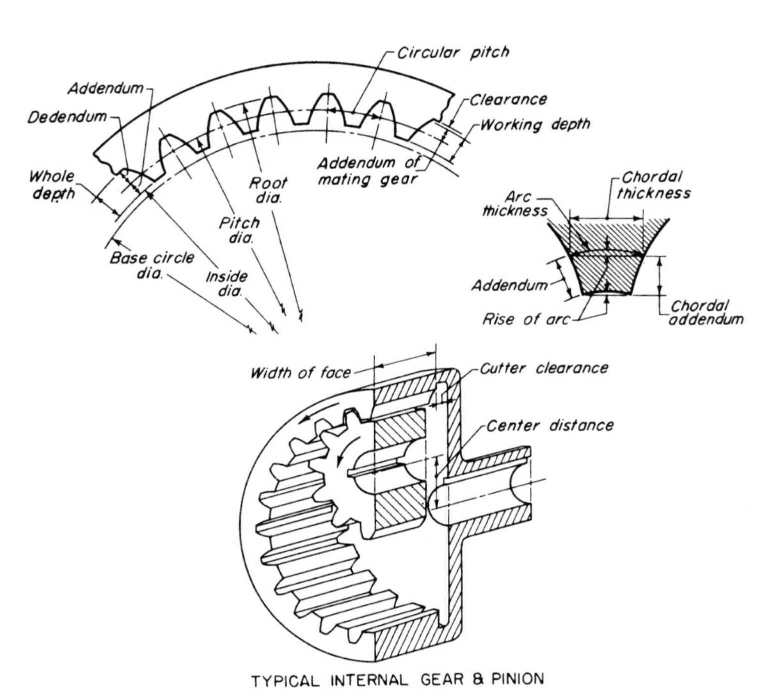

FIG. 1.20 Internal-gear terminology.

a natural guard over the meshing gear teeth. This is very advantageous for some types of machinery.

Internal gears have the disadvantage that fewer types of machine tools can produce them. Internal gears cannot be hobbed.* They can be shaped, milled, or cast. In small sizes they can be broached. Both helical and spur internals can be finished by shaving, grinding, lapping, or burnishing.

An internal gear has the same helix angle in degrees and the same hand as its mating pinion (right-hand pinion meshes with right-hand gear and vice versa).

Figure 1.20 shows the terminology used for a spur internal gear. All the previously given formulas apply to internal gearing except those involving center distance [Eqs. (1.4), (1.5), and (1.6) do not hold for internals]. Formulas for internal-gear center distance are

$$\text{Center distance} = \frac{\text{pitch dia. of gear} - \text{pitch dia. of pinion}}{2} \tag{1.13}$$

*Some very special hobs and hobbing machines have been used—to a rather limited extent—to hob internal gears.

$$\text{Pitch dia. (operating) of pinion} = \frac{2 \times \text{op. cent. dist.}}{\text{ratio} - 1} \tag{1.14}$$

$$\text{Pitch dia. (operating) of gear} = \frac{2 \times \text{op. cent. dist.} \times \text{ratio}}{\text{ratio} - 1} \tag{1.15}$$

1.15 Straight Bevel Gears

Bevel gear blanks are conical in shape. The teeth are tapered in both tooth thickness and tooth height. At one end the tooth is large, while at the other end it is small. The tooth dimensions are usually specified for the *large* end of the tooth. However, in calculating bearing loads, the central-section dimensions and forces are used.

The simplest type of bevel gear is the *straight* bevel gear. These gears are commonly used for transmitting power between intersecting shafts. Usually the shaft angle is 90°, but it may be almost any angle. The gears impose both radial and thrust load on their bearings.

Bevel gears must be mounted on axes whose shaft angle is almost exactly the same as the design shaft angle. Also, the axes on which they are mounted must either intersect or come very close to intersecting. In addition to the accuracy required of the axes, bevel gears must be mounted at the right distance from the cone center. The complications involved in mounting bevel gears make it difficult to use sleeve bearings with large clearances (which is often done on high-speed, high-power spur and helical gears). Ball and roller bearings are the kinds commonly used for bevel gears. The limitations of these bearings' speed and load-carrying capacity indirectly limit the capacity of bevel gears in some high-speed applications.

Straight bevel teeth are usually cut on bevel-gear generators. In some cases, where accuracy is not too important, bevel-gear teeth are milled. Bevel teeth may also be cast. Lapping is the process often used to finish straight bevel teeth. Shaving is not practical for straight bevel gears, but straight bevels may be ground.

The size of bevel-gear teeth is defined in module for the metric system and in diametral pitch for the English system. The specified size dimensions are given for the large end of the tooth. A bevel gear tooth which is 12 module at the large end may be only around 10 module at the small end. The commonly used modules (or diametral pitches) are the same as those used for spur gears (see Sec. 1.12). There is no particular advantage to using standard tooth sizes for bevel gears. A set of cutting tools will cut more than a single pitch.

The two views of bevel gears in Fig. 1.21 show bevel-gear terminology. Bevel-gear teeth have profiles which closely resemble an involute curve. The

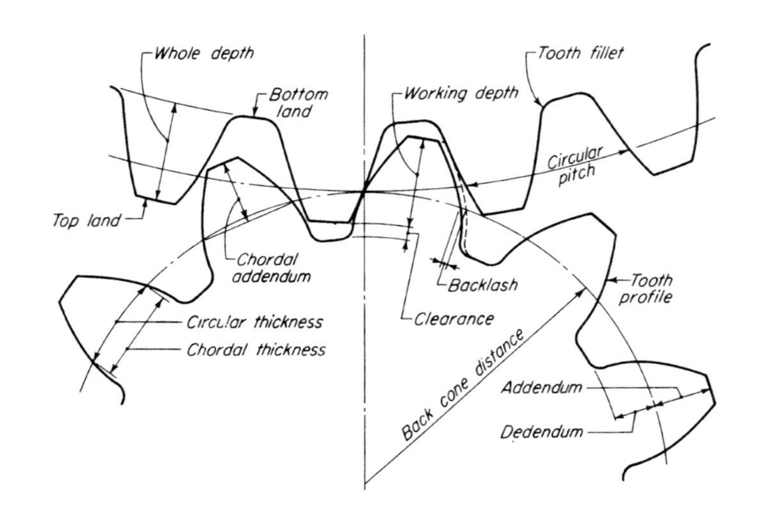

FIG. 1.21 Bevel-gear terminology.

shape of a straight bevel-gear tooth (in a section *normal* to the tooth) closely approximates that of an involute spur gear with a larger number of teeth. This larger number of teeth, called the *virtual number of teeth*, is equal to the actual number of teeth divided by the cosine of the pitch angle.

Straight bevel-gear teeth have been commonly made with $14\frac{1}{2}$, $17\frac{1}{2}$, and 20° pressure angles. The 20° design is the most popular.

The pitch angle of a bevel gear is the angle of the pitch cone. It is a measure of the amount of taper in the gear. For instance, as the taper is reduced, the pitch angle approaches zero and the bevel gear approaches a spur gear.

The pitch angles in a set of bevel gears are defined by lines meeting at the cone center. The root and face angles are defined by lines which *do not* hit the cone center (or apex). In old-style designs these angles did meet at the apex, but modern designs make the outside cone of one gear parallel to the root cone of its mate. This gives a constant clearance and permits a better cutting-tool design and gear-tooth design than the old-style design with its tapering clearance.

The circular pitch and the pitch diameters of bevel gears are calculated the same as for spur gears. [See Eqs. (1.1) to (1.3).] The pitch-cone angles may be calculated by one of the following sets of equations:

$$\text{Tan pitch angle, pinion} = \frac{\text{no. teeth in pinion}}{\text{no. teeth in gear}} \tag{1.16}$$

$$\text{Tan pitch angle, gear} = \frac{\text{no. teeth in gear}}{\text{no. teeth in pinion}} \tag{1.17}$$

When the shaft angle is less than 90°,

$$\text{Tan pitch angle, pinion} = \frac{\text{sin shaft angle}}{\text{ratio} + \text{cos shaft angle}} \tag{1.18}$$

$$\text{Tan pitch angle, gear} = \frac{\text{sin shaft angle}}{1/\text{ratio} + \text{cos shaft angle}} \tag{1.19}$$

When the shaft angle is over 90°,

$$\text{Tan pitch angle, pinion} = \frac{\sin\,(180° - \text{shaft angle})}{\text{ratio} - \cos\,(180° - \text{shaft angle})} \tag{1.20}$$

$$\text{Tan pitch angle, gear} = \frac{\sin\,(180° - \text{shaft angle})}{1/\text{ratio} - \cos\,(180° - \text{shaft angle})} \tag{1.21}$$

In all the above cases,

$$\text{Pitch angle, pinion} + \text{pitch angle, gear} = \text{shaft angle} \tag{1.22}$$

FIG. 1.22 A pair of Zerol bevel gears. (*Courtesy of the Gleason Works, Rochester, NY, U.S.A.*)

1.16 Zerol Bevel Gears

Zerol* bevel gears are similar to straight bevel gears except that they have a curved tooth in the lengthwise direction. See Fig. 1.22. Zerol bevel gears have 0° spiral angle. They are made in a different kind of machine from that used to make straight bevel gears. The straight-bevel-gear-generating machine has a cutting tool which moves back and forth in a straight line. The Zerol is generated by a rotary cutter that is like a face mill. It is the curvature of this cutter that makes the lengthwise curvature in the Zerol tooth.

The Zerol gear has a profile which somewhat resembles an involute curve. The pressure angle of the tooth varies slightly in going across the face width. This is caused by the lengthwise curvature of the tooth.

Zerol gear teeth may be finished by grinding or lapping. Since the Zerol gear can be ground, it is favored over straight bevel gears in applications in which both high accuracy and full hardness are required. Even in applications in which cut gears of machinable hardness can be used, the Zerol may be the best choice if speeds are high. Because of its lengthwise curvature, the Zerol tooth has a slight overlapping action. This tends to make it run more smoothly than the straight-bevel-gear tooth. A cut Zerol bevel gear is usually more accurate than a straight bevel gear.

In making a set of Zerol gears, one member is made first, using theoretical machine settings. Then a second gear is finished in such a way that its

*Zerol is a registered trademark of the Gleason Works, Rochester, NY, U.S.A.

profile and lengthwise curvature will give satisfactory contact with the first gear. Several trial cuts and adjustments to machine settings may be required to develop a set of gears which will conjugate properly. If a number of identical sets of gears are required, a matching set of test gears is made. Then each production gear is machined so that it will mesh satisfactorily with one or the other of the test gears. In this way a number of sets of interchangeable gears may be made.

Zerol gears are usually made to a 20° pressure angle. In a few ratios where pinion and gear have small numbers of teeth, $22\frac{1}{2}°$ or 25° is used.

The calculations for pitch diameter and pitch-cone angle are the same for Zerol bevel gears as for straight bevel gears.

1.17 Spiral Bevel Gears

Spiral bevel gears have a lengthwise curvature like Zerol gears. However, they differ from Zerol gears in that they have an appreciable angle with the axis of the gear. See Fig. 1.23. Although spiral bevel teeth do not have a true helical spiral, a spiral bevel gear looks somewhat like a helical bevel gear.

Spiral bevel gears are generated by the same machines that cut or grind Zerol gears. The only difference is that the cutting tool is set at an angle to the axis of the gear instead of being set essentially parallel to the gear axis.

Spiral bevel gears are made in matched sets like Zerol bevel gears. Different sets of the same design are not interchangeable unless they have been purposely built to match a common set of test gears.

FIG. 1.23 A pair of spiral bevel gears. (*Courtesy of the Gleason Works, Rochester, NY, U.S.A.*)

Generating types of machines are ordinarily used to cut or grind spiral-bevel-gear teeth. In some high-production jobs, a special kind of machine is used which cuts the teeth without going through a generating motion. Spiral-bevel-gear teeth are frequently given a lapping operation to finish the teeth and obtain the desired tooth bearing.

In high-speed gear work, the spiral bevel is preferred over the Zerol bevel because its spiral angle tends to give the teeth a considerable amount of overlap. This makes the gear run more smoothly, and the load is distributed over more tooth surface. However, the spiral bevel gear imposes much more thrust load on its bearings than does a Zerol bevel gear.

Spiral bevel gears are commonly made to 16°, $17\frac{1}{2}$°, 20°, and $22\frac{1}{2}$° pressure angles. The 20° angle has become the most popular. It is the only angle used on aircraft and instrument gears. The most common spiral angle is 35°.

1.18 Hypoid Gears

Hypoid gears resemble bevel gears in some respects. They are used on crossed-axis shafts, and there is a tendency for the parts to taper as do bevel gears. They differ from true bevel gears in that their axes do not intersect. The distance between a hypoid pinion axis and the axis of a hypoid gear is called the *offset*. This distance is measured along the perpendicular common to the two axes. If a set of hypoid gears had *no offset*, they would simply be spiral bevel gears. See Fig. 1.24 for *offset* and other terms.

Hypoid pinions may have as few as five teeth in a high ratio. Since the various kinds of bevel gears do not often go below 10 teeth in a pinion, it can be seen that it is easier to get high ratios with hypoid gears.

Contrary to the general rule with spur, helical, and bevel gears, hypoid pinions and gears *do not* have pitch diameters which are in proportion to their numbers of teeth. This makes it possible to use a large and strong pinion even with a high ratio and only a few pinion teeth. See Fig. 1.25.

Hypoid teeth have unequal pressure angles and unequal profile curvatures on the two sides of the teeth. This results from the unusual geometry of the hypoid gear rather than from a nonsymmetrical cutting tool.

Hypoid gearsets are matched to run together, just as Zerol or spiral gearsets are matched. Interchangeability is obtained by making production gears fit with test masters.

Hypoid gears and pinions are usually cut on a generating type of machine. They may be finished by either grinding or lapping.

The hypoid gears for passenger cars and for industrial drives usually have a basic pressure angle of 21°15′. For tractors and trucks the average pressure angle is 22°30′. Pinions are frequently made with a spiral angle of 45° or 50°.

In hypoid gearing, module and diametral pitch are used for the gear *only*.

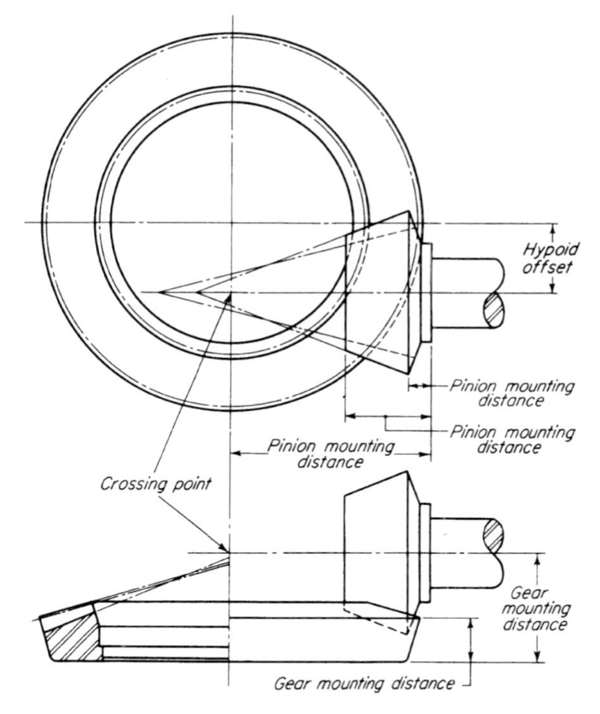

FIG. 1.24 Hypoid-gear arrangement.

FIG. 1.25 A pair of hypoid gears (*Courtesy of the Gleason Works, Rochester, NY, U.S.A.*)

Likewise, the pitch diameter and the pitch angle are figured for the gear only. If a pitch were used for the pinion, it would be smaller than that of the gear. The size of a hypoid pinion is established by its outside diameter and its number of teeth. The geometry of hypoid teeth is defined by the various dimensions used to set up the machines to cut the teeth.

1.19 Face Gears

Face gears have teeth cut on the end face of a gear, just as the name "face" implies. They are not ordinarily thought of as bevel gears, but functionally they are more akin to bevel gears than to any other kind.

A spur pinion and a face gear are mounted—like bevel gears—on shafts that intersect and have a shaft angle (usually 90°). The pinion bearings carry mostly radial load, while the gear bearings have both thrust and radial load. The mounting distance of the pinion from the pitch-cone apex is not critical, as it is in bevel or hypoid gears. Figure 1.26 shows the terminology used with face gears.

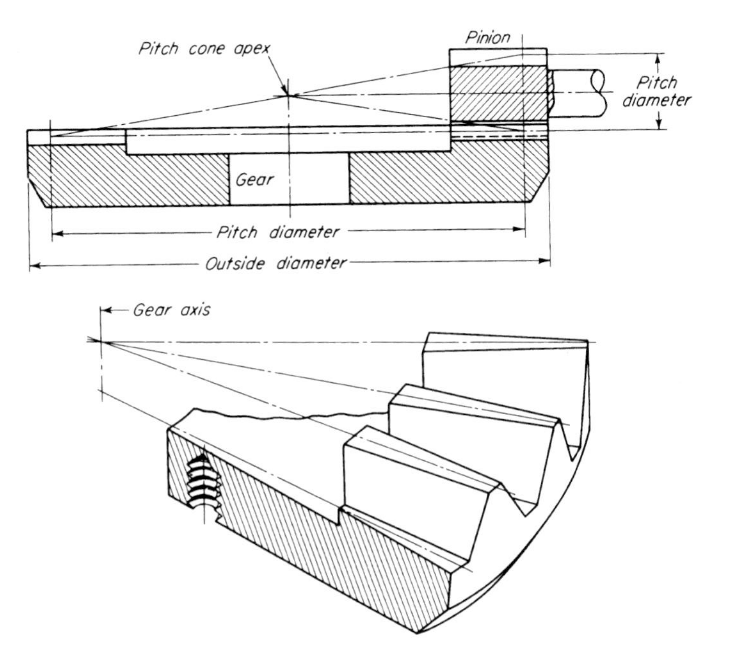

FIG. 1.26 Face-gear terminology.

The pinion that goes with a face gear is usually made spur, but it can be made helical if necessary. The formulas for determining the dimensions of a pinion to run with a face gear are no different from those for the dimensions of a pinion to run with a mating gear on parallel axes. The pressure angles and pitches used are similar to spur-gear (or helical-gear) practice.

The pinion may be finished or cut by all the methods previously mentioned for spur and helical pinions. The gear, however, must be finished with a shaper-cutter which is almost the same size as the pinion. Equipment to grind face gears is not available. The teeth can be lapped, and they might be shaved without too much difficulty, although ordinarily they are not shaved.

The face-gear tooth changes shape from one end of the tooth to the other. The face width of the gear is limited at the outside end by the radius at which the tooth becomes pointed. At the inside end, the limit is the radius at which undercut becomes excessive. Practical considerations usually make it desirable to make the face width somewhat short of these limits.

The pinion to go with a face gear is usually made with 20° pressure angle.

1.20 Crossed-Helical Gears (Nonenveloping Worm Gears)

The word "spiral" is rather loosely used in the gear trade. The word may be applied to both helical and bevel gears. In this section we shall consider the special kind of worm gear that is often called a "spiral gear." More correctly, though, it is a *crossed-helical* gear.

Crossed-helical gears are essentially *nonenveloping* worm gears. Both members are cylindrically shaped. (See Fig. 1.27.) In comparison, the *single-enveloping* worm gearset has a cylindrical worm, but the gear is throated so that it tends to wrap around the worm. The *double-enveloping* worm gearset goes still further; both members are throated, and both members wrap around each other.

Crossed-helical gears are mounted on axes that do not intersect and that are at an angle to each other. Frequently the angle between the axes is 90°. The bearings for crossed-helical gears have both thrust and radial load.

A *point contact* is made between two spiral gear teeth in mesh with each other. As the gears revolve, this point travels across the tooth in a sloping line. After the gears have worn in for a period of time, a shallow, sloping line of contact is worn into each member. This makes the original point contact increase to a line as long as the width of the sloping band of contact. The load-carrying capacity of crossed-helical gears is quite small when they are new, but if they are worn in carefully, it increases quite appreciably.

Crossed-helical gearsets are able to stand small changes in center distance and small changes in shaft angle without any impairment in the accuracy

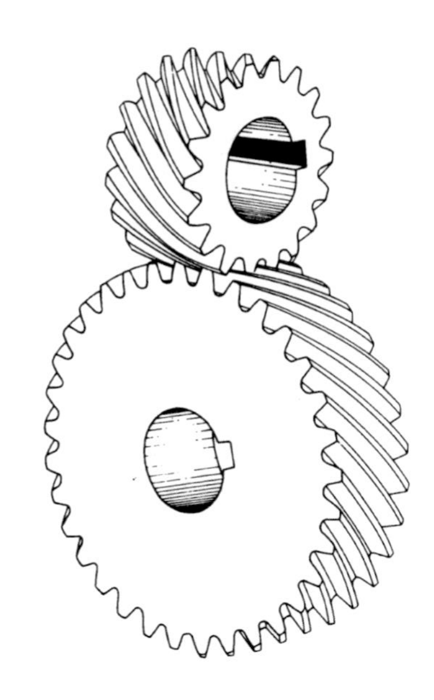

FIG. 1.27 Crossed-helical-gear drive.

with which the set transmits motion. This fact, and the fact that shifting either member endwise makes no difference in the amount of contact obtained, makes this the easiest of all gears to mount. There is no need to get close accuracy in center distance, shaft alignment, or axial position—provided the teeth are cut with reasonably generous face width and backlash.

Crossed-helical gears may be made by any of the processes used to make single helical gears. Up to the point of mounting the gear in a gearbox, there is no difference between a crossed-helical gear and a helical gear.

Usually a crossed-helical gear of one hand is meshed with a crossed-helical gear of the same hand. It is not necessary, though, to do this. If the shaft angle is properly set, it is possible to mesh opposite hands together. Thus the range of possibilities is

RH driver with RH driven
LH driver with LH driven
RH driver with LH driven
LH driver with RH driven

The pitch diameters of crossed-helical gears—like those of hypoid gears—are not in proportion to the tooth ratio. This makes the use of the word

"pinion" for the smaller member of the pair inappropriate. In a crossed-helical gearset, the small *pinion* might easily have more teeth than the *gear*! The two members of a crossed-helical set are described as *driver* and *driven*.

The same helix angle in degrees does not have to be used for each member. Whenever different helix angles are used, the module (or diametral pitch) for two crossed-helical gears that mesh with each other is not the same. The thing that is the same in all cases is the normal module (and the normal circular pitch). This makes the normal module (or normal diametral pitch) the most appropriate measure of tooth size.

Designers of crossed-helical gears usually get the best results when there is a contact ratio in the normal section of at least 2. This means that in all positions of tooth engagement, the load will be shared by at least two pair of teeth. To get this high contact ratio, a low normal pressure angle and a deep tooth depth are needed. When the helix is 45°, a normal pressure angle of $14\frac{1}{2}°$ gives good results.

Some of the basic formulas for crossed-helical gears are

$$\text{Shaft angle} = \text{helix angle of driver} \pm \text{helix angle of driven} \tag{1.23}$$

$$\text{Normal module} = \text{normal circ. pitch} \div \text{pi} \tag{1.24}$$

$$\text{Normal diam. pitch} = \frac{\text{pi}}{\text{normal circ. pitch}} \tag{1.25}$$

$$\text{Pitch dia.} = \frac{\text{no. of teeth} \times \text{normal module}}{\text{cosine of helix angle}} \tag{1.26}$$

$$\text{Center distance} = \frac{\text{pitch dia. driver} + \text{pitch dia. driven}}{2} \tag{1.27}$$

$$\text{Cosine of helix angle} = \frac{\text{no. teeth} \times \text{normal circ. pitch}}{\text{pi} \times \text{pitch dia.}} \tag{1.28}$$

1.21 Single-Enveloping Worm Gears

Figure 1.28 shows a single-enveloping worm gear. Worm gears are characterized by one member having a screw thread. Frequently the thread angle (lead angle) is only a few degrees. The worm in this case has the outward appearance of the thread on a bolt, greatly enlarged. When a worm has multiple threads and a lead angle approaching 45°, it may be (if it has an involute profile) geometrically just the same as a helical pinion of the same lead angle. In this case the only difference between a worm and a helical pinion would be in their usage.

Worm gears are usually mounted on nonintersecting shafts which are at a 90° shaft angle. Worm bearings usually have a high thrust load. The worm-

FIG. 1.28 Single-enveloping worm gearset. (*Courtesy of Transamerica Delaval, Delroyd Worm Gear Div., Trenton, NJ, U.S.A.*)

gear bearings have a high radial load and a low thrust load (unless the lead angle is high).

The single-enveloping worm gear has a line contact which extends either across the face width or across the part of the tooth that is in the zone of action. As the gear revolves, this line sweeps across the whole width and height of the tooth. The meshing action is quite similar to that of helical gears on parallel shafts, except that much higher sliding velocity is obtained for the same pitch-line velocity. In a helical gearset, the sliding velocity at the tooth tips is usually not more than about one-fourth the pitch-line velocity. In a high-ratio worm gearset, the sliding velocity is greater than the pitch-line velocity of the worm.

Worm gearsets have considerably more load-carrying capacity than crossed-helical gearsets. This results from the fact that they have *line* contact instead of *point* contact. Worm gearsets must be mounted on shafts that are

very close to being correctly aligned and at the correct center distance. The axial position of a single-enveloping worm is not critical, but the worm gear must be in just the right axial position so that it can wrap around the worm properly.

Several different kinds of worm-thread shapes are in common use. These are

Worm thread produced by straight-sided conical milling or grinding wheel

Worm thread straight-sided in the axial section

Worm thread straight-sided in the normal section

Worm thread an involute helicoid shape

The shape of the worm thread defines the worm-gear tooth shape. The worm gear is simply a gear element formed to be "conjugate" to a specified worm thread.

A worm and a worm gear have the same hand of helix. A RH worm, for example, meshes with a RH gear. The helix angles are usually very different for a worm and a worm gear. Usually the worm has a more than 45° helix angle, and the worm gear has a less than 45° helix angle. Customarily the *lead angle*—which is the complement of the helix angle—is used to specify the angle of the worm thread. See Fig. 1.29 for a diagram of the worm gear and its terminology. When the worm gearset has a 90° shaft angle, the worm lead angle is numerically equal to the worm-gear helix angle.

The *axial pitch* is the dimension that is used to specify the size of worm threads. It is the distance from thread to thread measured in an axial plane. When the shaft angle is 90°, the axial pitch of the worm is numerically equal to the worm-gear circular pitch. In the metric system, popular axial pitch values are 5, $7\frac{1}{2}$, 10, 15, 20, 30, and 40 mm. In the English system, commonly used values have been 0.250, 0.375, 0.500, 0.750, 1.000, 1.250, and 1.500 in. Fine-pitch worm gears are often designed to standard lead- and pitch-diameter values so as to obtain even lead-angle values (see AGMA* standard 374.04).

Worm threads are usually milled or cut with a single-point lathe tool. In fine pitches, some designs can be formed by rolling. Grinding is often employed as a finishing process for high-hardness worms. In fine pitches, worm threads are sometimes ground from the solid.

Worm-gear teeth are usually hobbed. The cutting tool is essentially a duplicate of the mating worm in size and thread design. New worm designs should be based on available hobs wherever possible to avoid the need for procuring a special hob for each worm design.

*The abbreviation AGMA stands for the American Gear Manufacturers Association.

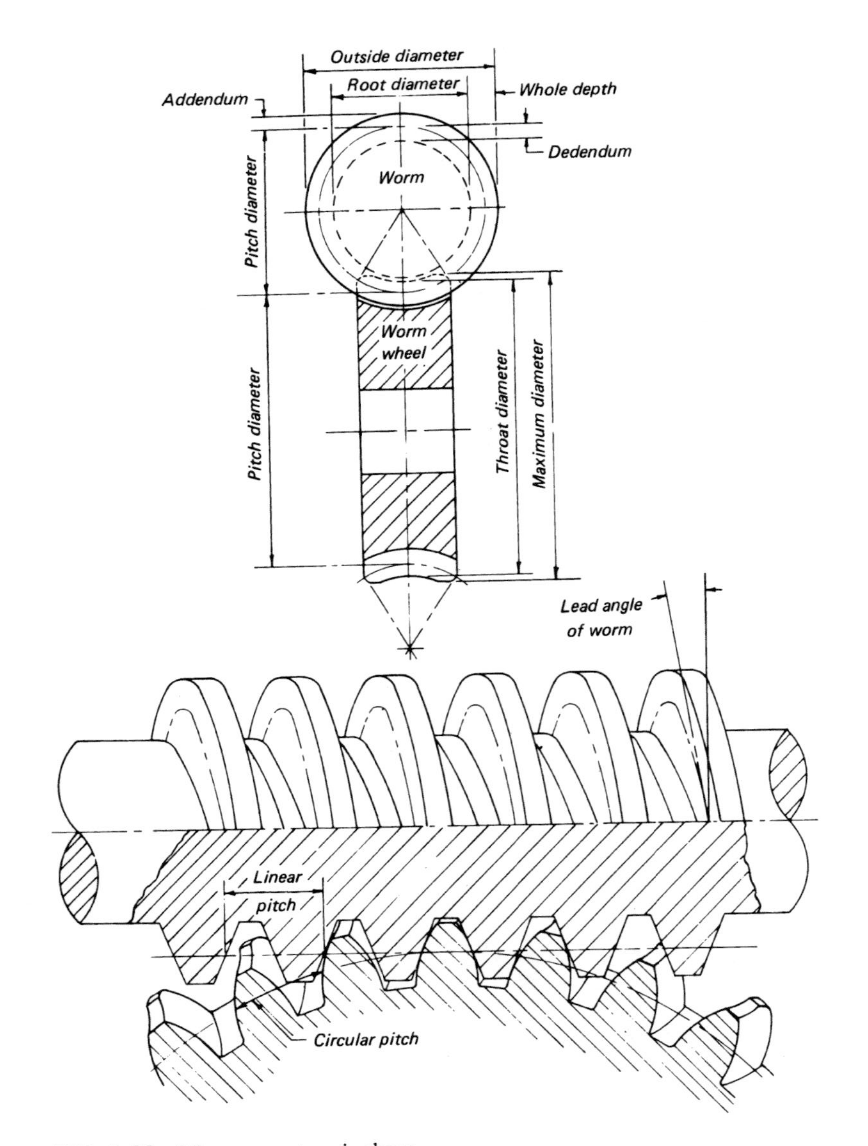

FIG. 1.29 Worm-gear terminology.

A variety of pressure angles are used for worms. Single-thread worms used for indexing purposes frequently have low axial pressure angles, like $14\frac{1}{2}°$. Multiple-thread worms with high lead angles like 30 or 40° are often designed with about 30° axial pressure angles.

The following formulas apply to worm gears which are designed to run on 90° axes:

$$\text{Axial pitch of worm} = \text{circ. pitch of worm gear} \quad (1.29)$$

$$\text{Pitch dia. of gear} = \frac{\text{no. of teeth} \times \text{circ. pitch}}{\text{pi}} \quad (1.30)$$

$$\text{Pitch dia. of worm} = 2 \times \text{center distance} - \text{pitch dia. of gear} \quad (1.31)$$

$$\text{Lead of worm} = \text{axial pitch} \times \text{no. of threads} \quad (1.32)$$

$$\text{Tan lead angle} = \frac{\text{lead of worm}}{\text{pitch dia.} \times \text{pi}} \quad (1.33)$$

$$\text{Lead angle of worm} = \text{helix angle of gear} \quad (1.34)$$

1.22 Double-Enveloping Worm Gears

The double-enveloping worm gear is like the single-enveloping worm gear except that the worm envelops the worm gear. Thus both members are throated. See Fig. 1.30.

Double-enveloping worm gears are used to transmit power between nonintersecting shafts, usually those at a 90° angle. Double-enveloping worm gears load their bearings with thrust and radial loads the same as single-enveloping worm gears do.

Double-enveloping worm gears should be accurately located on all mounting dimensions. Shafts should be at the right shaft angle and at the right center distance.

The double-enveloping type of worm gear has more tooth surface in contact than a single-enveloping worm gear. Instead of line contact, it has *area* contact at any one instant in time. The larger contact area of the

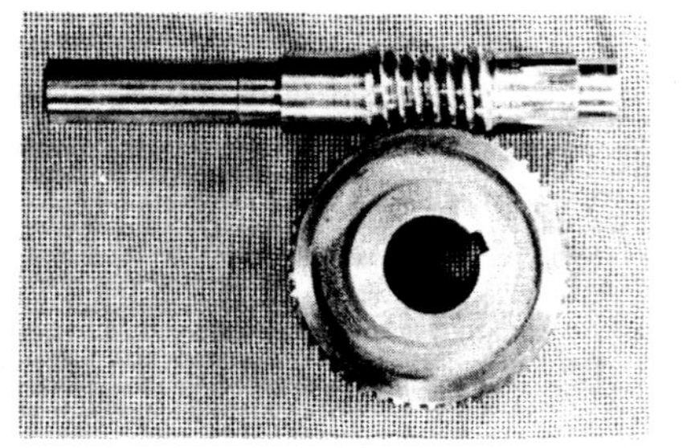

FIG. 1.30 Double-enveloping worm gearset. *(Courtesy of Ex-Cell-O Corp., Cone Drive Div., Traverse City, MI, U.S.A.)*

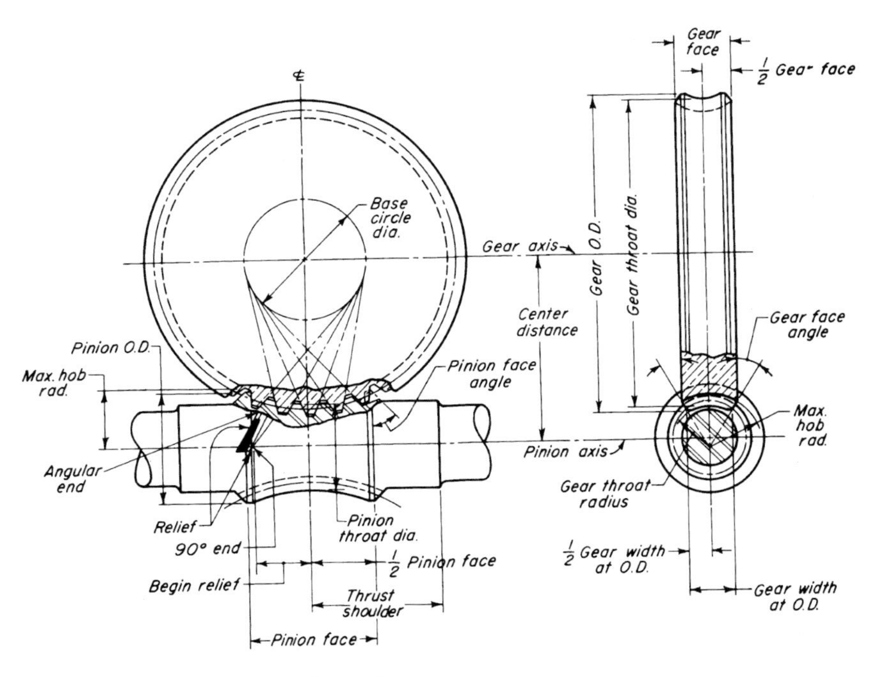

FIG. 1.31 Terminology of Cone-Drive worm gears.

double-enveloping worm gear increases the load-carrying capacity. On most double-enveloping worm gearsets, the worm rubbing speed is below 10 m/s (2000 fpm). Above 10 m/s, it is possible to get good results with oil lubrication, using a circulating system and coolers. The lubrication system must be good enough to prevent scoring and overheating.

The only double-enveloping worm gear that is in widespread use today is the Cone-Drive* design. Figure 1.31 shows the terminology used with Cone-Drive worm gears.

The Cone-Drive worm has a straight-sided profile in the axial section, but this profile changes its inclination as you move along the thread. At any one position, this slope is determined by a line which is tangent to the *base cylinder* of the gear. The base cylinder of a Cone-Drive gear is like an involute base circle in that it is an imaginary circle used to define a profile. Geometrically, though, the base circle of a Cone-Drive gear is not used in the same way as the base circle of an involute gear.

*Cone-Drive is a registered trademark of the Ex-Cell-O Corp., Cone-Drive Div., Traverse City, Michigan, U.S.A.

The size of Cone-Drive gear teeth is measured by the circular pitch of the gear.

The normal pressure angle is ordinarily 20° or 22°. The Cone-Drive worm and gear diameters are not in proportion to the ratio. With low ratios, it is possible (although not recommended) to have a worm which is larger than the gear!

Equations (1.30) and (1.31) apply to Cone-Drive gears as well as to regular worm gears. The other formulas for single-enveloping worm gears apply only to the center of the Cone-Drive worm, since it does not have a fixed axial pitch and lead like a cylindrical worm.

In both single-enveloping worm gears and Cone-Drive gears, it is generally recommended that the worm or pinion diameter be made a function of the center distance (see AGMA 342.02). Thus

$$\text{Pitch dia. of worm} = \frac{(\text{center distance})^{0.875}}{2.2} \tag{1.35}$$

Following the recommendation of Eq. (1.35) is, of course, not necessary. This formula merely recommends a good proportion of worm to gear diameter for best power capacity. In instrument and control work, the designer may not be interested in power transmission at all. In such cases, it may be desirable to depart considerably from Eq. (1.35) in picking the size of a worm or pinion. In fact, AGMA 374.04 shows a whole series of worm diameters for fine-pitch work which do not agree with Eq. (1.35).

When a worm diameter is picked in accordance with Eq. (1.35), the gear diameter and the circular pitch may be obtained by working backward through Eqs. (1.31), (1.30), and (1.29).

The helix angle of a worm gear or a Cone-Drive gear may be obtained from the following general formula:

$$\text{Tan center helix angle of gear} = \frac{\text{pitch dia. of gear}}{\text{pitch dia. of worm} \times \text{ratio}} \tag{1.36}$$

1.23 Spiroid Gears

The Spiroid family of gears operates on nonintersecting, nonparallel axes. The most famous family member is called Spiroid. It involves a tapered pinion that somewhat resembles a worm (see Fig. 1.32). The gear member is a face gear with teeth curved in a lengthwise direction; the inclination to the tooth is like a helix angle—but not a true helical spiral.

FIG. 1.32 Double-reduction Spiroid linear actuator unit. (*Courtesy of Spiroid Div. of Illinois Tool Works, Inc., Chicago, IL, U.S.A.*)

Figure 1.33 shows the schematic relation of the Spiroid type of gear to worm gears, hypoid gears, and bevel gears.

The Spiroid family has Helicon* and Planoid* types as well as the Spiroid type. The Helicon is essentially a Spiroid with no taper in the pinion. The Planoid is used for lower ratios than the Spiroid, and its offset is lower—more in the range of the hypoid gear.

The Spiroid pinions may be made by hobbing, milling, rolling, or thread chasing. Spiroid gears are typically made by hobbing, using a specially built (or modified) hobbing machine and special hobs. The gear may be made with molded or sintered gear teeth using tools (dies or punches) that have teeth resembling hobbed gears. Shaping and milling are not practical to use in making Spiroid gears.

The Spiroid gears may be lapped as a final finishing process. Special "shaving" type hobs may also be used in a finishing operation.

Spiroid gears are used in a wide variety of applications, ranging from aerospace actuators to automotive and appliance uses. The combination of a high ratio in compact arrangements, low cost when mass-produced, and good load-carrying capacity makes the Spiroid-type gear attractive in many situations. The fact that the gearing can be made with lower-cost machine tools and manufacturing processes is also an important consideration.

*Helicon and Planoid are registered trademarks of Illinois Tool Works Inc., Chicago, IL, U.S.A., as is the term Spiroid.

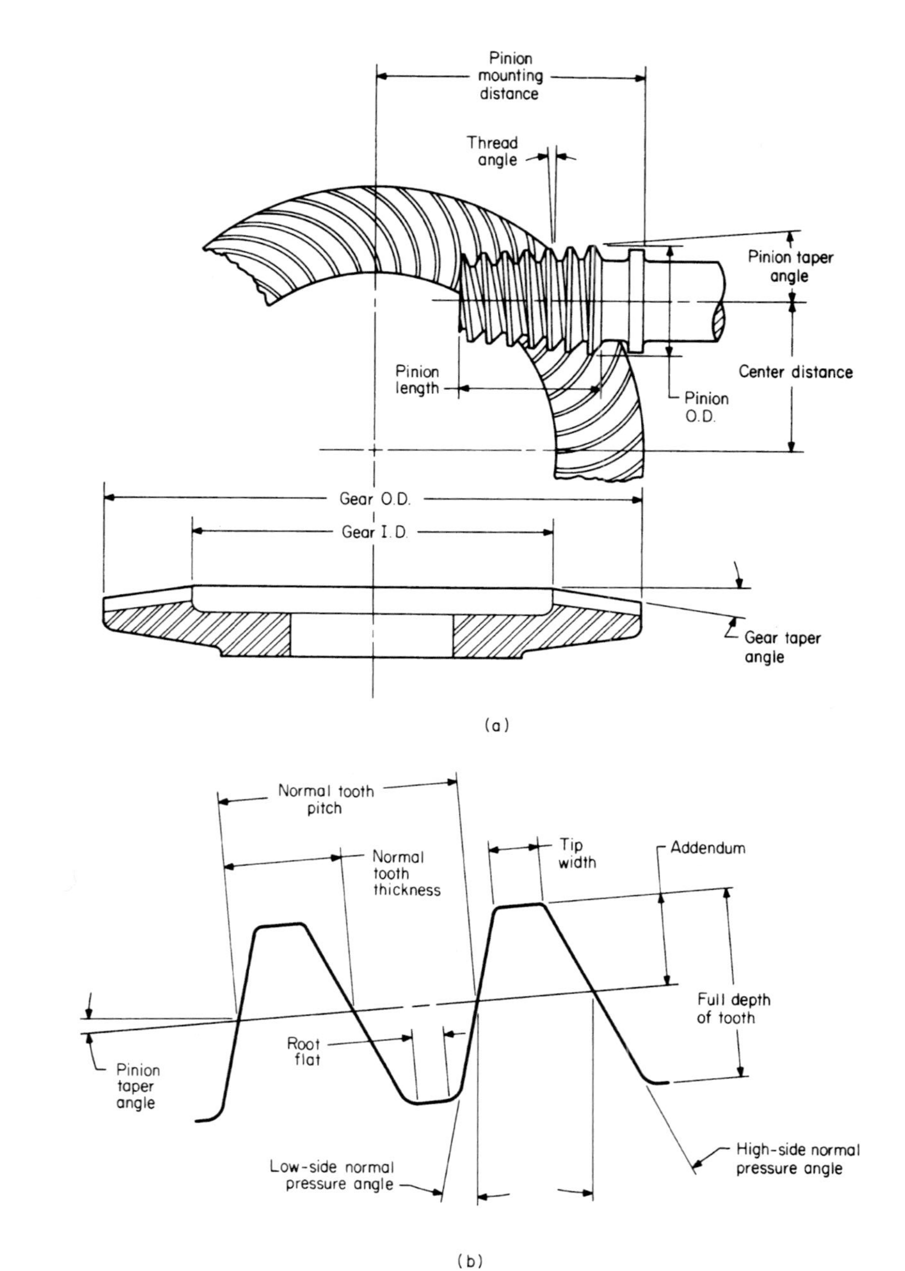

FIG. 1.33 Spiroid-gear terminology.

Lieu and Sorby, 2nd Edition – Chapter 17

Textbook: Visualization, Modelling, and Graphics for Engineering Design
Chapter Title: Advanced Visualization Techniques

Chapter 17

Advanced Visualization Techniques

Chapter Introduction

Objectives

After completing this chapter, you should be able to

- Increase your visualization skills
- Move seamlessly between 2-D and 3-D representations of objects

- Use basic strategies for visualizing complex objects

17.01 Introduction

Imagine trying to communicate in a foreign country without knowing one word of the native language. You might be able to pantomime adequately to take care of your basic needs—food and drink, for example—but your level of communication would be superficial. You would not be able to exchange views on your philosophy, your feelings, or your thoughts. There would be no chance for a meaningful dialogue. Trying to communicate with other engineers without well-developed graphics and visualization skills is similar to trying to communicate with a French person when you do not know the language. Without graphics skills, you can engage in a superficial conversation about basics; however, you cannot participate in a meaningful discussion about design details. That is why this chapter is included in the text—to help you move beyond the basics in your graphics and visualization skills so you can be a full-fledged member of an engineering design team.

17.02 Basic Concepts and Terminology in Visualization

In previous chapters, you learned about some of the basic building blocks of graphical communication. You learned how to make simple sketches. You developed basic skills for visualizing simple objects, and you learned about pictorial representation of objects. You learned how engineers use multiview drawings to accurately portray 3-D objects on a 2-D sheet of paper. In this chapter, you will bring all of your skills together to tackle increasingly difficult problems in visualization. Mostly through examples, this chapter will illustrate techniques you can use to hone your visualization skills, which will improve your effectiveness in graphical communication.

Before continuing with the chapter, you need to become familiar with some fundamental definitions and concepts. As you learned in a previous chapter, four basic types of surfaces make up 3-D objects. Table 17.01 outlines the basic surface types and explains how they appear in a standard drawing that shows top, front, and right-side views.

Table 17.01

Basic Surface Types for Objects

Surface Type	Surface Characteristics	Appearance

NORMAL	PARALLEL TO ONE OF THE PRIMARY VIEWING PLANES; PERPENDICULAR TO THE OTHER TWO PLANES	SURFACE IN ONE VIEW; EDGE IN TWO REMAINING VIEWS
INCLINED	PERPENDICULAR TO ONE PRIMARY VIEW; ANGLED WITH RESPECT TO THE OTHER TWO VIEWS	SURFACE IN TWO VIEWS; EDGE (ANGLED LINE) IN ONE VIEW
OBLIQUE	NEITHER PARALLEL NOR PERPENDICULAR TO ANY OF THE PRIMARY VIEWS	SURFACE IN ALL THREE VIEWS
CURVED	PERPENDICULAR TO ONE VIEW	CURVED EDGE IN ONE VIEW; EXTENTS OF SURFACE VISIBLE IN TWO REMAINING VIEWS

The definition of surface types can be further refined based on their characteristics. **normal surface** (A surface on an object being viewed that is parallel to one of the primary viewing planes.) are defined by the primary view to which they are parallel, and they appear in true shape and size within this primary view. When normal surfaces are parallel to the front view, they are referred to as **frontal surface** (A surface on an object being viewed that is parallel to the front viewing plane.) ; when normal surfaces are parallel to the top view, they are referred to as **horizontal surface** (A surface on an object being viewed that is parallel to the top viewing plane.) ; and when normal surfaces are parallel to the side view, they are referred to as **profile surface** (A surface on an object being viewed that is parallel to a side viewing plane.) . Figure 17.01 illustrates the three types of normal surfaces found on objects. Notice that each normal surface shown is viewed as a surface in one view and as edges in the remaining two views.

Figure 17.01

An object made up of normal surfaces.

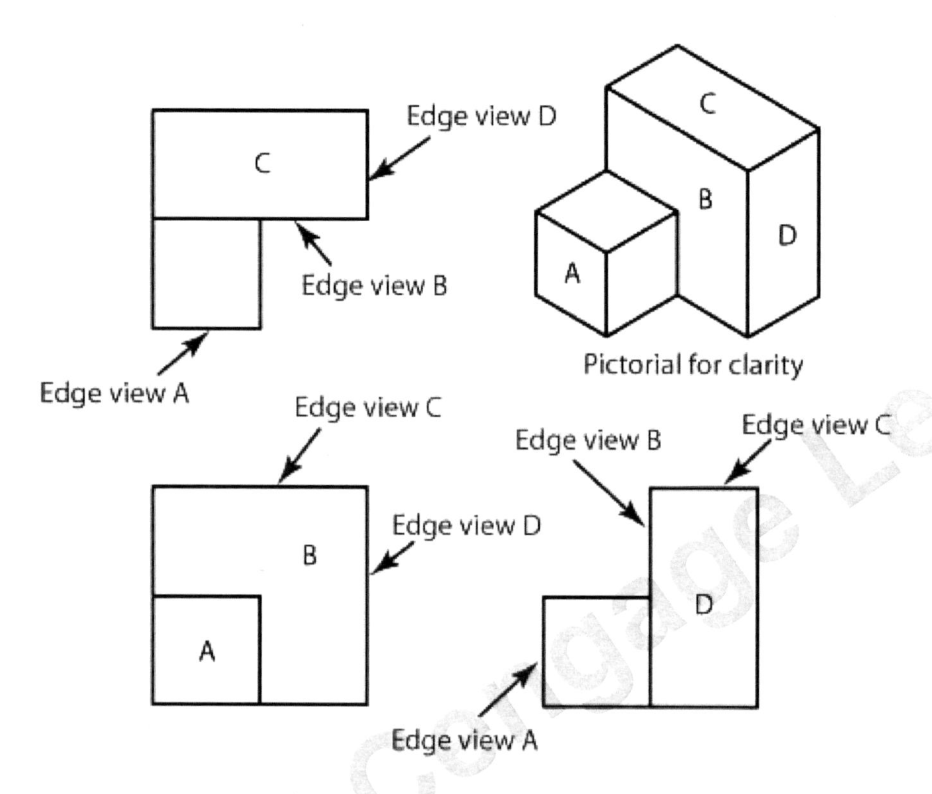

inclined surface (A flat surface on an object being viewed that is perpendicular to one primary view and angled with respect to the other two views; in other words, a plane that appears as an edge view in one primary view but is not parallel to any of the principal views.) are defined by the primary view to which they are perpendicular—in other words, they are defined by the view in which they appear as an edge. Thus, an inclined frontal plane appears as an edge in the front view and as a surface in the top and side views; an inclined horizontal plane appears as an edge in the top view and as a surface in the front and side views; and an inclined profile surface appears as an edge in the side view and as a surface in the front and top views. Figure 17.02 illustrates the three classifications for inclined surfaces that are found on objects.

Figure 17.02

Inclined surface definitions.

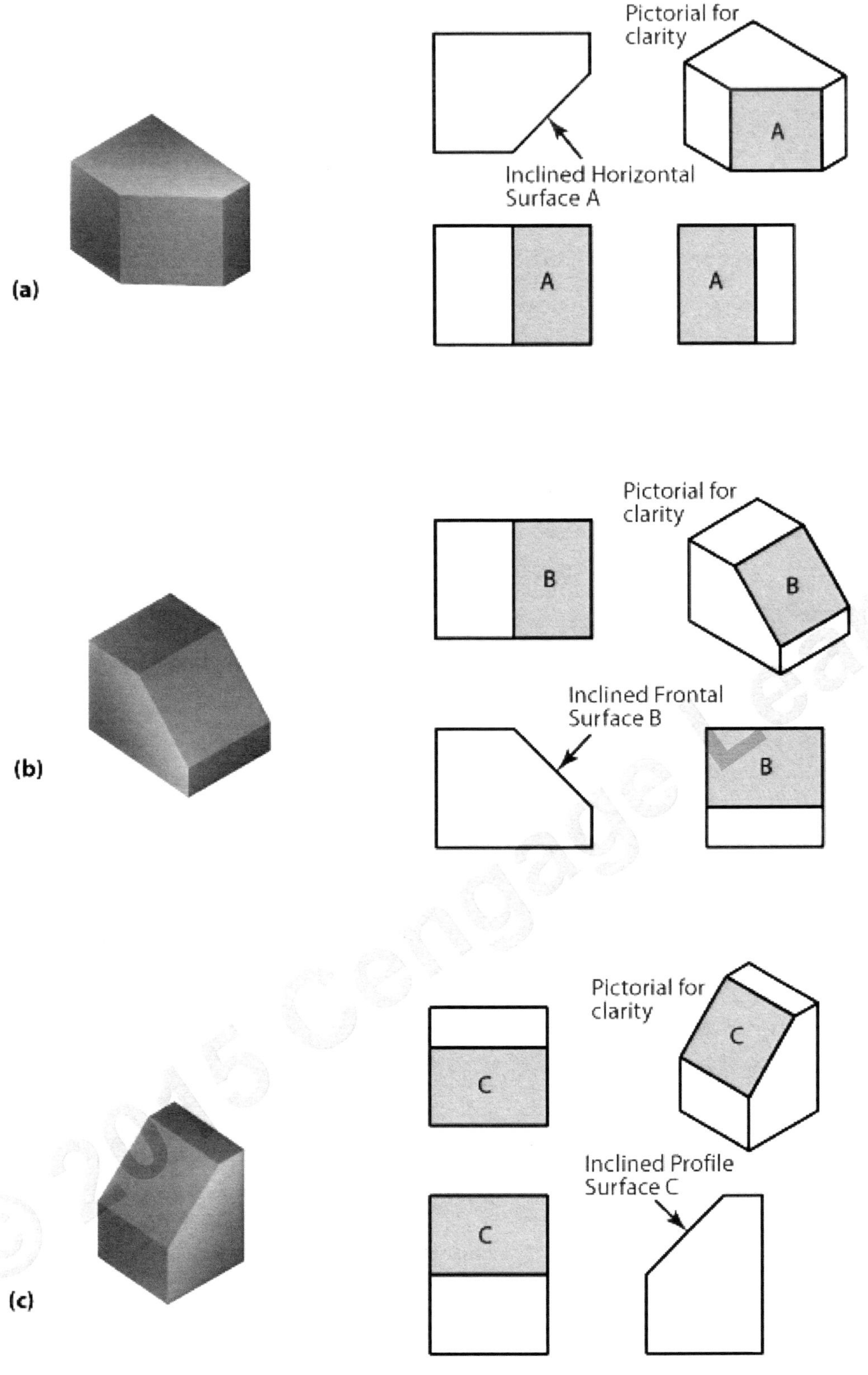

oblique surface (A flat surface on an object being viewed that is neither parallel nor perpendicular to any of the primary views.) do not have any special classification since they appear neither as an edge nor in their true shape and size in any of the primary views. Figure 17.03 illustrates the basic characteristic of an oblique surface.

Figure 17.03

An object with oblique surface A.

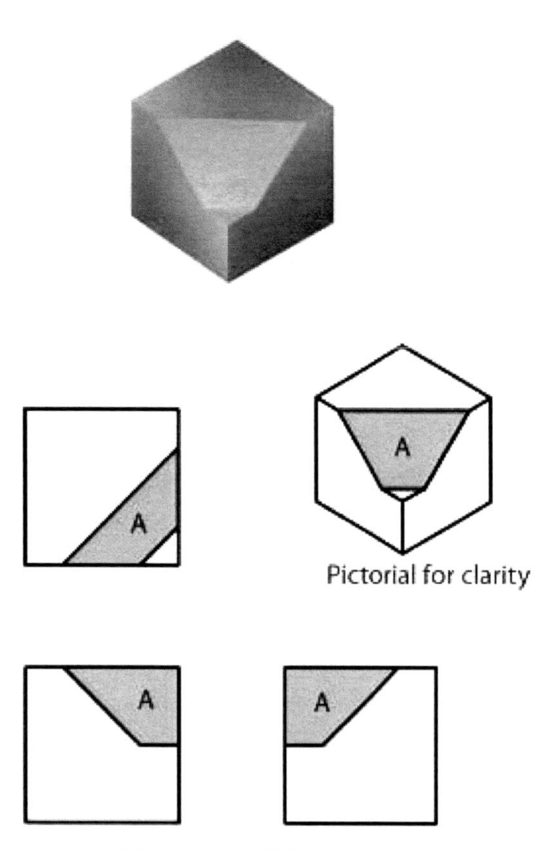

There is no special classification for **curved surface** (Any nonflat surface on an object.) , although they can be perpendicular to any of the primary views, as shown in Figure 17.04.

Figure 17.04

Various orientations for curved surfaces.

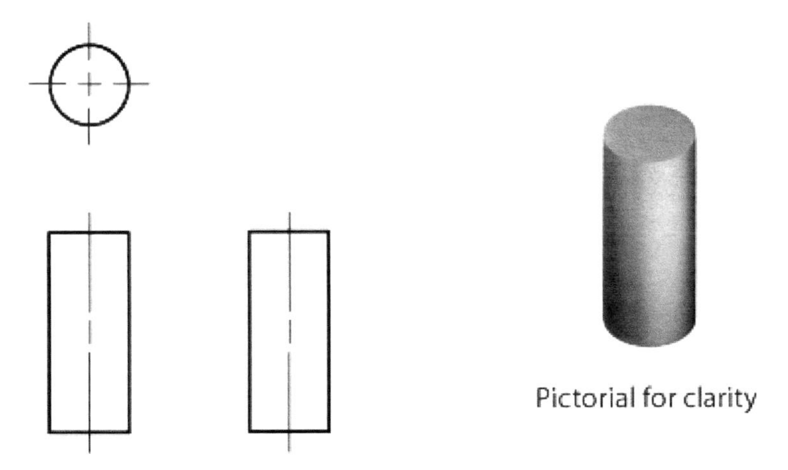

(a) Curved surface perpendicular to top view

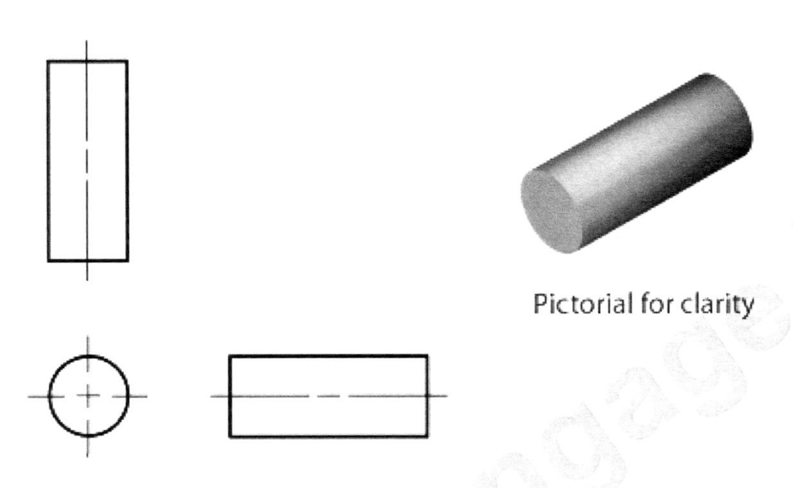

(b) Curved surface perpendicular to front view

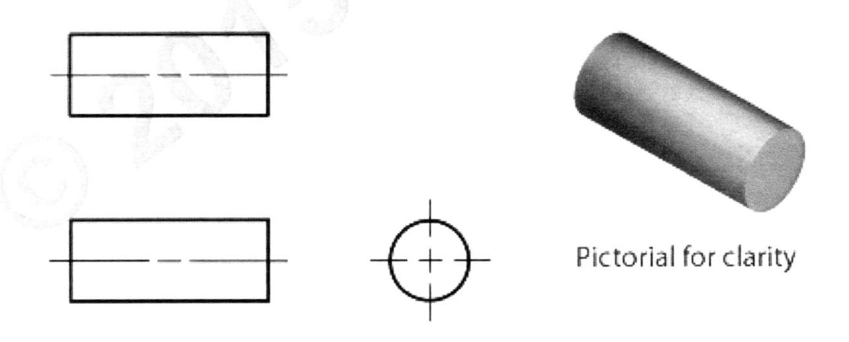

(c) Curved surface perpendicular to side view

Engineering drawings are typically set up in the familiar L-shaped pattern that shows the top, front, and right-side views. With this view layout, the top is projected vertically from the front and the right side is projected horizontally from the front with the panels folded out of the glass box and hinged at the edges of the front view. Sometimes you may want to include the top, front, and left-side views, resulting in a backward *L* view pattern. In this case, the top hinges on the front view as before, with the left view also hinging on the front, similar to the way the right view was constructed. Other times, you will want the top view as the stationary view and the side and front views hinging from it. This is particularly true when the edge view of a curved surface is visible in the top view. Figure 17.05 shows the alternative ways the glass box can be unfolded to produce varying view layouts. Figure 17.06 shows a comparison of two projection systems used for sketching a cylinder. Notice that the view layout shown in Figure 17.06b more clearly defines the curved surface of the cylinder than the layout in Figure 17.06a. In Figure 17.06b, any two adjacent views (e.g., either the front and top views or the top and sides views) clearly define the surface as being curved. In Figure 17.06a, the front and side views together cannot, by themselves, define the curved surface. The addition of a top view is required.

Figure 17.05

One way to present the drawing of an object: showing the front, top, and right-side views on a hypothetical glass box in (a), the bow panels opening in (b), to present all three views on a single plane in (c), for the multiview drawing in (d). Another way to present the drawing of an object: showing the front, top, and right-side views on a hypothetical glass box in (e), the bow panels opening in (f), to present all three views on a single plane in (g), for the multiview drawing in (h). Still another way to present the drawing of an object; showing the front, top, and right-side views on a hypothetical glass box in (i), the bow panels opening in (j), to present all three views on a single plane in (k), for the multiview drawing in (l).

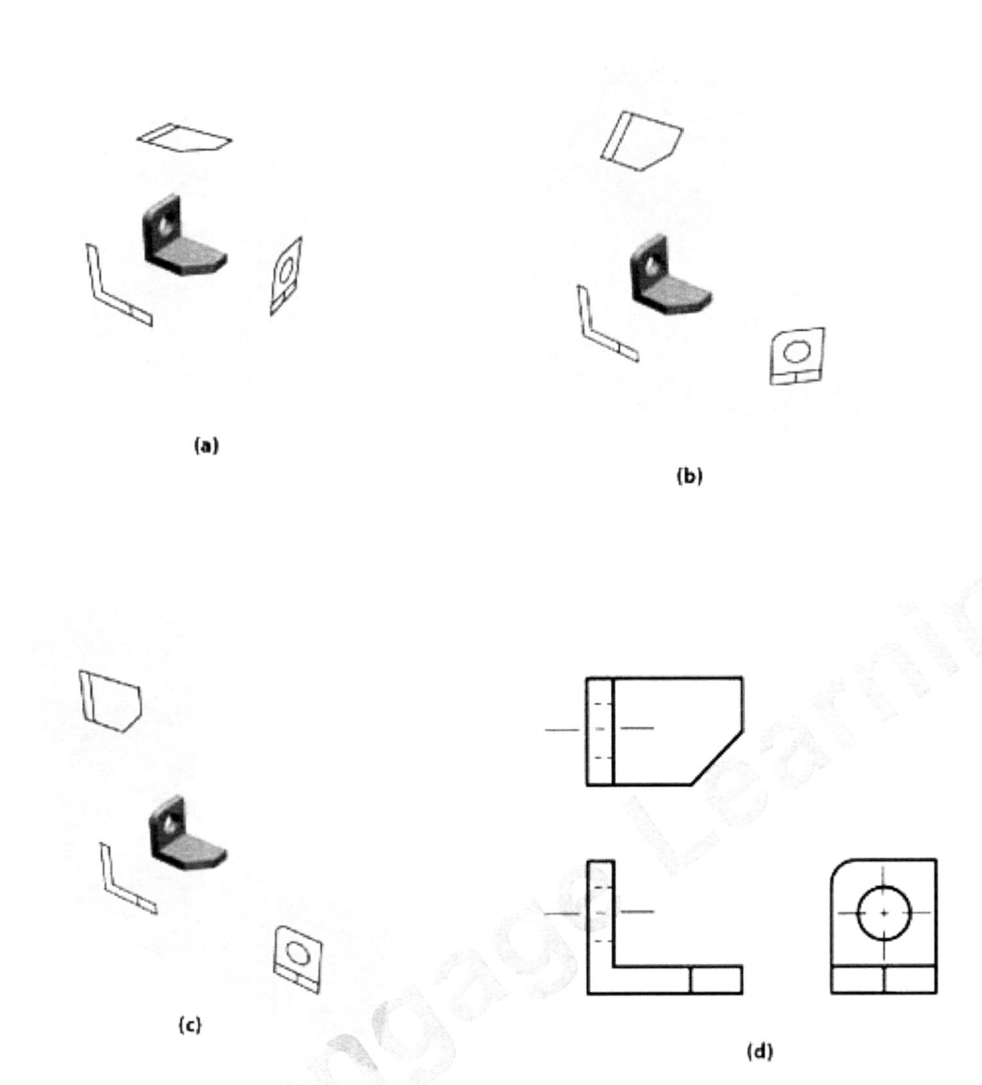
(a)
(b)
(c)
(d)

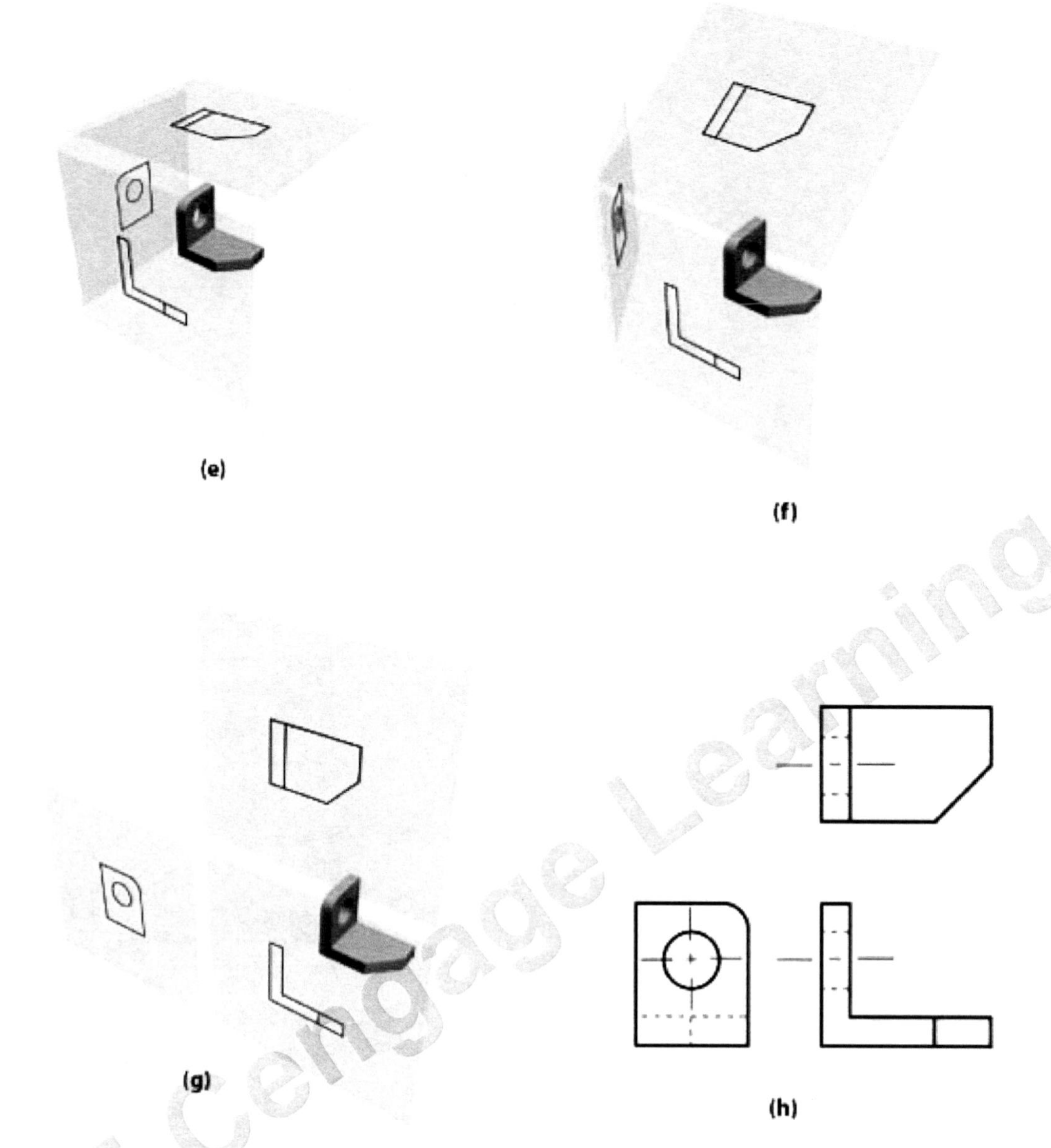

© Cengage Learning® Courtesy of D. K. Lieu

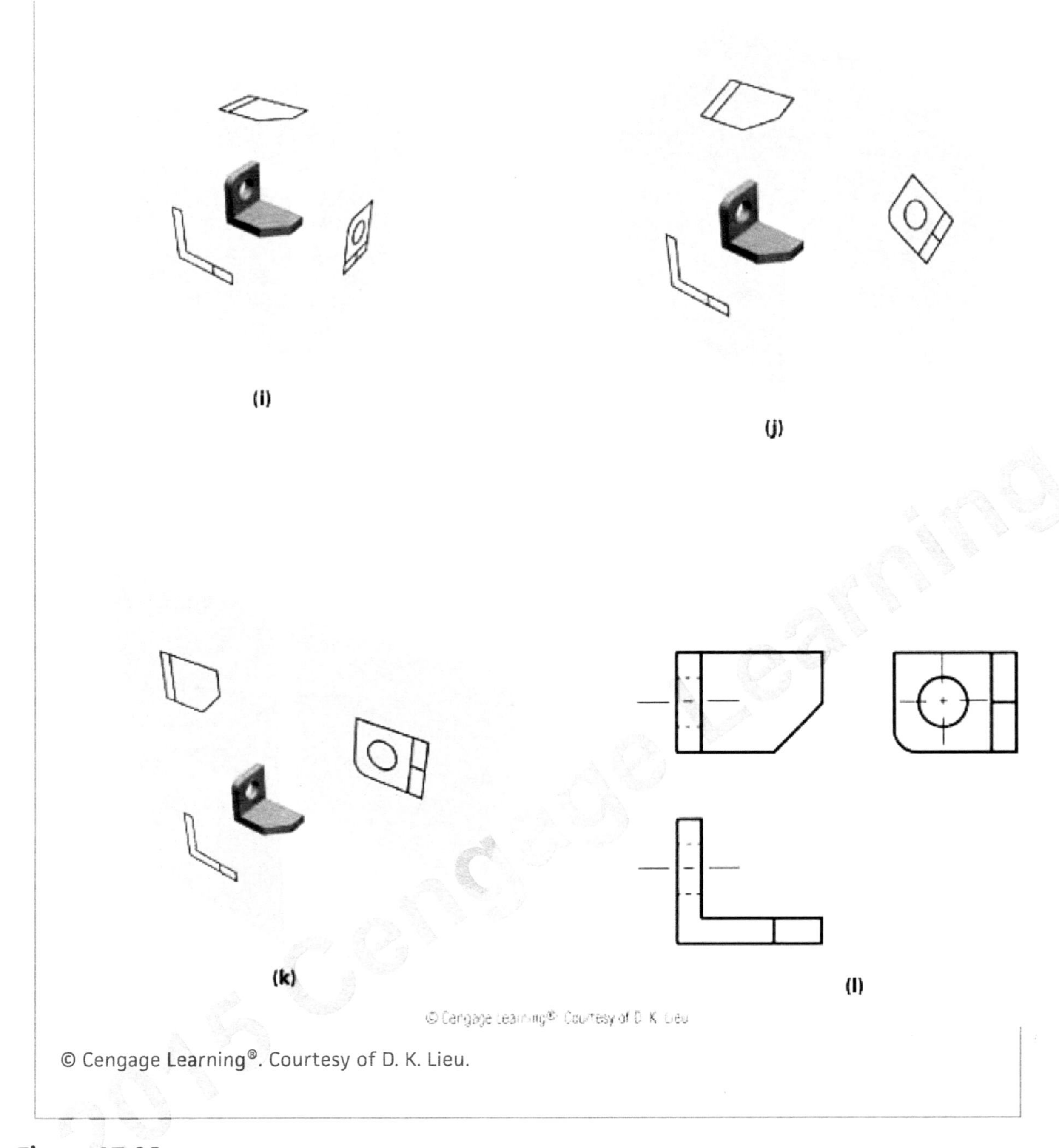

Figure 17.06

The side view projected from the top view.

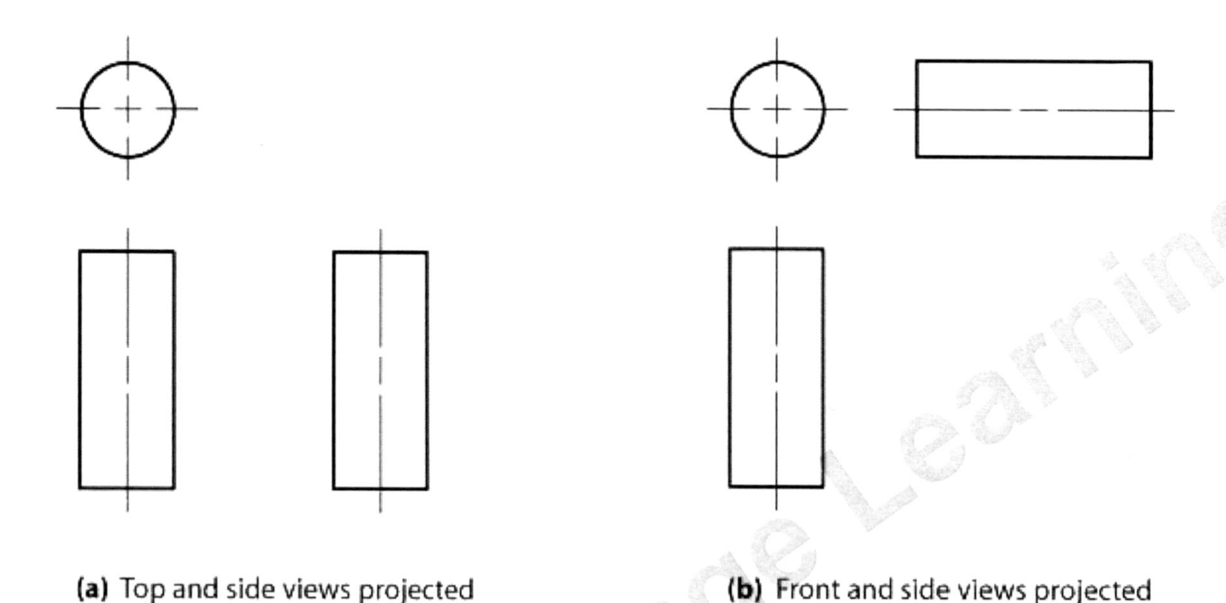

(a) Top and side views projected from the front view.

(b) Front and side views projected from the top view

One important visualization concept you must understand is that surfaces retain their basic shapes from view to view. If a surface has four edges and four vertices in one view, it will have four edges and four vertices in the next view. If the surface has a basic L shape in one view, it will have a basic L shape in all views. The exception, of course, is that a normal surface appears as a single edge in at least one of the primary views and merely looks like a line in that view—you will not be able to determine how many edges and vertices it has. Figure 17.07 shows a multiview drawing, including an isometric pictorial, of an object that has an L-shaped surface on it. Notice that this surface (labeled *A* in all views) maintains its basic shape from view to view, except in the top view where it appears as a single edge.

Figure 17.07

An object with an L-shaped inclined horizontal surface.

Pictorial for clarity

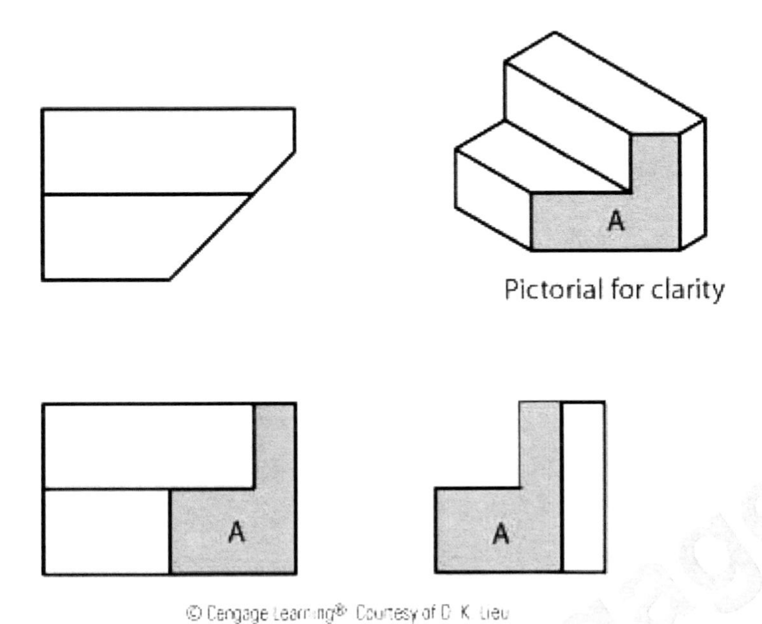

Another basic concept to keep in mind when visualizing objects from multiview drawings is that the limits of surfaces should correspond from one view to the next. Because you have a system of orthographic projection, features and vertices on the object will be aligned between views. This principle holds true for surfaces as well. If the height of a surface goes from bottom to top in the front view of the object, it also will go from bottom to top in the side view or in an isometric pictorial view. Note that because the top view does not show changes in object height, those changes will not be apparent in that view. Similarly, the width and depth limits of a given surface are maintained from one view to the next. Figure 17.08 shows a multiview drawing of an object along with an isometric pictorial. The height, width, and depth limits of surface B are labeled in each view. Notice that these dimensions are maintained between views.

Figure 17.08

Height, width, and depth measurements maintained between views.

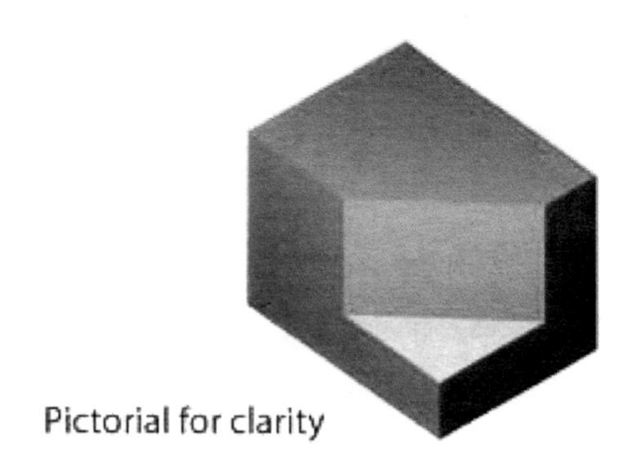

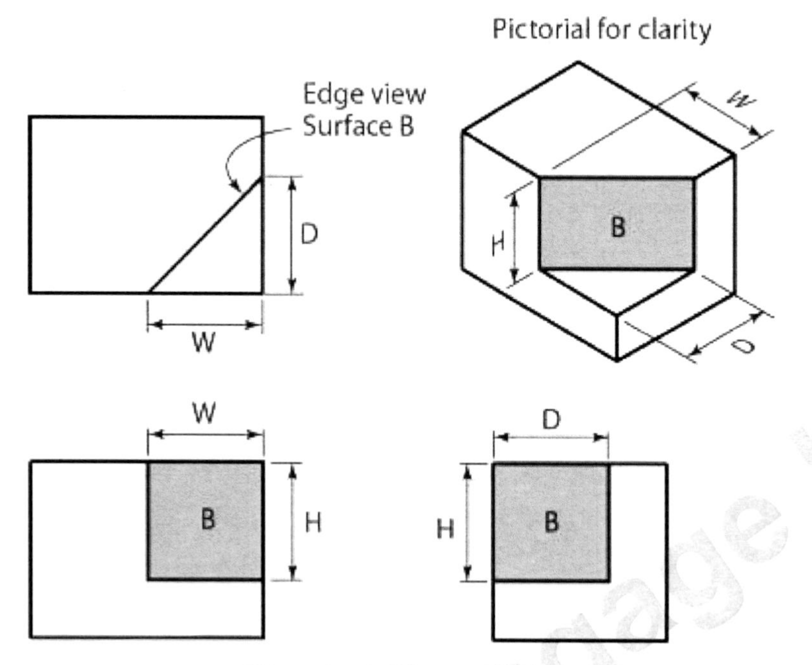

A final basic concept to keep in mind when looking at multiview drawings is that for normal and inclined surfaces, right angles on surfaces are maintained from view to view. In fact, one way to identify an oblique surface on an object is to observe that an angle appears to change from perpendicular to something other than perpendicular from one view to the next. Figure 17.09 shows an object with a T-shaped inclined surface on it. Notice that in each view of the surface, all angles appear to be right angles (and that the surface has the same basic shape and same number of edges and vertices in each view). Figure 17.10 shows a multiview drawing of an object with an oblique surface (A) on it. Notice that the angle at the vertex labeled *X* appears to be a right angle in the front view but about 40 degrees in the top view. This change in the relative size of the angle at vertex X is a clear indication that surface A is an oblique surface and, as such, appears as a surface in all views.

Figure 17.09

An object with inclined T-shaped surface A.

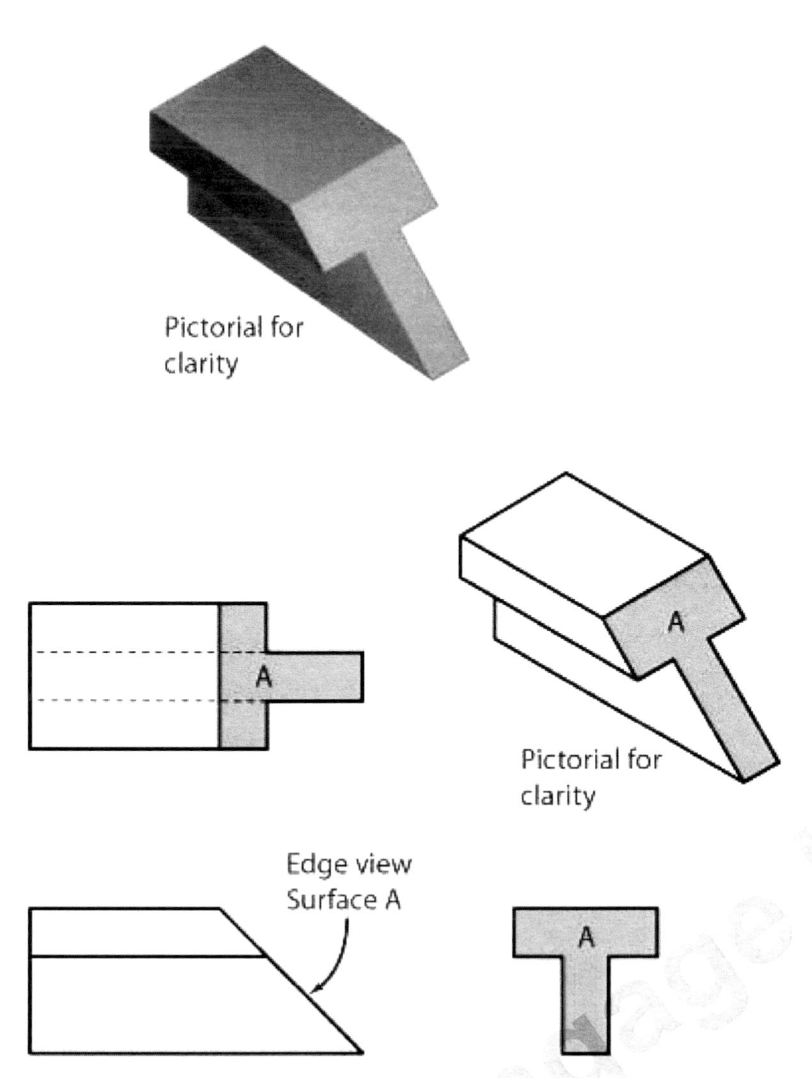

Figure 17.10

An apparent change in the angle at vertex X for oblique surface A.

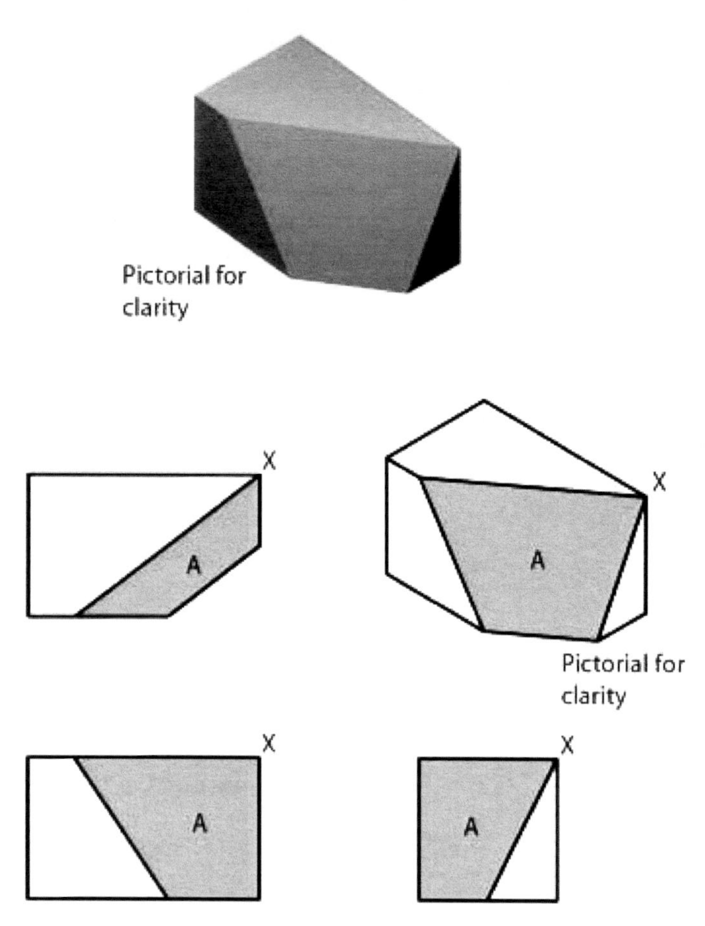

17.03 The Possibilities for a Feature Representation

To understand a drawing or a view completely, you must be able to imagine all of the possibilities inherent to a given feature. Then you can reject the possibilities that do not match the information provided. Figure 17.11 shows the top view of a simple object.

Figure 17.11

The top view of an object.

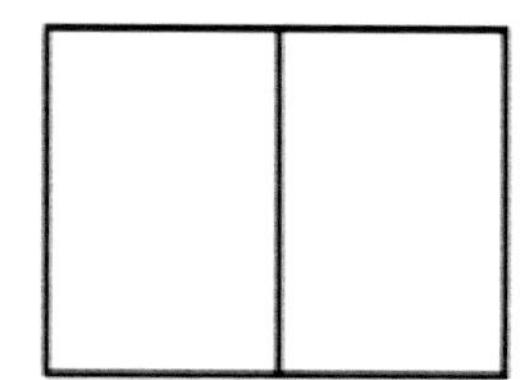

Can you tell what the object looks like from this view alone? Figure 17.12 shows several pictorial sketches of objects that could produce this single top view. Notice that the possible objects include normal, inclined, and curved surfaces as defined in this chapter. Several other possibilities produce this top view, including some that contain oblique surfaces. Can you imagine others? Which one shows the correct object? Without more information, it is impossible to determine the correct object that corresponds to this top view.

Figure 17.12

Possible objects from a given top view.

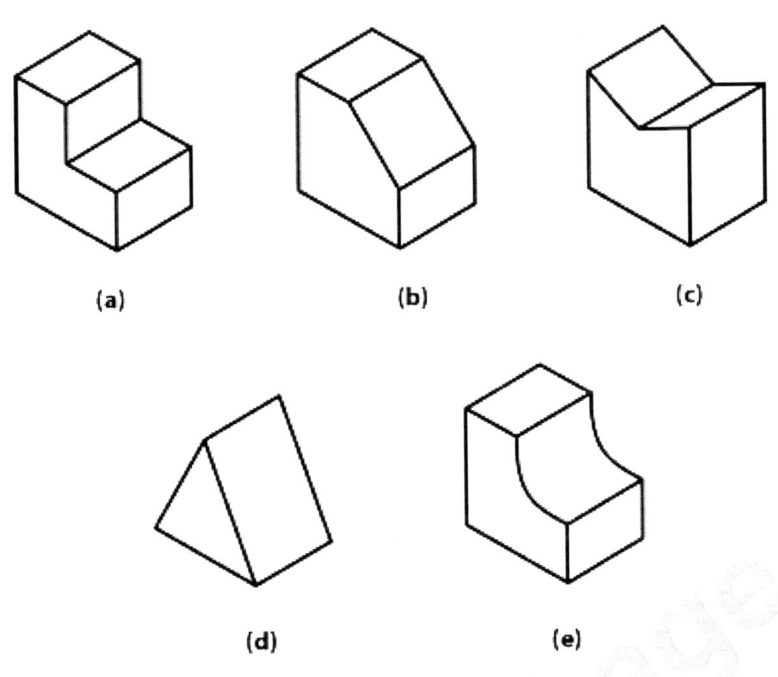

Figure 17.13 shows the top and front views of the object. By examining this additional view, you can rule out all of the possibilities from Figure 17.12 except choice (d).

Figure 17.13

Top and front views of an object.

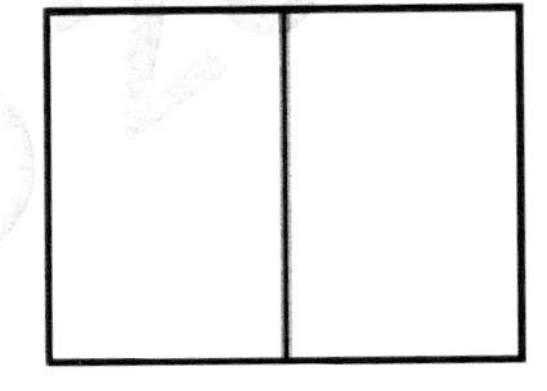

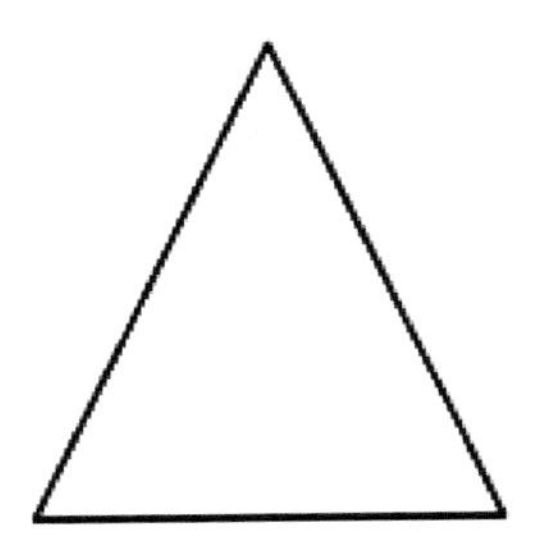

Figure 17.14 shows the top view of an object, and Figure 17.15 shows multiple possibilities for objects that correspond to this view. Thinking about all of the possibilities that exist will help you develop your visualization abilities. Again, which of the objects is correct? It is impossible to tell from this limited information.

Figure 17.14

The top view of an object.

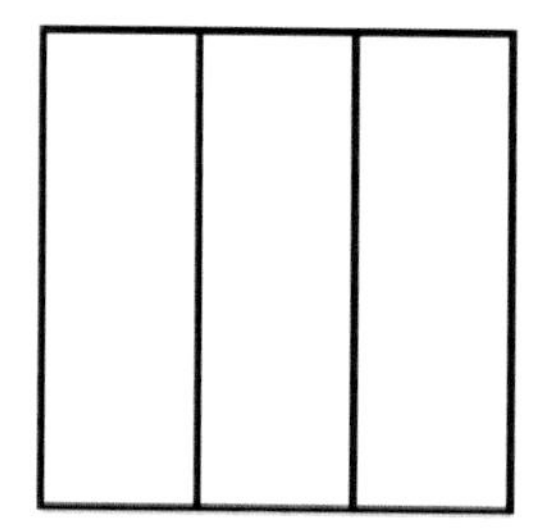

Figure 17.15

Possible objects based on the top view given.

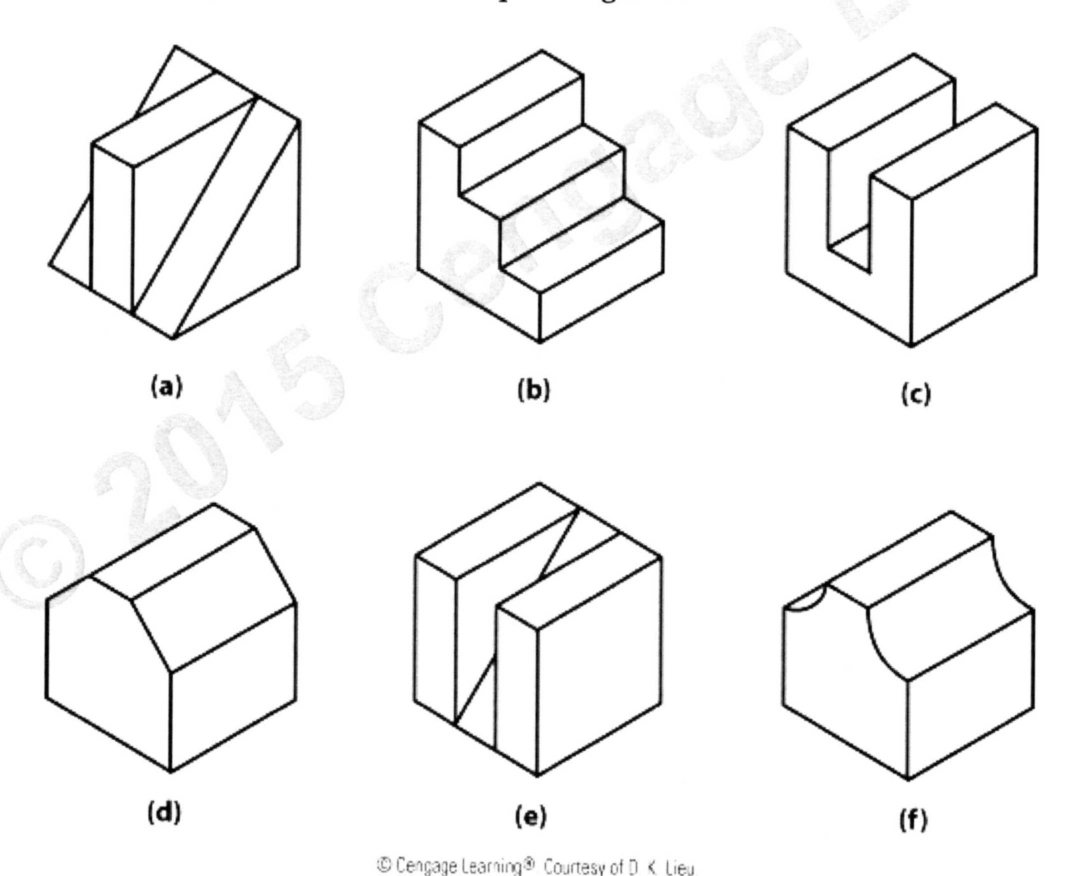

Figure 17.16 shows the top and front views of the object. Based on this new information, you can rule out several of the possibilities; only two remain: choices (a) and (e).

Figure 17.16

Top and front views of an object.

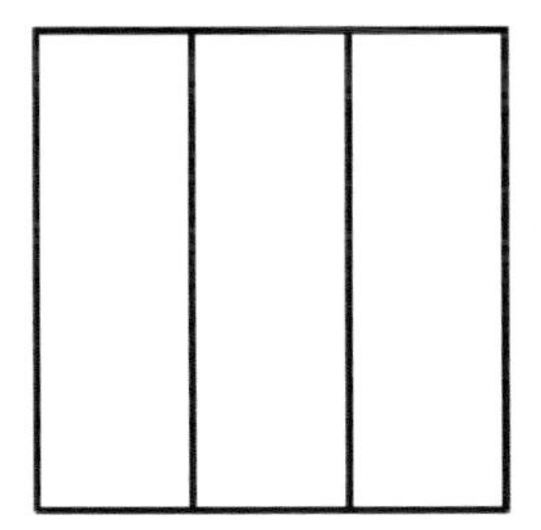

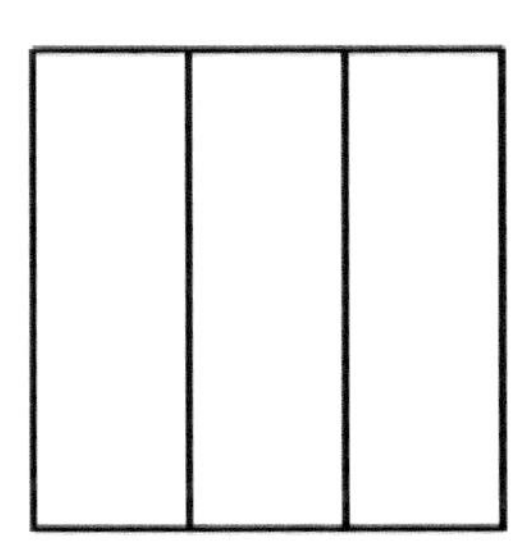

To determine which of these is correct, a third view of the object is required. Figure 17.17 shows the top, front, and right-side views of the object. Based on this new information, it is clear that choice (a) is the correct interpretation of the object.

Figure 17.17

A multiview drawing of an object.

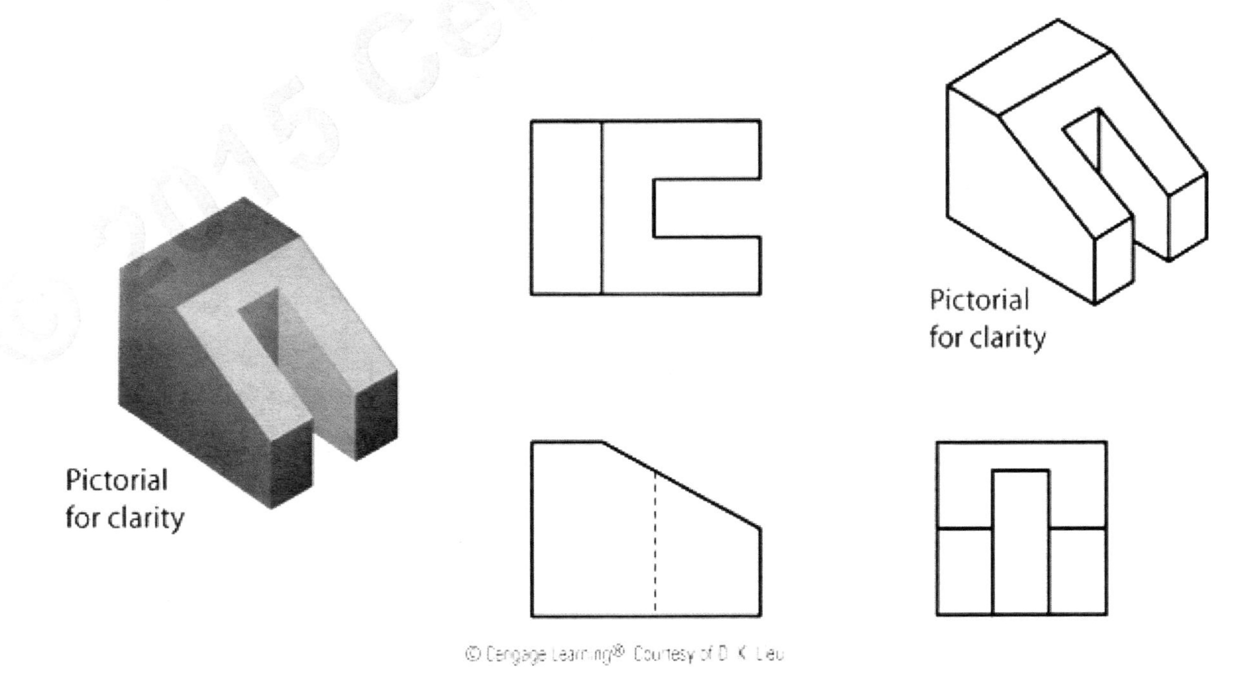

17.04 Other Viewpoints

In a previous chapter, you learned about coded plans and the way they are used to define an object. You also learned how the object can be viewed from any one of the corners defined by the coded plan. Figure 17.18 shows a coded plan and the four corner views defined by it.

Figure 17.18

A coded plan and four corner views from above.

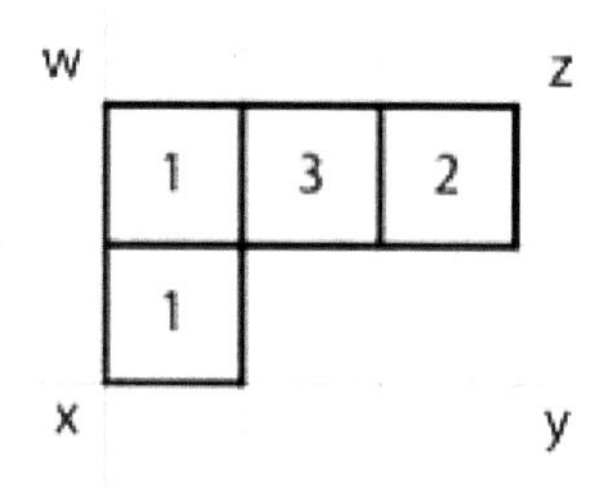

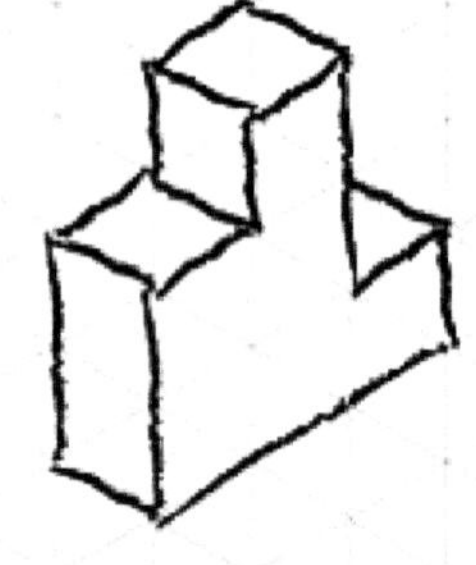

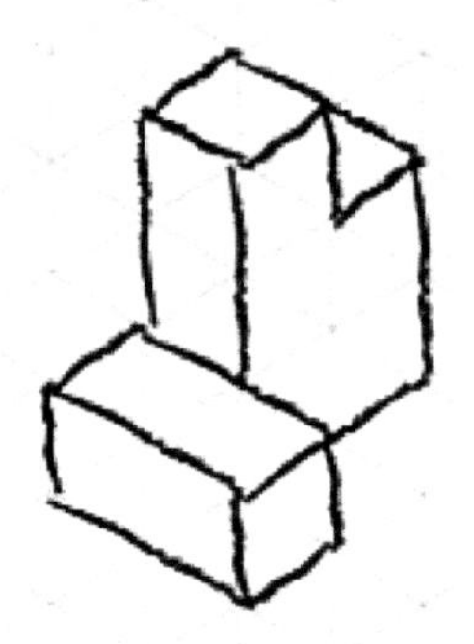

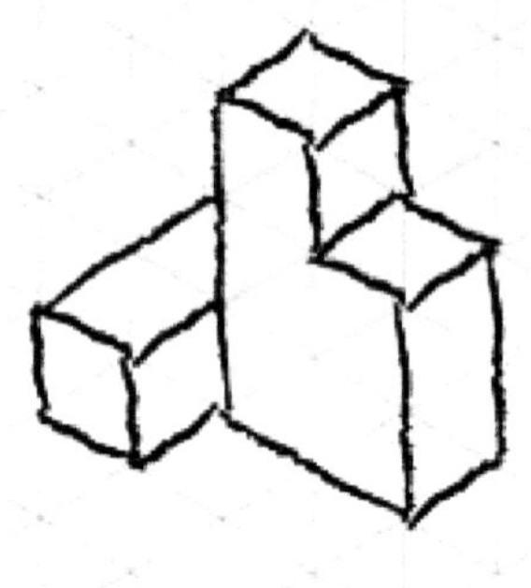

In the previous discussion of coded plans and corner views, you learned to look at an object from the indicated corner by imagining that your eye was located *above* the object. Because you are looking at the object from above, this view is sometimes referred to as the bird's-eye view. What if you wanted to look at the object from beneath it? Figure 17.19 shows the same object in Figure 17.18 except that the corner views are defined by locating your eye beneath the object, looking upward toward it. These views are referred to as worm's-eye views. (Think of yourself as a worm burrowing underground and looking up at the object rather than a bird soaring overhead and looking down on the object.)

Figure 17.19

A coded plan and four corner views from below.

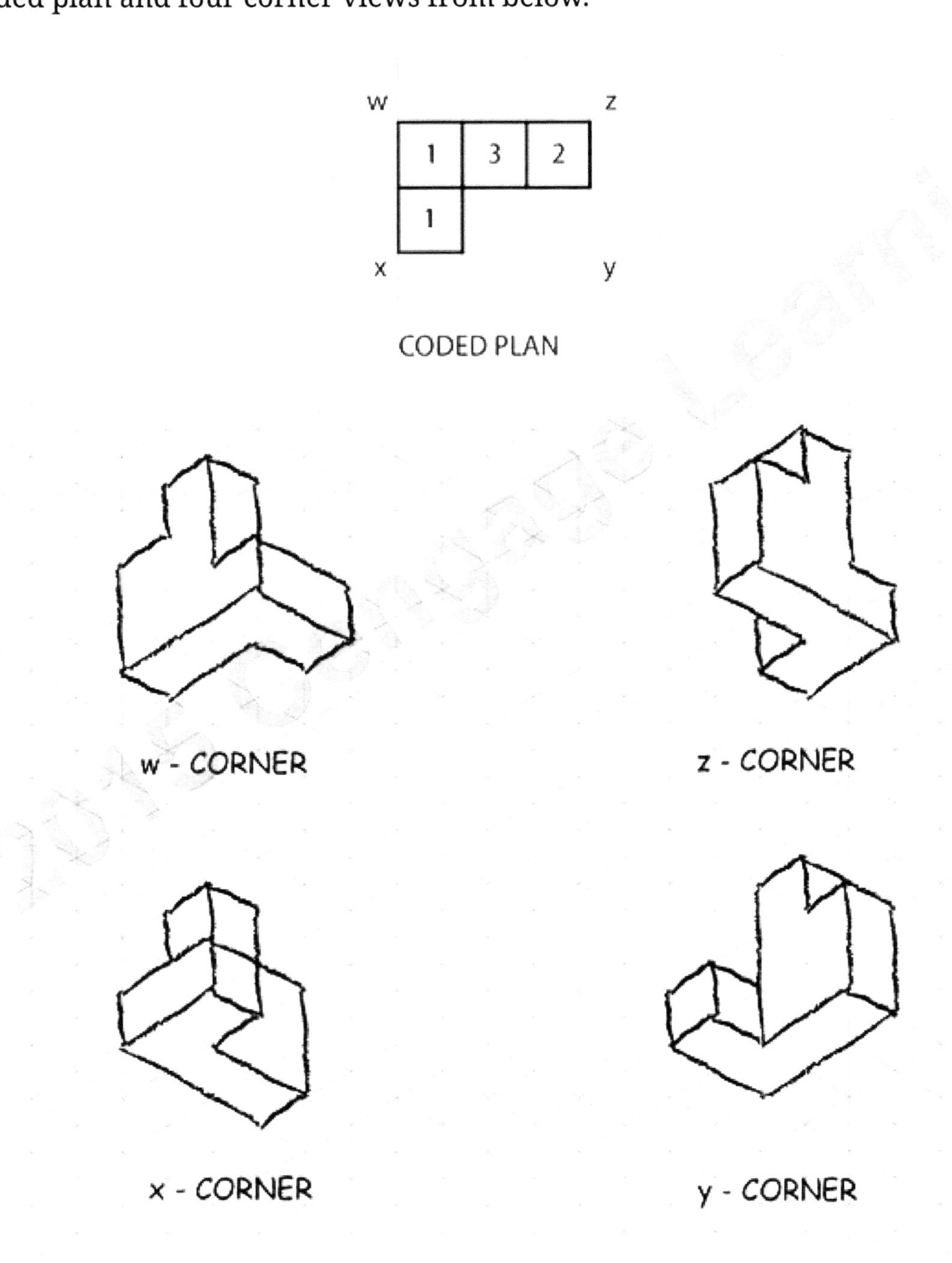

If you compare the corner views from these two viewpoints, you will notice that with a bird's-eye view, you see all of the top surfaces of the object (except for those that are partially or completely hidden by projections on the object). But with the worm's-eye view, you see the bottom surfaces of the object and none of the top surfaces are shown. Figure 17.20 shows a bird's-eye and worm's-eye view for an object, with the top and bottom surfaces labeled.

Figure 17.20

Bird's-eye view versus worm's-eye view of the same object from the same corner.

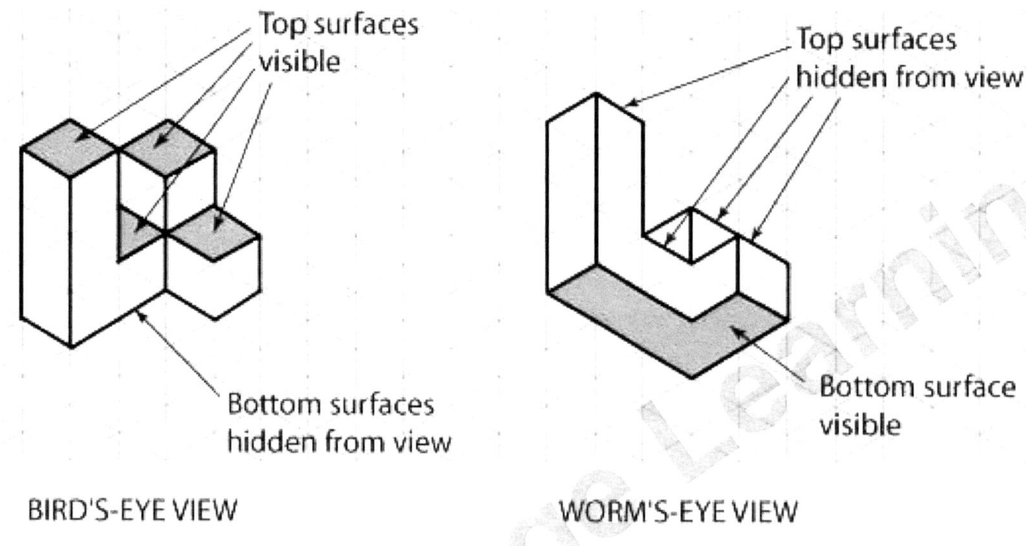

Recall that for isometric pictorials, you learned that the axes are set up such that you are looking down the diagonal of a cube. The difference between a bird's-eye view and a worm's-eye view is that you are looking down two different diagonals of the cube. Figure 17.21 shows two cubes with the diagonals for both viewpoints labeled.

Figure 17.21

Cube diagonals for bird's-eye (a) and worm's-eye (b) views.

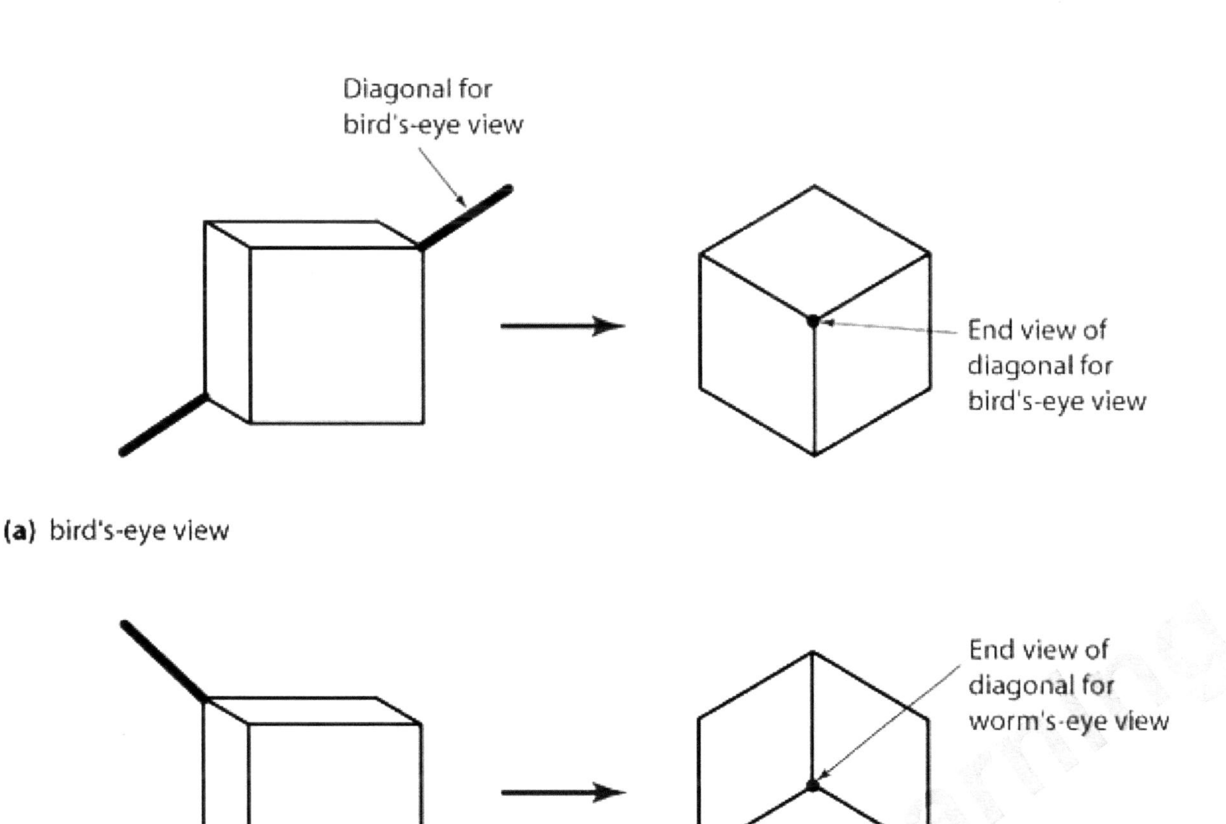

17.05 Advanced Visualization Techniques

In the next sections, you will learn some systematic strategies for visualizing objects and for seamlessly moving between multiview and pictorial representations. Many of these techniques have been tried and tested with engineering students over the years. Although you may find some of the techniques to be difficult in the beginning, with practice and with improvement in your visualization skills, you will be able to tackle increasingly difficult problems. Some of these techniques may be easier for you than others. You should find the methods that work for you and stick with them as you complete the exercises at the end of the chapter. Good luck!

17.05.01 Visualization with Basic Concepts

Figure 17.22 shows the top and front views of an object, along with three possible right-side views. Using what you know about basic shapes, which of the possible right-side views is correct?

Figure 17.22

Top and front views, with three possible side views.

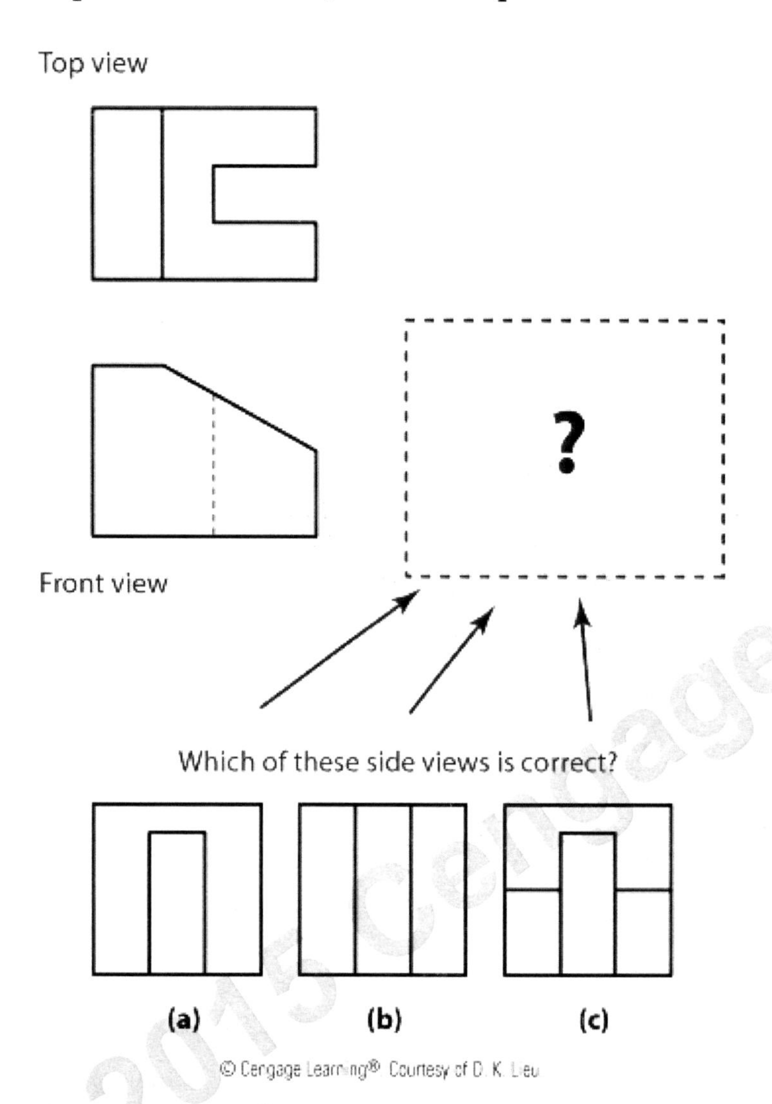

Begin by looking at the angled edge of the object in the front view labeled *A*. The most probable explanation for this angled edge is that it is an inclined frontal surface—seen as an edge in the front and as surfaces in the top and right side. This edge does not extend all the way across the object in the front view, but goes about three-quarters of the way from right to left. You know that this surface must be visible in the top view, and the only surface there that extends from the right side to about three-quarters of the way to the left side is the U-shaped surface labeled *A* in Figure 17.23.

Figure 17.23

The width dimension of an inclined surface.

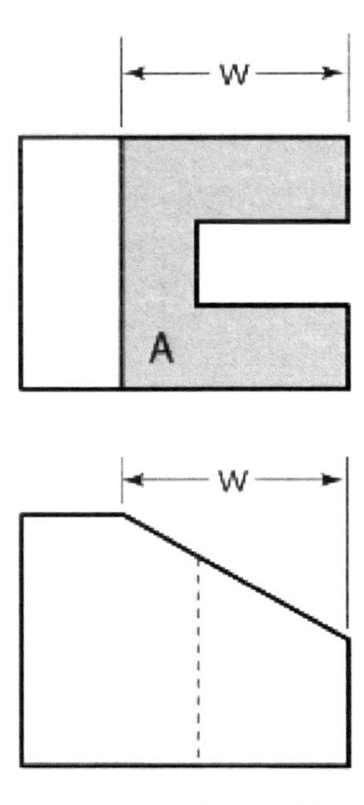

© Cengage Learning®. Courtesy of D. K. Lieu.

You know that this surface also must appear as a U-shaped surface in the correct side view, since the overall shape of the surface cannot change from view to view. There are two possibilities from the choices given for the correct right-side view—choices (a) and (c). Look again at the inclined line that represents the surface in the front view. Notice that it does not extend all the way from the bottom to the top in that view. The inclined line starts at a point about halfway up from the bottom and extends from there to the top of the object, which is illustrated in Figure 17.24.

Figure 17.24

The height dimension of an inclined surface.

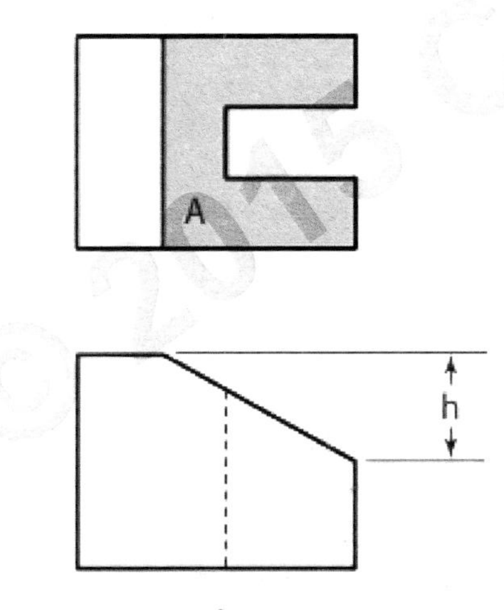

© Cengage Learning®. Courtesy of D. K. Lieu.

Of the choices given, only choice © has the correct overall height for the U-shaped surface; the U-shaped surface in choice (a) goes all the way from the bottom to the top of the object. Figure 17.25 shows the correct multiview sketch of the object. An isometric pictorial sketch also is shown in the figure for clarity.

Figure 17.25

A multiview drawing of an object.

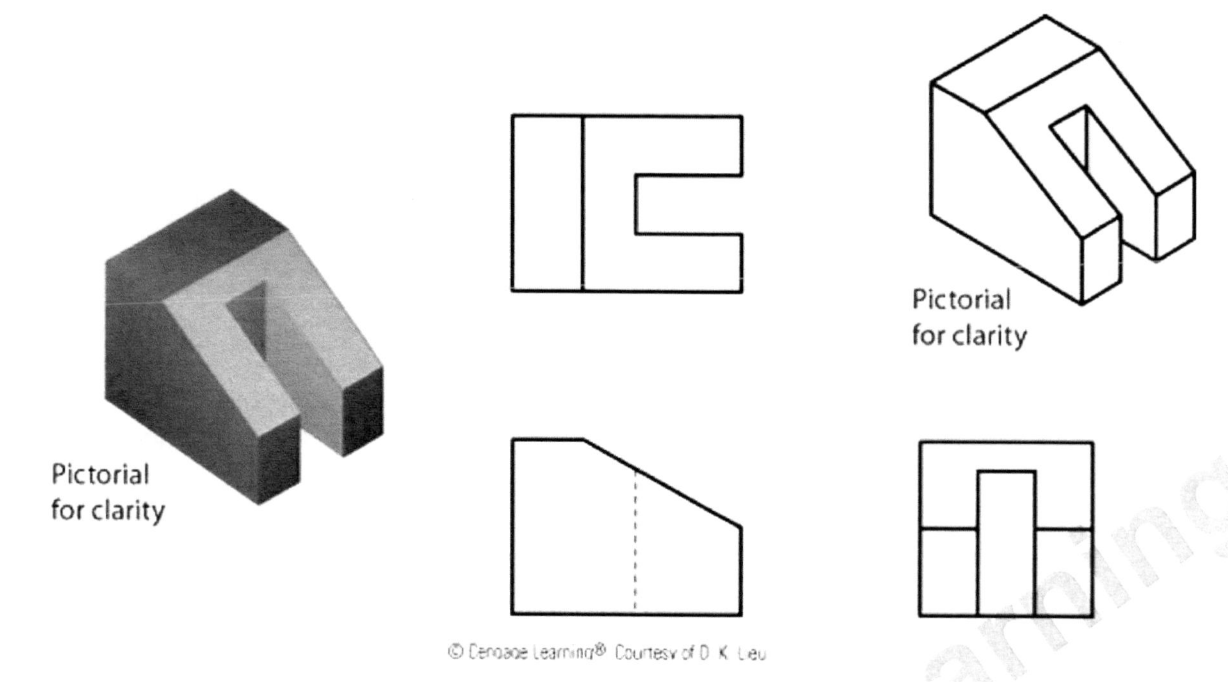

© Cengage Learning®. Courtesy of D. K. Lieu.

What about the object shown in Figure 17.26? For this particular object, the front and side views are given and you are to select the top from the three choices given.

Figure 17.26

Front and side views of an object, with three possible top views.

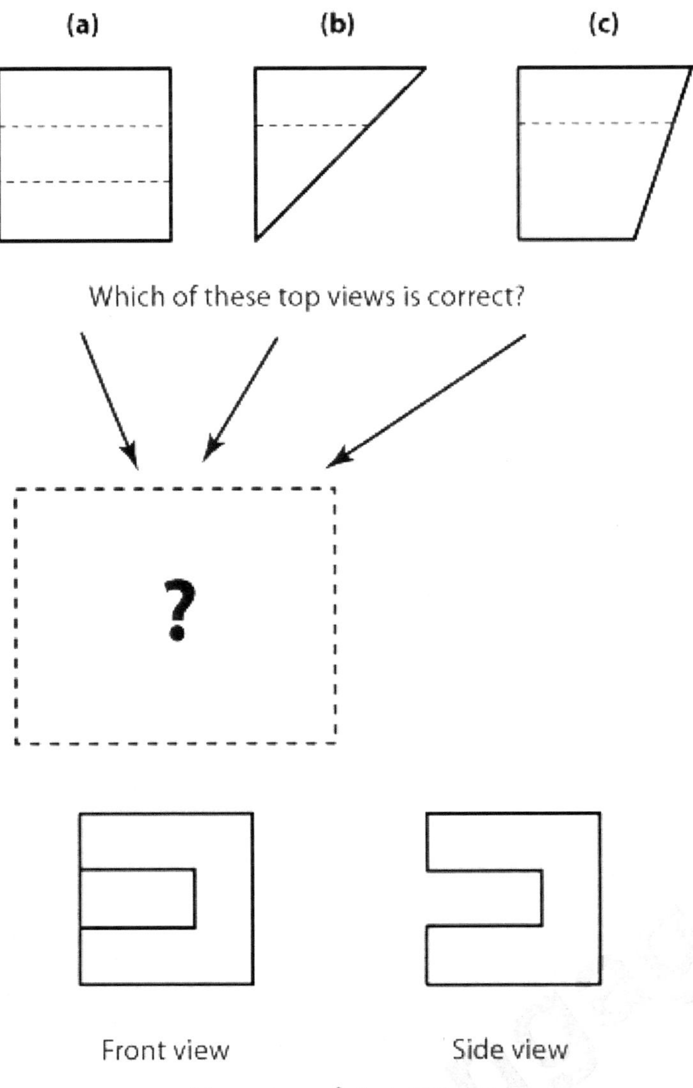

A U-shaped surface in both views for this object extends all the way from the bottom to the top. Chances are good that this is an inclined surface since it appears as a similarly shaped surface in two views. (It also could be an oblique surface; but for simplicity, assume it is inclined.) If it is an inclined surface, it will be classified as an inclined horizontal surface and, therefore, would be seen as an edge in the top view. Two of the choices, (b) and (c), show an angled edge in Figure 17.27.

Figure 17.27

Angled lines in choices (b) and (c) for the top view.

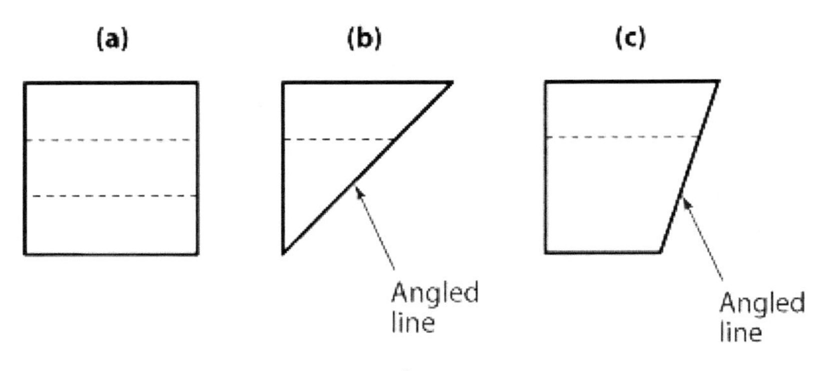

From the front view, the U-shaped surface goes all the way across the object from left to right. Only choice (b) shows the angled line extending all the way across the object, so it must be the correct top view. Figure 17.28 shows the correct multiview sketch for the object. Once again, a pictorial of the object has been included for clarity.

Figure 17.28

The correct multiview drawing of the object.

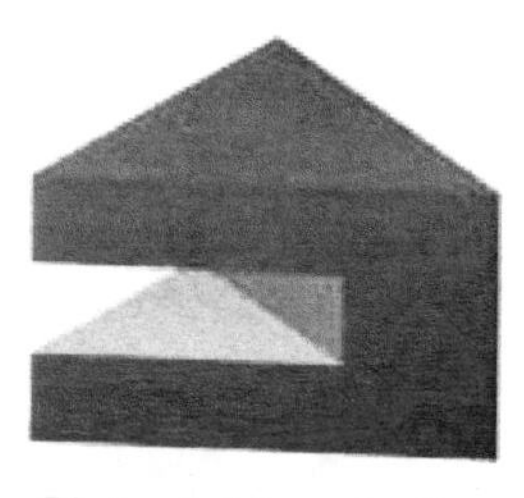

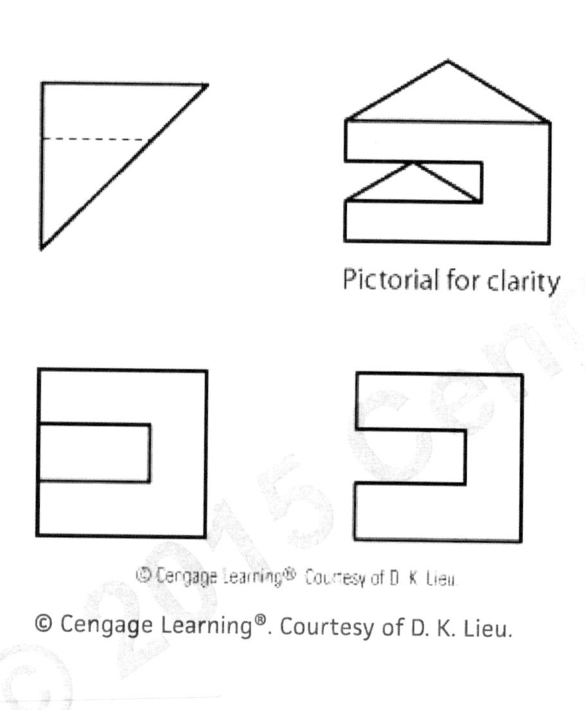

17.05.02 Strategy for a Holistic Approach to Constructing Pictorials from Multiview Drawings

The inverse tracking of points, edges, and surfaces covered in the later sections are detailed methods that can be used to create a pictorial from a multiview drawing. A different, more holistic approach also can be employed to achieve the same result, as outlined in the following paragraphs.

A few ground rules must be established before you can begin development of this approach:

- The objects you devise must be solid.
- You cannot have a single, infinitely thin plane on the resulting object.
- You cannot have two objects that are next to each other in the result.
- You should pick one view as the "base" and work from there until all views match the object.

An illustration of this approach will begin with a simple object. Figure 17.29 shows the front and right-side views of an object from which you want to create a pictorial. This example will use the front view as the base view.

Figure 17.29

The front and right-side views of an object.

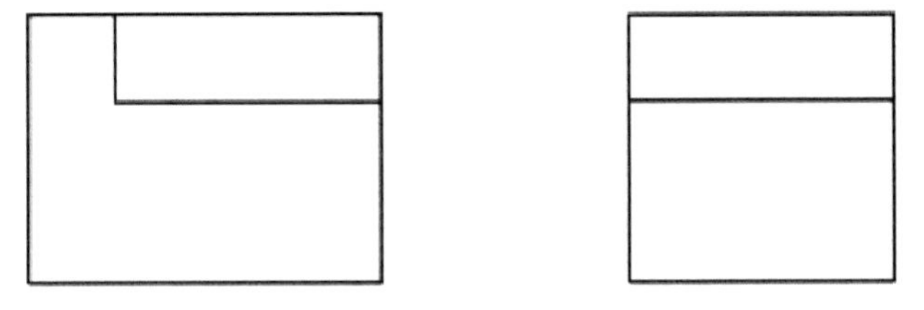

The front view shows two surfaces—one is L-shaped; the other, rectangular. When you start with the L-shaped surface, you can assume that it is a normal surface and that it is located all the way forward on the object. You will consider other cases of this surface later; but for this method, it is best to start with the simplest form of the object and then consider increasingly complex solutions. Normal surfaces are the simplest type, so it makes sense to start there. When you also consider the rectangular surface to be normal, it must be parallel to the front view as well. (Otherwise, it would appear as an edge in the front.) The three possibilities for this surface are that it appears all the way at the front of the object, all the way to the back of the object, or somewhere in between. These three possibilities are illustrated in Figure 17.30.

Figure 17.30

Three possible solutions when the rectangular area in Figure 17.29 is a normal surface.

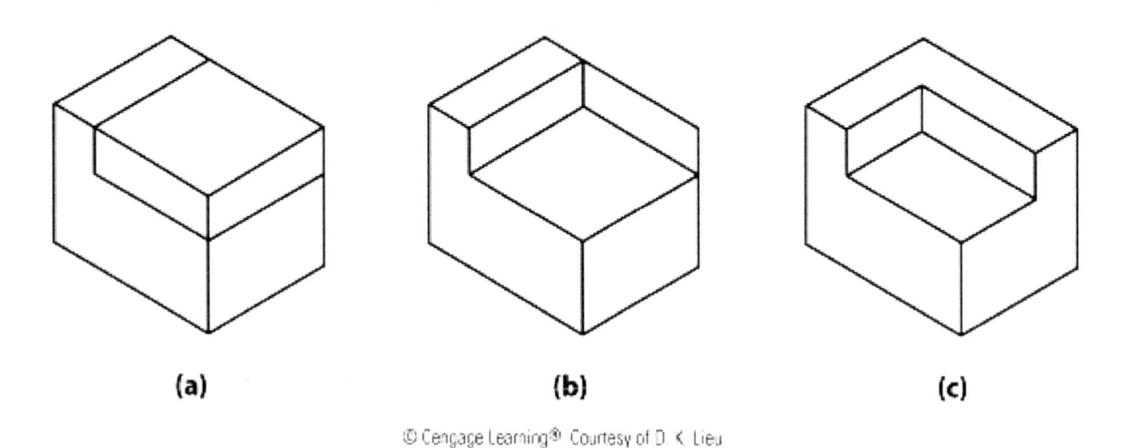

The first of these choices is not a valid solution because if both surfaces were all the way to the front of the object, there would be no need for the edges separating the two areas—lines are not included on a single plane of an object. The second choice produces front and side views that match those that were given initially; however, this choice is not valid since it results in an infinitely thin plane at the back of the object. The third choice also can be eliminated, since it results in an object that does not match the right-side view. It does not matter where the surface is located with respect to the object depth—there is no way to include a normal surface, such as this, on the object to produce a correct right-side view.

When the L-shaped surface is located correctly, there is no possible way the rectangular area represents a normal surface. Now you will assume that the rectangular area represents an inclined surface. In this case, you have to consider only an inclined horizontal or an inclined profile surface, since an inclined frontal would appear as an angled line in the front view. (No such lines are there.) Figure 17.31 shows two possible inclined horizontal surfaces. Notice that the first choice results in an object that does not correspond to the right-side view that is given, while the second choice does. This means that an inclined horizontal surface, such as the one shown, is one possible solution to this problem.

Figure 17.31

Two possible inclined horizontal surfaces for an object.

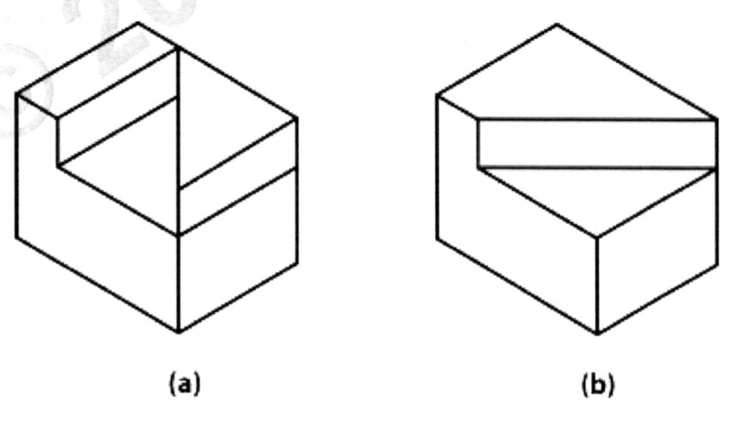

Now consider the possible inclined profile surface shown in Figure 17.32. Notice that this inclined profile surface is not a possible solution, since there are no angled lines in the side view.

Figure 17.32

An inclined profile surface for an object.

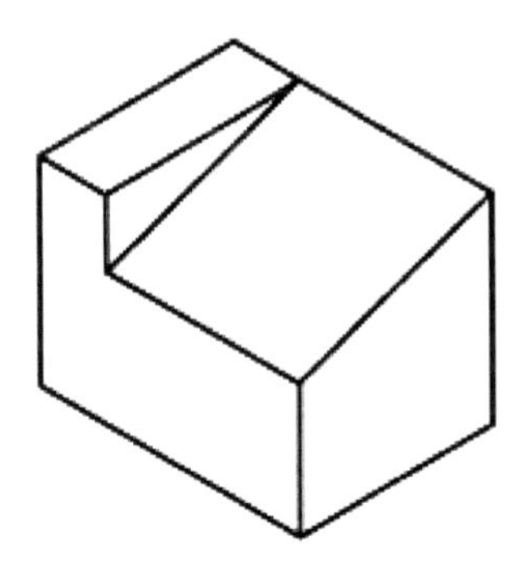

Are there other possible solutions? Are curved surfaces a possibility? Figure 17.33 shows two possible solutions where the rectangle seen in the front view represents a curved surface [(b) and (d)] and several solutions that include curved surfaces that do not result in valid solutions [(a), (c), and (e)]. The invalid solutions result in a front or side view that does not match those given.

Figure 17.33

Possible curved surfaces for an object.

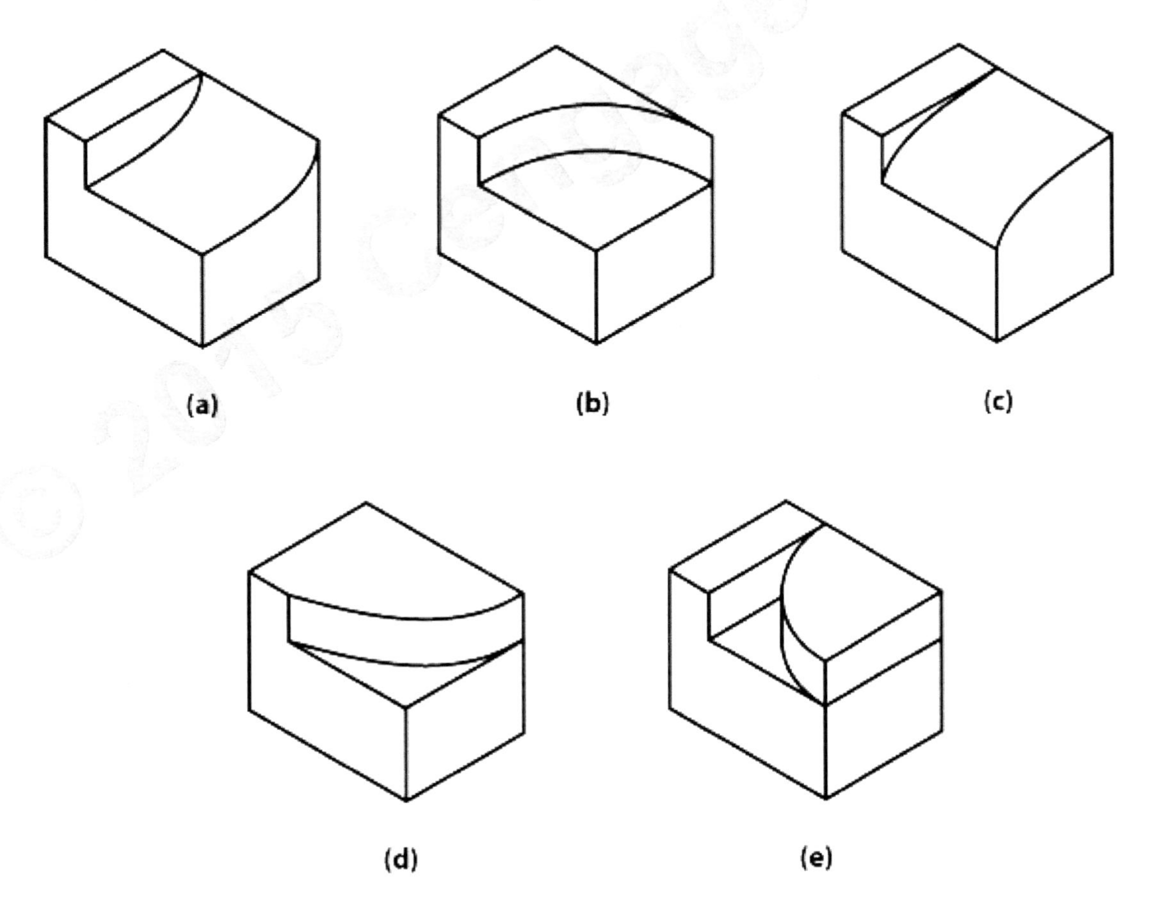

What about an oblique surface for the rectangular area in the front view? Figure 17.34 shows two possible oblique surfaces that satisfy the front view, but notice that the side view produced by these objects is not valid. You could try any number of different oblique surfaces and not find one that satisfies the two views given.

Figure 17.34

Two possible oblique surfaces.

Now consider choices where the rectangular area in the front view is a normal surface that is located all the way to the front of the object. This time you should again consider the L-shaped surface as a normal surface all the way to the front, all the way to the back, or somewhere in between. These three possibilities are shown in Figure 17.35. Once again in this case, the front view of the object is okay for each possibility, but either the side view is incorrect or the result is not permissible.

Figure 17.35

Three possibilities with an L-shaped normal surface.

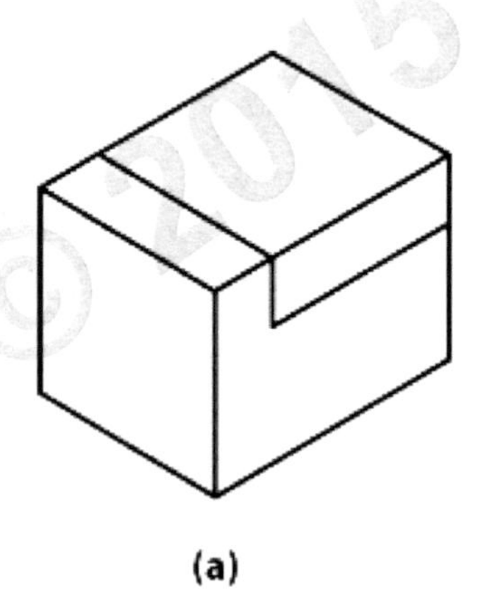

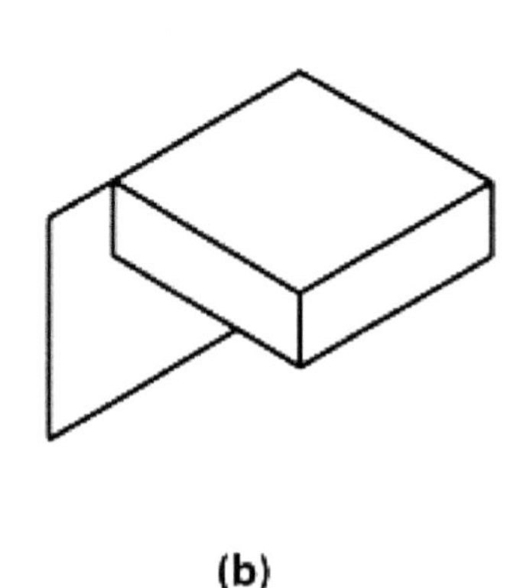

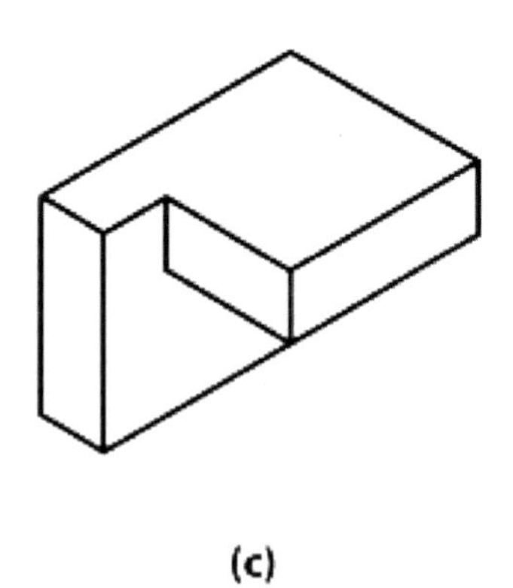

Next, consider the L-shaped surface to be an inclined profile. The two possibilities are shown in Figure 17.36. Notice that neither of these produces the correct side view.

Figure 17.36

Two possible solutions with inclined profile L-shaped surfaces.

Figure 17.37 shows the two possibilities for the L-shaped area to represent an inclined horizontal surface. Once again, neither of these produces a side view that corresponds to the one that is given. An inclined frontal surface is not possible, since there are no angled lines in the front view.

Figure 17.37

Two possible solutions with inclined horizontal L-shaped surfaces.

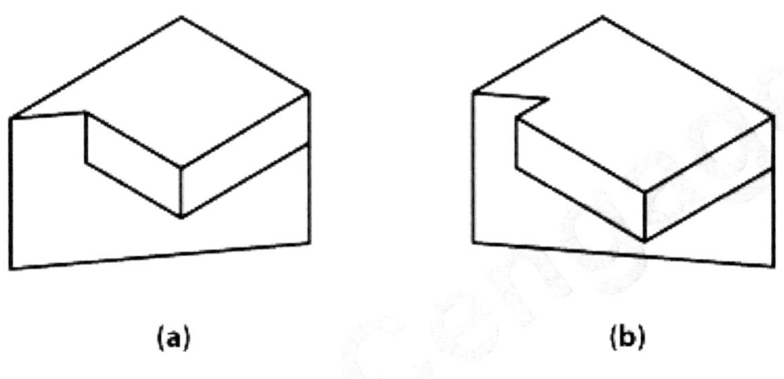

You also can consider curved surfaces for the L-shaped area as before, as shown in Figure 17.38. In this case, only one of the possible solutions with a curved surface matches the two views given.

Figure 17.38

Two possible curved surfaces for the object.

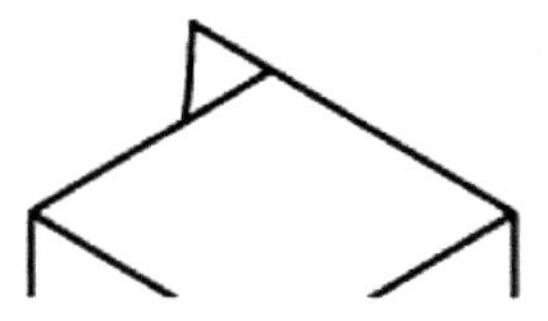 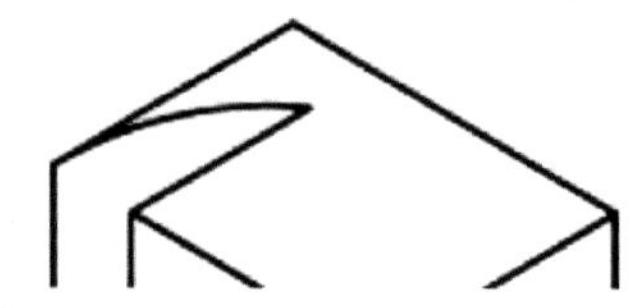

Can the L-shaped surface be an oblique surface? Figure 17.39 shows one possible oblique surface that satisfies the front view. Notice that when the L-shaped area is this oblique surface, once again the side view does not correspond to the one that is given.

Figure 17.39

A possible L-shaped oblique surface.

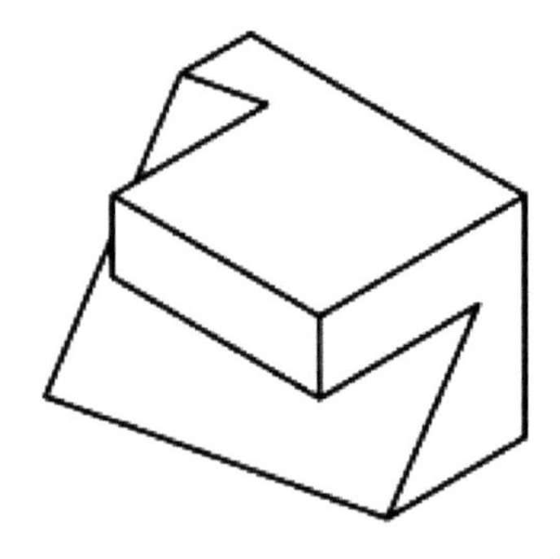

© Cengage Learning®. Courtesy of D. K. Lieu.

After sketching multiple possible pictorials that match the two views that are given, you have identified three possibilities—one that includes an inclined surface and two that contain a curved surface. All of the correct pictorials identified in this exercise are shown in Figure 17.40. You may find this method of creating pictorials from multiview drawings tedious at first; however, as you practice, you will find it increasingly easy. In time, you will probably be able to skip certain pictorials altogether.

Figure 17.40

Possible correct solutions for the object.

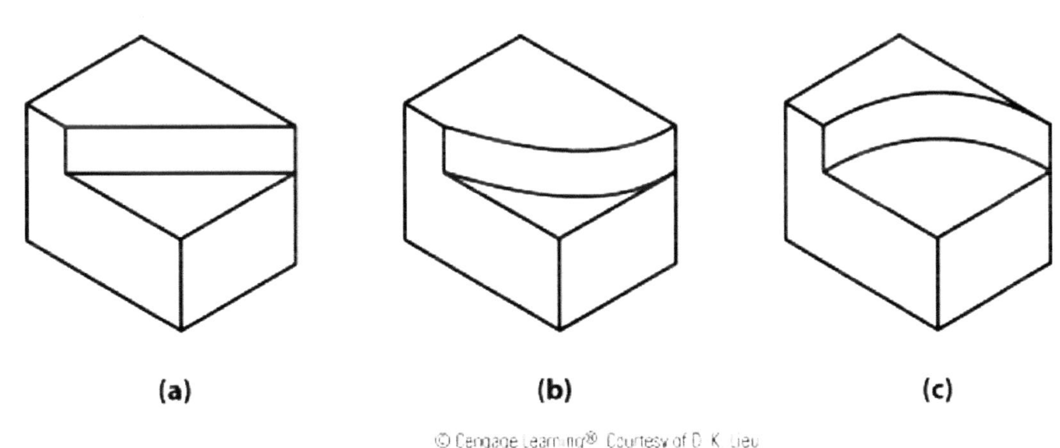

17.05.03 Strategy for Constructing Pictorials by Inverse Tracking of Edges and Vertices

The process of creating a multiview drawing from a pictorial (or from an idea or image in your mind) is quickly mastered with practice. However, the inverse process of creating a mental image, or pictorial drawing, of an object from its multiview drawing is considerably more challenging. With some practice, though, you can master this skill as well. People who deal with multiview drawings on a daily basis, such as professional design engineers, drafters, and machinists, usually can create a mental image of a part quickly after they see its multiview drawing.

To develop this skill, it may be best to start with a technique that is well-defined and methodical, as well as an object of fairly simple geometry, such as that shown in Figure 17.41. For this example, sketching techniques will once again be used as a method for developing visualization skills. Later examples will use more formal graphics for presentation clarity.

Figure 17.41

To create an isometric pictorial of the object shown in this multiview drawing, define the anchor point and coordinate axes in each view, then the same point and axes in the pictorial grid.

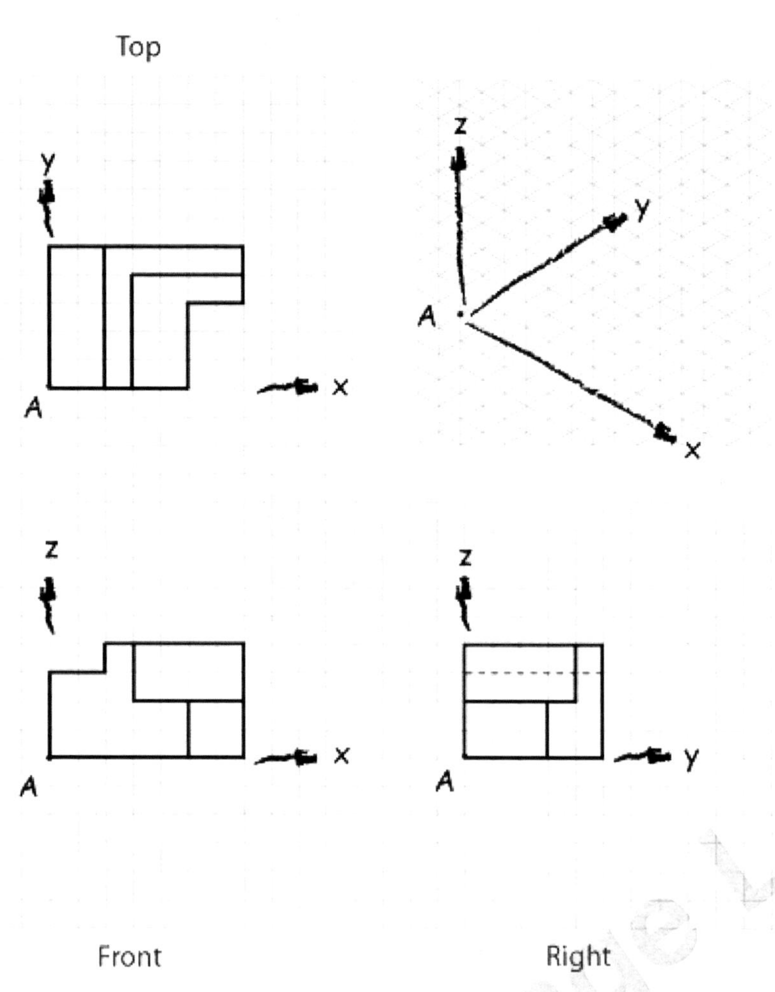

© Cengage Learning®. Courtesy of D. K. Lieu.

Inverse Tracking with Edges and Vertices for Normal Surfaces

An eight-step process can be used, but this time the process is inverted compared to the process presented in an earlier chapter where a multiview drawing was created from a pictorial. The steps are as follows:

Step 1:

Define the location and directions of a coordinate system consistent in all views.

Step 2:

Define an anchor point on the object.

Step 3:

Mark the limits of the foundation volume.

Step 4:

Locate a vertex or an edge adjacent to the anchor point and draw that edge.

Step 5:

Successively locate other vertices and edges and draw those edges.

Step 6:

Convert hidden lines.

Step 7:

Add internal features.

Step 8:

Check model validity.

The first step (Step 1) is to create a set of 3-D axes on an isometric grid and define the viewing directions. Next, select an anchor point for the object (Step 2). The **anchor point** (The same point, usually a vertex, which can be located easily and confidently on multiple views for an object.) should be a point on the part that you can easily and confidently locate on each of the orthogonal views. A good selection for this example would be the vertex on the lower-left front of the part. Called point A, it is identified in each view in Figure 17.41; and a set of axes is drawn at that point in all of the views, including the pictorial grid. Make sure the directions of the coordinate axes are consistent in every view.

The **foundation space** (The rectilinear volume that represents the limits of the volume occupied by an object.) (Step 3) is then outlined with respect to the anchor point on the grid to show the limits of the volume that is occupied by the object in space, as shown in Figure 17.42. To find the foundation space, note that the width and height limits of the part can be seen in the front view. The top view shows the width of the part as well as its depth. The side view shows the height and depth of the part. Thus, the height, width, and depth of the foundation space can be established with only two views as long as those view planes are orthogonal to each other. Any two adjacent views, such as the front and top views or the front and side views, will be orthogonal to each other. For this example, the width of the part is 7 units in the x-direction, the height of the part is 4 units in the z-direction, and the depth of the part is 5 units in the y-direction. A rectangular volume with these x, y, and z measurements is created on the isometric grid. Although most objects probably will not be brick-shaped, a foundation shape is easy to draw and easy to visualize. It is typically much easier to locate the points, edges, and surfaces of the object with respect to its foundation than to a set of axes. Also, if you start creating points and lines on the isometric grid that are outside the foundation space, you will know you are doing something wrong.

Figure 17.42

Coordinate axes and viewing direction are defined on an isometric grid. An anchor point A is selected, and the foundation space is outlined.

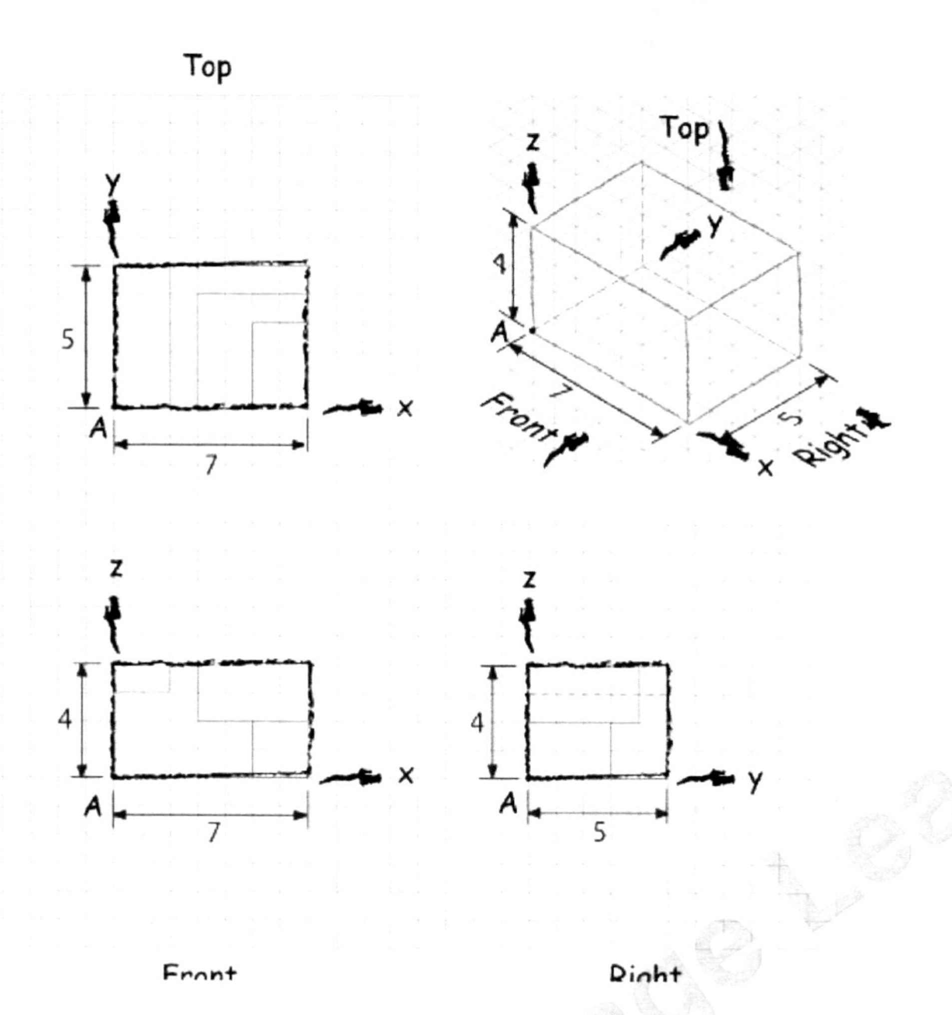

Next, locate a vertex or an edge adjacent to the anchor point (Step 4). Remember that a vertex can be defined as a point that connects at least three edges and any corner is automatically a vertex. To locate an edge on the pictorial, that edge must be defined in at least two orthogonal views. For parts that contain surfaces that are parallel or perpendicular to the viewing planes, it is likely that two or more edges of the part will appear on top of each other in the orthogonal views. Thus, when you are tracking an edge from a drawing view to the pictorial, the edge needs to be carefully specified in another view. **point tracking** (A procedure by which successive vertices on an object are simultaneously located on a pictorial image and a multiview image of that object.) and **edge tracking** (A procedure by which successive edges on an object are simultaneously located on a pictorial image and on a multiview image of that object.) are done by selecting vertices on the part in the orthogonal views and then locating those vertices on the pictorial by moving along the edges of the part. Moving along the part edges in a multiview drawing is more difficult than doing so on a pictorial because you need to track the motion on different orthogonal views simultaneously. Also, on a multiview drawing, many edges and vertices lie directly above or below each other. Identification of the individual vertices may be confusing, but it can be done easily when you exercise some care. The trick is to define the location of the vertices in at least two views simultaneously. Tracking is made easier by the fact that vertices in adjacent orthogonal views must align either horizontally or vertically. For example, as shown in Figure 17.43, as you move 5 units to the right along the **x− axis** from point A to reach point B in the front view along the edge indicated, you also must identify this motion in the top and side views and on the pictorial simultaneously. To find point B in the top view, draw a vertical alignment line through point B in the front view and see which vertices on the object lie on this line in the top view. The top view shows a few possible vertex locations on the alignment line for point B; but only one of the locations is accessible by moving along a single edge (also in this case, the x− axis from point A). This edge also is parallel to the front viewing plane. In the side view, the location of point B is directly in front of point A, thus appearing coincident with it. With the location of point B identified in each of the orthogonal views, this location and the edge between points A and B are transferred to the pictorial by moving 5 units along the x− axis .

Figure 17.43

Identify and track the edges of the part on the orthogonal views and locate these edges on the pictorial.

Next, another vertex on the front view and on the same surface as point A is selected on the orthogonal views. Calling this point C, you can easily see the edge formed by points A and C on the side view. It is located 3 units above point A along the z− axis . Even though the horizontal view alignment line goes through four vertices in the side view, only one of these locations is accessible by moving along an edge from point A. This edge is parallel to the front viewing plane. In the top view, point C is directly above point A; so they are seen as coincident in this view. The location of point C on the pictorial is placed on the z− axis 3 units above point A.

Additional vertices are located by noting their x, y, and z positions on the multiview drawing and then transferring those locations to the pictorial view (Step 5). Before moving to the next surface, you should select the points and edges to outline an entire surface of the object. In this manner, instead of seeing connected edges that appear in a variety of directions on the pictorial, you see that the surfaces of the part appear one at a time. Surfaces or parts of surfaces on the part can be identified easily on the orthogonal views as simple closed loops, as shown in Figure 17.44. Once all of the edges in all of the orthogonal views have been accounted for and placed on the pictorial, the object pictorial should be complete, as shown in Figure 17.45. Note that in the final pictorial, one of the surface loops is partially obscured by some of the other surfaces. In this case, the obscured edges should be removed or shown as hidden lines (Step 6). Since isometric views show only three sides of the glass box, the front, side, and top views are generally sufficient to create the isometric view of most objects. If only two of these views are available, you should create the third view before attempting to construct the pictorial.

Figure 17.44

Surfaces are loops on the orthogonal views. Complete the edges of an entire surface before proceeding to the next surface.

Figure 17.45

When all surface loops on the orthogonal views are accounted for, the isometric view should be complete.

The final steps are to add any internal features (Step 7), of which there are none in this example, and then check to ensure that each vertex on the pictorial is connected to at least three edges (Step 8) for the pictorial to be valid. For this pictorial, all vertices are connected to three edges (some of the edges are hidden); so it appears to be valid.

Inverse Tracking with Edges and Vertices for Inclined Surfaces

Any time there is an edge on a multiview drawing that appears at an angle (i.e., it is not horizontal or vertical), an inclined surface will exist on the part. The drawing of such a part is shown in Figure 17.46. Creating an isometric pictorial of a part with inclined surfaces is only slightly more complicated than creating a part with horizontal or vertical surfaces. The same procedure of vertex and edge tracking is used for both types of parts.

Figure 17.46

It is desired to create an isometric pictorial of the object shown in this multiview drawing.

First, an anchor point, called point A, is selected and identified in each view. The anchor point, coordinate axes, and view directions are defined on an isometric grid (Steps 1 and 2). A foundation space is created on the grid (Step 3) based on the limits of the object extracted from the orthogonal views. This space is shown in Figure 17.47. Note that in this particular example, the anchor point is not located at one of the extreme limits of the foundation space. The anchor point can be located just about anywhere on the object as long as it is convenient to use and you are confident of its location in all of the orthogonal views and in the pictorial.

Figure 17.47

Coordinate axes and viewing directions are defined on an isometric grid. An anchor point, A, is selected, and the foundation space is outlined.

Start with vertices near the anchor point (Step 4). By tracking points and identifying the edges of the part in the orthogonal view, build the edges surrounding one of the surfaces on the part. Note that as you build these edges, the direction of travel for any edge at an angle will be two-dimensional; and you must keep track on the pictorial of the distance traveled in each direction. This is shown in Figure 17.48.

Figure 17.48

Identify and track the edges of a surface at the anchor point on the orthogonal views and locate these edges on the pictorial.

Track the edges of one surface at a time (Step 5). Surfaces or parts of surfaces in each view are seen as simple loops, as shown in Figure 17.49. As long as you keep careful track of the location of each vertex on each surface, the inclined surfaces defined by their vertices should appear on the pictorial.

Figure 17.49

Surfaces appear as simple loops on the orthogonal views. Complete the entire surface and its edges before proceeding to the next surface.

Once all edges in all of the orthogonal views are accounted for, the object should be complete, as shown in Figure 17.50. Note that in this case, one of the surfaces on the vertical slot seen in the pictorial is hidden in the multiview drawing; that is, the surface loop on the multiview drawing contains a hidden line. Surfaces must always be connected on all of their edges by other surfaces. Also note for this example that the hidden edges for a partially obscure surface, such as the ledge on the rear of the object, are not shown in the pictorial (Step 6).

Figure 17.50

When all surface loops on the orthogonal views are accounted for, the isometric view should be complete.

If so desired, the full object with its hidden edges also can be created, as shown in Figure 17.51. For the full model with hidden edges to be created, all of the surfaces of the object need to be created on the pictorial, even the hidden surfaces. The hidden surfaces are identified by creating the remaining three of the six standard views or by carefully extracting the required information from the existing three views. When there are many hidden edges, it is often better to remove the hidden edges and offer a second pictorial presentation from another viewing position, as shown in Figure 17.51.

Figure 17.51

The full pictorial presentation with visible and hidden edges may be confusing. It is sometimes better to offer a presentation using two separate pictorial views.

Once again, the pictorial is complete after any internal features are added and its model validity is checked (Steps 7 and 8).

Inverse Tracking with Edges and Vertices for Oblique Surfaces and Hidden Features

The next problem is more complicated because of the existence of a hidden feature and an oblique surface. It is desired to create an isometric pictorial of the object shown on the multiview drawing in Figure 17.52. This object is an approximate rectangular solid with one corner that has been cut off obliquely and a cubical cutout at the opposite corner. A five-view presentation would probably have been clearer, but the luxury of additional views is not always offered. Although you can create the additional views yourself, you will proceed with the existing information.

Figure 17.52

It is desired to create an isometric pictorial of the object shown in this multiview drawing.

An anchor point, called point A, is selected and identified in each view. The anchor point, coordinate axes, and view directions are defined on an isometric grid. A foundation space is created on the grid based on the limits of the object extracted from the orthogonal views. This space is shown in Figure 17.53.

Figure 17.53

Coordinate axes and viewing directions are defined on an isometric grid. An anchor point, A, is selected, and the foundation space is outlined.

Start with the anchor point. By tracking points and identifying the edges of the part in the orthogonal view, build the edges surrounding one of the surfaces on the part. For the first surface you create, you should select a surface created by loops made of visible edges only. In fact, it is a good idea to create all of the surfaces made only with visible edges (i.e., temporarily ignore the hidden edges). This is shown in Figure 17.54.

Figure 17.54

Consider the visible edges first. Use point and edge tracking to create these edges on the pictorial.

For this object, the isometric pictorial does not give a clear representation of the object. The hidden lines in the multiview drawing indicate a hidden feature that must somehow be indicated in the pictorial. The hidden edges are then tracked and added to the pictorial (Step 7), as shown in Figure 17.55, until the pictorial is complete, as shown in Figure 17.56. A check of model validity shows that this pictorial is valid (Step 8).

Figure 17.55

Track the vertices of the hidden edges. Surfaces are loops formed by the hidden edges. Complete the edges of the entire surface before proceeding to the next surface.

Figure 17.56

When all of the edges, visible and hidden, are accounted for, the pictorial should be complete.

Inverse Tracking with Edges and Vertices for Curved Surfaces

A curved edge on any view on a multiview drawing means that there is a curved surface on the part. The most common curved surface is cylindrical, such as a hole or an external round, which intersects another planar surface. When a cylindrical surface intersects a plane at a right angle, a circular edge is formed. As you learned in a previous chapter, in an isometric pictorial, a circular edge will appear as an ellipse, with its inclination angle and orientation dependent upon its orientation on the object. As shown in Figure 17.57, for an isometric pictorial, depending on the location of the circular edge, that edge will appear as an ellipse with an inclination angle of 0, 60, or –60 degrees. For a unit diameter circle, the minor axis of the isometric ellipse will have a length of 0.707 unit, and the major axis will have a length of 1.225 units. These orientations and sizes for the ellipses are true as long as the circle lies in a plane that is parallel to one of the standard orthogonal views. Also recall that a solid line on the drawing may not be an edge on the part at all, but rather a representation of the optical limit of the part seen along a curved surface. When an isometric pictorial of such an object is constructed, this physical limit of the part must be shown.

Figure 17.57

The size and orientation of a circle of unit diameter on the faces of an isometric cube.

Consider the part shown in Figure 17.58, which has a cylindrical round on its external surface as well as a hole, which is an internal feature.

Figure 17.58

It is desired to create an isometric pictorial of the object shown in this multiview drawing.

Start the creation of the isometric view the same way you began the previous cases. An anchor point, point A, is selected and identified in each view. The anchor point, coordinate axes, and view directions are defined on an isometric grid. A foundation space is created on the grid, based on the limits of the object extracted from the orthogonal views. This space is shown in Figure 17.59.

Figure 17.59

Coordinate axes and viewing directions are defined on an isometric grid. An anchor point, A, is selected, and the foundation space is created.

When dealing with curved surfaces, one common strategy you can use is to ignore the curved surfaces temporarily, create the remaining planar surfaces as extended surfaces, and then go back to add the curved surfaces. This approach has the advantage of allowing you to quickly draw the rough frame of the part and visualize it. The extended planar surfaces are shown in Figure 17.60.

Figure 17.60

Create a rough frame by using linear edges instead of curved edges. Use point and edge tracking to create these edges on the pictorial.

The circular edges are then added to the appropriate planar surfaces to create the curved surfaces. In the isometric pictorial, the circular edges appear as segments of ellipses. Linear edges that are tangent to circular edges on the orthogonal views also are tangent to their corresponding elliptical edges on the pictorial. This process is shown in Figure 17.61.

Figure 17.61

Add the circular edges using elliptical segments in the pictorial.

For holes, make sure you include the far edge of the hole when it can be seen through the thickness of the part. For external curves, make sure you include the physical limit of the part as seen on the curved surface, as shown in Figure 17.62, even though a real edge does not exist there. In this example, notice that some surfaces that were hidden or shown on edge in the multiview drawing become partially visible on the pictorial. The continuity of the surfaces must be maintained.

Figure 17.62

Add the optical limits of the curved surfaces, seen as lines tangent to the curved edge. Remove portions of edges that are hidden by the curved surfaces.

As a final check, make sure each vertex has at least three edges connected to it. Note that tangent edges, created by planes tangent to cylinders or cones, still count as edges, even though they usually are not shown.

17.05.04 Strategy for Constructing Pictorials by Inverse Tracking of Surfaces

The process of tracking vertices and edges to create a mental image, or a pictorial drawing, from a multiview drawing is a slow but reliable method. However, after gaining experience doing this, some people may find it faster and easier to skip this process and go directly to the **surface tracking** (A procedure by which successive surfaces on an object are simultaneously located on a pictorial image and a multiview image of that object.) . The eight-step model creation process is the same as that used with points and edges except for some slight modifications.

Step 1:

Define the location and directions of a coordinate system consistent in all views.

Step 2:

Define an anchor surface.

Step 3:

Mark the limits of the foundation volume.

Step 4:

Locate a surface adjacent to the anchor surface and draw its boundary.

Step 5:

Successively locate other adjacent surfaces and draw those boundaries.

Step 6:

Convert hidden lines.

Step 7:

Add internal features.

Step 8:

Check model validity.

For example, the multiview drawing shown in Figure 17.63 is composed of multiple loops (representing surfaces) in each view. The orientation and viewing directions (Step 1) for the model are shown in Figure 17.64. Also shown in this figure is an **anchor surface** (The same surface that can be located easily and confidently on multiple views for an object.) , surface A, which can be easily located in each view and on the pictorial (Step 2). The rectangular extent of the outer loop in each view gives the size of the foundation space as previously described for edges and vertices (Step 3).

Figure 17.63

It is desired to create an isometric pictorial of the object shown in this multiview drawing.

Figure 17.64

Viewing directions are defined on an isometric grid. The foundation space is created, and an anchor surface is selected.

The next step involves identifying a surface adjacent to the anchor surface, locating the adjacent surface in each view, and then adding this surface to the pictorial view (Step 4), as shown in Figure 17.65. Continue to identify surfaces on the object and transfer them to the pictorial, until all of the surfaces are accounted for. The object shown in Figure 17.66 should now be complete (Step 5).

Figure 17.65

Begin the process of locating each surface. Note that in this case, a surface in the right-side view is partially obscured in that view.

Figure 17.66

When all surface loops in the orthogonal views are accounted for, the pictorial should be complete. Remove hidden edges or show them as hidden lines.

© Cengage Learning®. Courtesy of D. K. Lieu.

Note on the completed pictorial in Figure 17.66 that some of the surfaces that are fully visible in the multiview drawing may be partially obscured in the pictorial representation. The best way to tell is to look at the view that is adjacent to the view that contains the loop and see if another surface is in "front" of it. If so, you will see one or more hidden lines on the orthogonal view that form the edges of the plane you are trying to create on the pictorial (Step 6). There are no internal features (Step 7) in this example, but remember to check for model validity (Step 8).

Of course, the ultimate goal is to inspect any multiview drawing and create a mental image of the object with very little effort or thought. However, developing this skill takes a great deal of practice. Creating pictorials from the multiview drawing is an exercise that will help you develop this skill. The more you practice, the easier it will become.

17.05.05 Strategy for Improving Spatial Skills through Imagining Successive Cuts to Objects

One last set of exercises to hone your visualization skills is presented next. With this method, you start with a basic shape and remove parts of it through **successive cuts** (A method of forming an object with a complex shape by starting with a basic shape and removing parts of it through subtraction of other basic shapes.) . With each new cut, you remove different portions of the original object and sketch the result.

You will begin with the basic L-shaped object shown in Figure 17.67. Since this object is made up of only normal surfaces, sketching the missing top view and the isometric pictorial is relatively easy, as shown in Figure 17.68.

Figure 17.67

Front and side views of a basic L-shaped object.

Figure 17.68

A multiview drawing of an L-shaped object, with a sketched top view and isometric pictorial.

What happens if you use a block to remove the upper-right portion of the original object? Figure 17.69 shows the front and side views of the new object. What do the top and isometric views look like?

Figure 17.69

Front and side views of an object with the upper-right portion removed.

The cutting block goes all the way through the object, exposing features as it continues. Since the cutting block cuts through the right side of the object, the left part is undisturbed.

You can start by sketching the isometric pictorial of the original object, then superimposing the cutting block on top of this sketch, as shown in Figure 17.70.

Figure 17.70

A sketched isometric pictorial with a sketched and shaded cutting block shown.

If you imagine the intersection between the smaller block and the original object, the result is shown in Figure 17.71a. After the small block has been removed from the original, the object shown in Figure 17.71b results.

Figure 17.71

The intersections on the cutting block to remove material from an object are shown in (a) on the isometric sketch. The final result (b) on the isometric sketch of using a cutting block to remove material from an object.

Using this isometric pictorial, now you can complete the top view. Recall that the left side of the object is undisturbed, so it will appear exactly as the left side of the top view of the original object. After this cut is made through the object, the multiview drawing shown in Figure 17.72 results.

Figure 17.72

A multiview drawing of the resulting object with the sketched top view.

What if the cutting block does not go all the way through the original object? Figure 17.73 shows the front and right-side views of the object in this case. Figure 17.74a shows the isometric with the cutting block superimposed on it, and Figure 17.74b shows the result of using the block to cut the original object.

Figure 17.73

Front and side views of an object when the cutting block does not go all the way through the object.

Figure 17.74

The intersection of the cutting block, which cuts partway through the object, is shown in (a) on the isometric sketch. The final result (b) on the isometric sketch of the block cutting partway through an object to remove material from it.

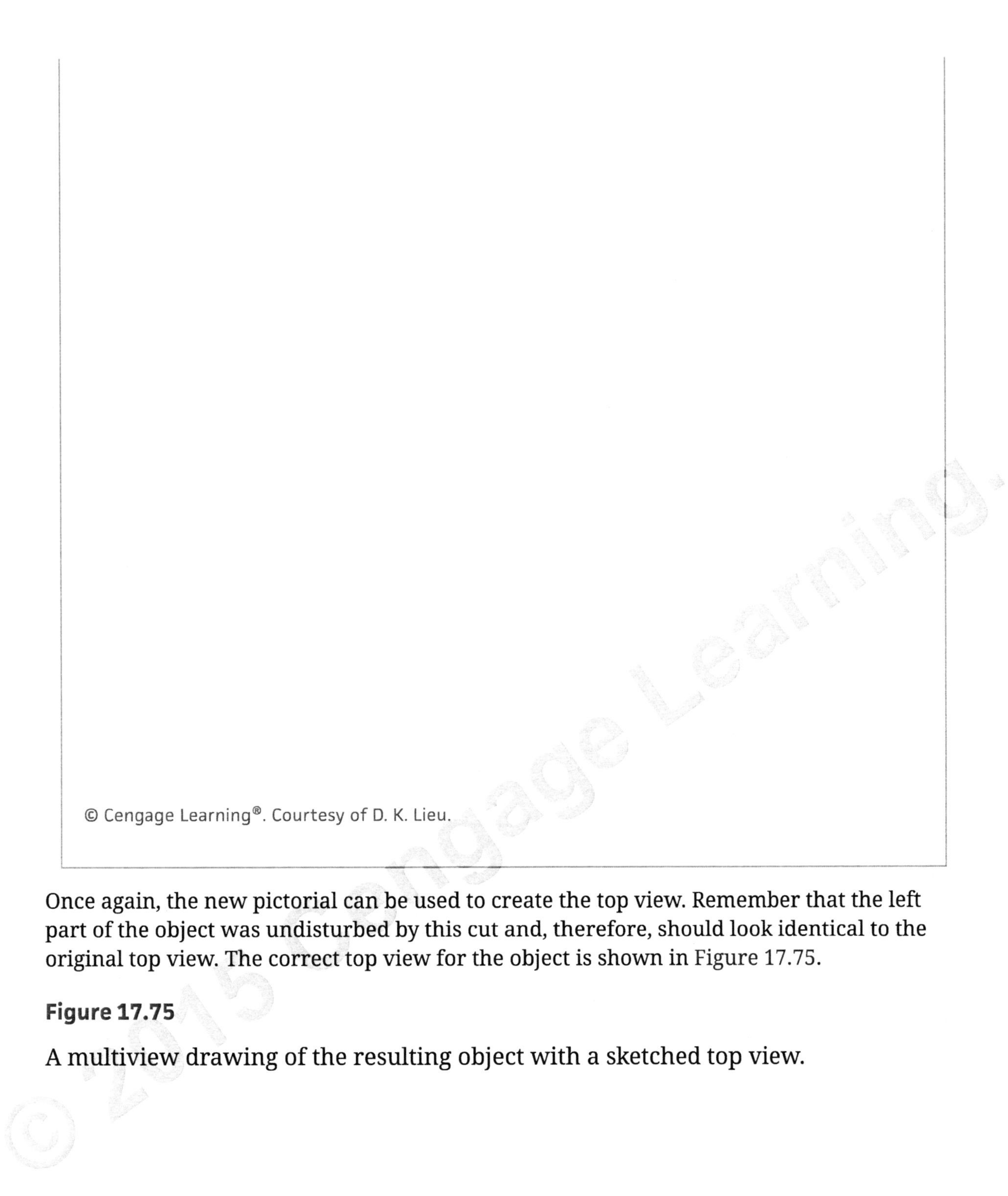

Once again, the new pictorial can be used to create the top view. Remember that the left part of the object was undisturbed by this cut and, therefore, should look identical to the original top view. The correct top view for the object is shown in Figure 17.75.

Figure 17.75

A multiview drawing of the resulting object with a sketched top view.

What if the block had been angled before it was used to cut through the object? Figure 17.76 shows the front and right-side views of the original L-shaped object after this angled cut has been made. Figure 17.77a shows the original object with the cutting block superimposed on it, and Figure 17.77b shows the result of the cut.

Figure 17.76

An L-shaped object with an angled cut.

Figure 17.77

The intersection of the angled cutting block on the original object is shown in (a) in the isometric sketch. The final result (b) on the isometric sketch of the original object cut with the angled cutting block.

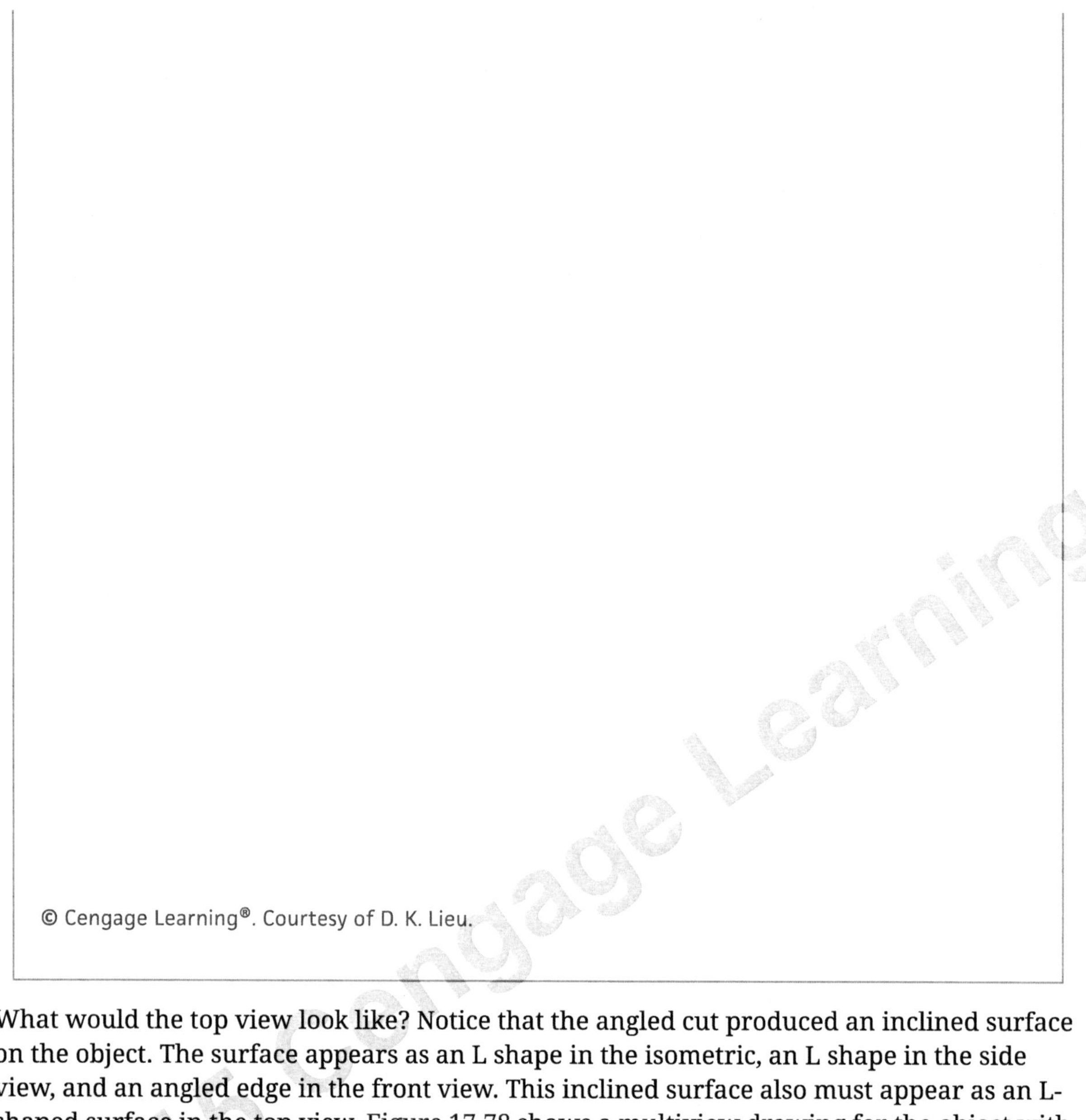

What would the top view look like? Notice that the angled cut produced an inclined surface on the object. The surface appears as an L shape in the isometric, an L shape in the side view, and an angled edge in the front view. This inclined surface also must appear as an L-shaped surface in the top view. Figure 17.78 shows a multiview drawing for the object with the correct top view included. Notice how the limits that define the L-shaped surface correspond between the front and top views.

Figure 17.78

A multiview drawing with a sketched top view of the object with the angled cut.

Now what if you take this angled cutting block and move it so that it cuts a *V* out of the middle of the original object? The front and side views that result from this cutting operation are shown in Figure 17.79. Figure 17.80a shows the cutting block superimposed on the original object, and Figure 17.80b shows the result of this cut.

Figure 17.79

An object with an angled wedge cut through the center.

Figure 17.80

The intersection of the wedge-shaped cutting block on the original object is shown in (a) in the isometric sketch. The final result (b) on the isometric sketch of the original object cut with the wedge-shaped cutting block.

Sketching the correct top view for this object is a bit more difficult since the result is two angled surfaces instead of one. Note that the two angled surfaces overlap each other in the side view—you see only one outline for the two surfaces. Also note that one of these surfaces is hidden from view in the isometric pictorial because the object is in the way. However, both inclined surfaces will be visible in the top view. To sketch these surfaces in the top view, use the front view as your guide regarding their limits. The correct multiview drawing, including a top view, is shown in Figure 17.81.

Figure 17.81

A multiview drawing with a sketched top view of the object with the angled wedge removed from it.

As the last step in this exercise, consider an oblique cut to the object. Figure 17.82 shows the resulting front and side views of the object after it experienced this type of cut. In this case, the cutting block has been oriented in space as shown in Figure 17.83a. The resulting isometric pictorial with this oblique cut is shown in Figure 17.83b.

Figure 17.82

Front and side views of an object with an oblique cut.

Figure 17.83

The intersection of the oblique surface cutting block on the original object is shown in the isometric sketch.

Recall that oblique surfaces will appear as areas in all views. The triangular surfaces that result from this oblique cut are seen as areas in the front, the side, and the isometric. They also will appear as triangular areas in the top. Use the other views to help you find the limits of the triangular areas in the top view to define the surfaces. The completed multiview drawing for this object is shown in Figure 17.84.

Figure 17.84

A multiview drawing with a sketched top view of the original object with the oblique cut.

Chapter Review

17.06 Chapter Summary

The ability to visualize objects in three dimensions is a key skill for engineering design. Physical models for proposed designs often do not exist, and an engineer must be able to envision, manipulate, and modify designs when they are presented in the form of 2-D drawings. You can develop visualization skills only through practice. The more you practice and the more geometrically complex the objects you practice with, the better your skills become. Mental reconstruction of a 3-D object from its 2-D drawing may, at first, seem like a daunting task. However, by keeping in mind some basic rules for the appearance of points, edges, and surfaces between the views on an engineering drawing, you can re-create a pictorial view of the object in a step-by-step process. Eventually, this visualization process becomes easier, quicker, and more natural. The ultimate goal is to be able to work seamlessly with 3-D computer models, pictorial presentations, and 2-D multiview presentations as tools for the development of engineering designs that are in your mind's eye.

Chapter Review

7.07 Glossary of Key Terms

anchor point (The same point, usually a vertex, which can be located easily and confidently on multiple views for an object.)

anchor surface (The same surface that can be located easily and confidently on multiple views for an object.)

curved surface (Any nonflat surface on an object.)

edge tracking (A procedure by which successive edges on an object are simultaneously located on a pictorial image and on a multiview image of that object.)

foundation space (The rectilinear volume that represents the limits of the volume occupied by an object.)

frontal surface (A surface on an object being viewed that is parallel to the front viewing plane.)

horizontal surface (A surface on an object being viewed that is parallel to the top viewing plane.)

inclined surface (A flat surface on an object being viewed that is perpendicular to one primary view and angled with respect to the other two views; in other words, a plane that appears as an edge view in one primary view but is not parallel to any of the principal views.)

normal surface (A surface on an object being viewed that is parallel to one of the primary viewing planes.)

oblique surface (A flat surface on an object being viewed that is neither parallel nor perpendicular to any of the primary views.)

point tracking (A procedure by which successive vertices on an object are simultaneously located on a pictorial image and a multiview image of that object.)

profile surface (A surface on an object being viewed that is parallel to a side viewing plane.)

successive cuts (A method of forming an object with a complex shape by starting with a basic shape and removing parts of it through subtraction of other basic shapes.)

surface tracking (A procedure by which successive surfaces on an object are simultaneously located on a pictorial image and a multiview image of that object.)

Uicker et al., 4th Edition – Chapter 7

Textbook: Theory of Machines and Mechanisms

Chapter Title: Spur Gears

7 Spur Gears

Gears are machine elements used to transmit rotary motion between two shafts, usually with a constant speed ratio. In this chapter we will discuss the case where the axes of the two shafts are parallel and the teeth are straight and parallel to the axes of rotation of the shafts; such gears are called *spur gears.*

7.1 TERMINOLOGY AND DEFINITIONS

A pair of spur gears in mesh is illustrated in Fig. 7.1. The *pinion* is a name given to the smaller of the two mating gears; the larger is often called the *gear* or the *wheel.* The pair of gears, chosen to work together, is often called a *gearset.*

The terminology of gear teeth is illustrated in Fig. 7.2, where most of the following definitions are given.

The *pitch circle* is a theoretical circle on which all calculations are based. The pitch circles of a pair of mating gears are tangent to each other, and it is these pitch circles that were pictured in earlier chapters as rolling against each other without slip.

The *diametral pitch* P is the ratio of the number of teeth on the gear to its pitch diameter, that is,

$$P = \frac{N}{2R}, \tag{7.1}$$

where N is the number of teeth and R is the pitch circle radius. Note that the diametral pitch cannot be directly measured on the gear itself. Also, note that, as the value of the diametral pitch becomes larger, the teeth become smaller; this is illustrated clearly in Fig. 7.3. In addition, a pair of mating gears have the same diametral pitch. The diametral pitch is used to indicate the tooth size in U.S. customary units and usually has units of teeth per inch.

Figure 7.1 Pair of spur gears in mesh. (Courtesy of Gleason Works, Rochester, NY.)

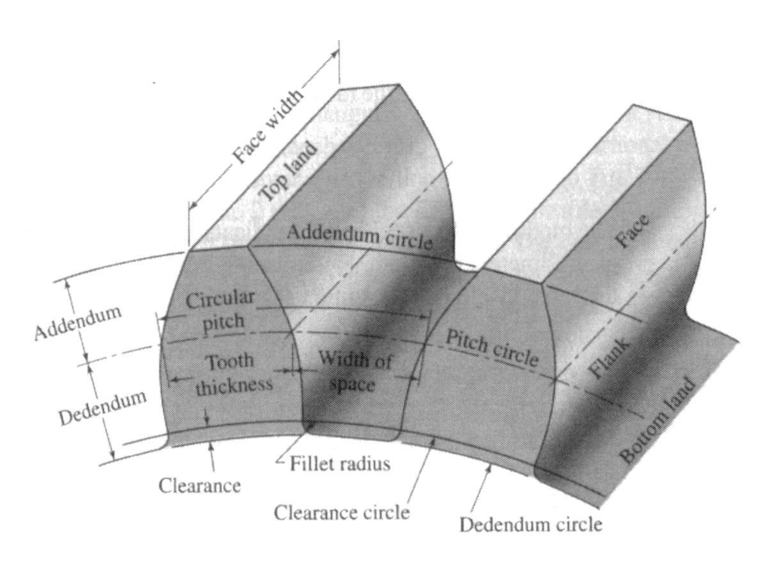

Figure 7.2 Gear tooth terminology.

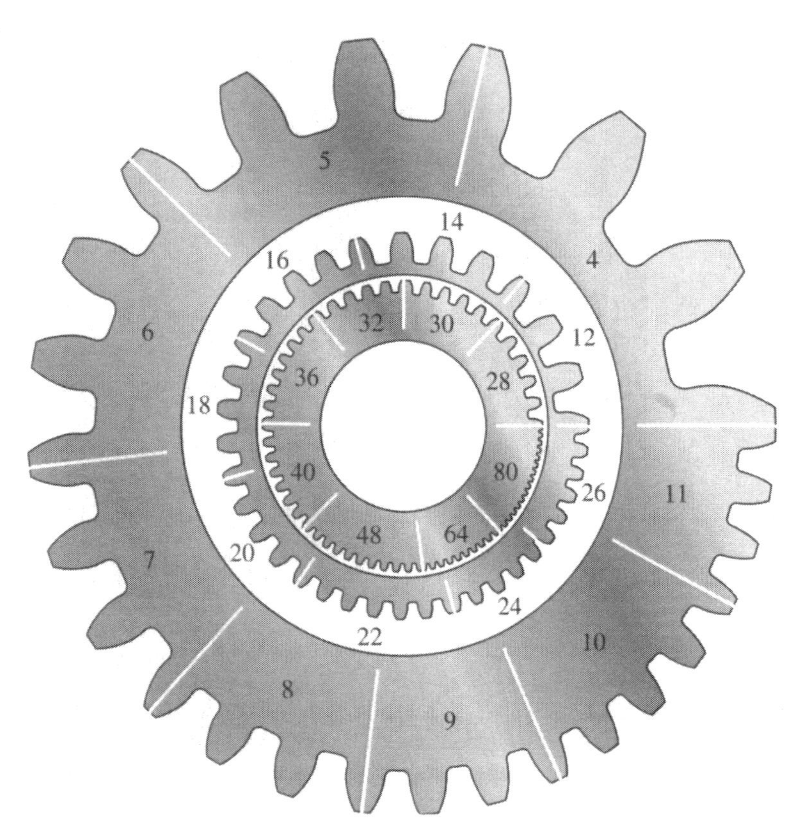

Figure 7.3 Tooth sizes for various diametral pitches in teeth per inch. (Courtesy of Gleason Cutting Tools Corp., Loves Park, IL.)

The *module m* is the ratio of the pitch diameter of the gear to its number of teeth, that is

$$m = \frac{2R}{N}. \tag{7.2}$$

The module is the usual unit for indicating tooth size in SI units and it customarily has units of millimeters per tooth. Note that the module is the reciprocal of the diametral pitch, and the relationship can be written as

$$m = \frac{25.4\ (\text{mm/in})}{P\ (\text{teeth/in})} = \frac{25.4}{P}\ \text{mm/tooth}.$$

Also note that metric gears should not be interchanged with U.S. gears because their standards for tooth sizes are not the same.

The *circular pitch p* is the distance from one tooth to the adjacent tooth measured along the pitch circle. Therefore, it can be determined from

$$p = \frac{2\pi R}{N}. \tag{7.3}$$

Circular pitch is related to the previous definitions, depending on the units, by

$$p = \frac{\pi}{P} = \pi m. \tag{7.4}$$

The *addendum* a is the radial distance between the pitch circle and the top land of each tooth.

The *dedendum* d is the radial distance from the pitch circle to the bottom land of each tooth.

The *whole depth* is the sum of the addendum and dedendum.

The *clearance* c is the amount by which the dedendum of a gear exceeds the addendum of the mating gear.

The *backlash* is the amount by which the width of a tooth space exceeds the thickness of the engaging tooth measured along the pitch circles.

7.2 FUNDAMENTAL LAW OF TOOTHED GEARING

Gear teeth mating with each other to produce rotary motion are similar to a cam and follower. When the tooth profiles (or cam and follower profiles) are shaped so as to produce a constant angular velocity ratio between the two shafts, then the two mating surfaces are said to be *conjugate*. It is possible to specify an arbitrary profile for one tooth and then to find a profile for the mating tooth so that the two surfaces are conjugate. One possible choice for such conjugate solutions is the *involute* profile, which, with few exceptions, is in universal use for gear teeth.

A single pair of mating gear teeth as they pass through their entire period of contact must be shaped such that the ratio of the angular velocity of the driven gear to that of the driving gear, that is, *the first-order kinematic coefficient, must remain constant*. This is the fundamental criterion that governs the choice of the tooth profiles. If this were not true in gearing, very serious vibration and impact problems would result, even at low speeds.

In Section 3.17 we learned that the angular velocity ratio theorem states that the first-order kinematic coefficient of any mechanism is inversely proportional to the segments into which the common instant center cuts the line of centers. In Fig. 7.4, two profiles are in contact at point T; let profile 2 represent the driver and profile 3 represent the driven. The normal to the surfaces CD is called the *line of action*. The normal to the profiles at the point of contact T intersects the line of centers O_2O_3 at the instant center of velocity. In gearing, this instant center is generally referred to as the *pitch point* and usually carries the label P.

Designating the pitch circle radii of the two gear profiles R_2 and R_3, from the angular velocity ratio theorem, Eq. (3.28), we see that

$$\frac{\omega_2}{\omega_3} = \frac{R_3}{R_2}. \tag{7.5}$$

This equation is frequently used to define what is called the fundamental law of gearing, which states that as gears go through their mesh, *the pitch point must remain stationary on the line of centers* so that the speed ratio remains constant. This means that the line of

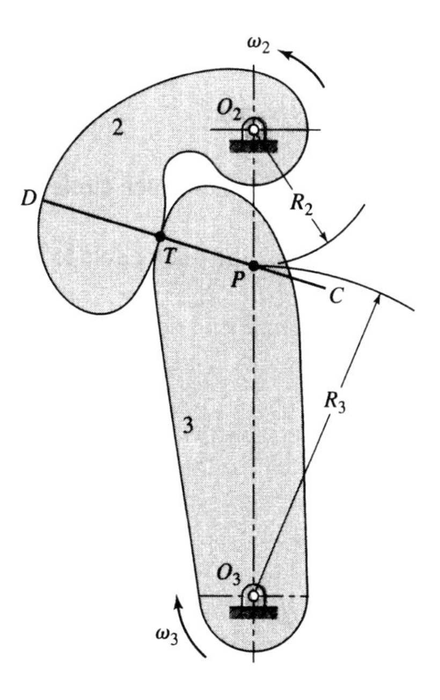

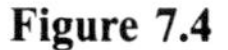
Figure 7.4

action for every new instantaneous point of contact must always pass through the stationary pitch point P. Thus, the problem of finding a conjugate profile for a given shape is to find a mating shape that satisfies the fundamental law of gearing.

It should not be assumed that any shape or profile is satisfactory just because a conjugate profile can be found. Although theoretically conjugate curves might be found, the practical problems of reproducing these curves from steel gear blanks or other materials while using existing machinery still exist. In addition, the sensitivity of the law of gearing to small dimensional changes of the shaft center distance caused either by misalignment or by large forces must also be considered. Finally, the tooth profile selected must be one that can be reproduced quickly and economically in very large quantities. A major portion of this chapter is devoted to illustrating how the involute curve profile fulfills these requirements.

7.3 INVOLUTE PROPERTIES

An *involute* curve is the path generated by a tracing point on a cord as the cord is unwrapped from a cylinder called the *base cylinder*. This is illustrated in Fig. 7.5, where T is the tracing point. Note that the cord AT is normal to the involute at T, and the distance AT is the instantaneous value of the radius of curvature. As the involute is generated from its origin T_0 to T_1, the radius of curvature varies continuously; it is zero at T_0 and increases continuously to T_1. Thus, the cord is the generating line, and it is always normal to the involute.

If the two mating tooth profiles both have the shapes of involute curves, the condition that the pitch point P remain stationary is satisfied. This is illustrated in Fig. 7.6, where two gear blanks with fixed centers O_2 and O_3 are illustrated having base cylinders with respective radii of O_2A and O_3B. We now imagine that a cord is wound clockwise around the base cylinder of gear 2, pulled tightly between points A and B, and wound counterclockwise

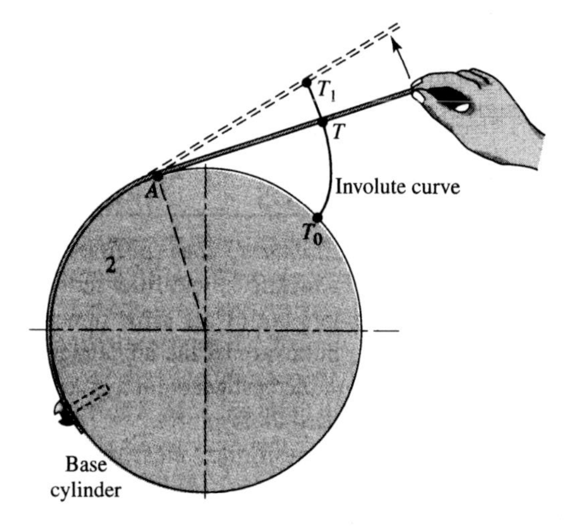

Figure 7.5 Involute curve.

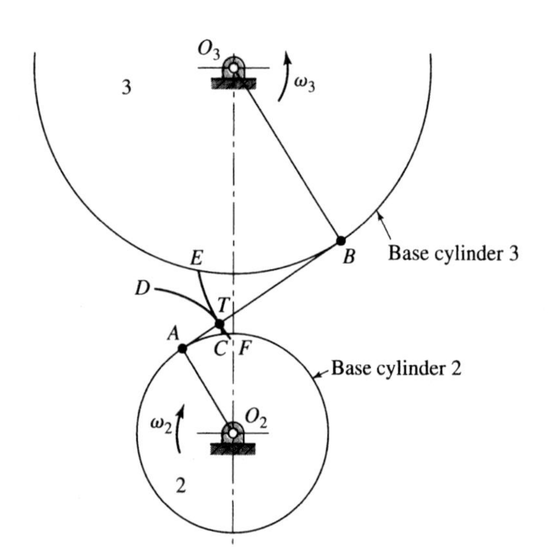

Figure 7.6 Conjugate involute curves.

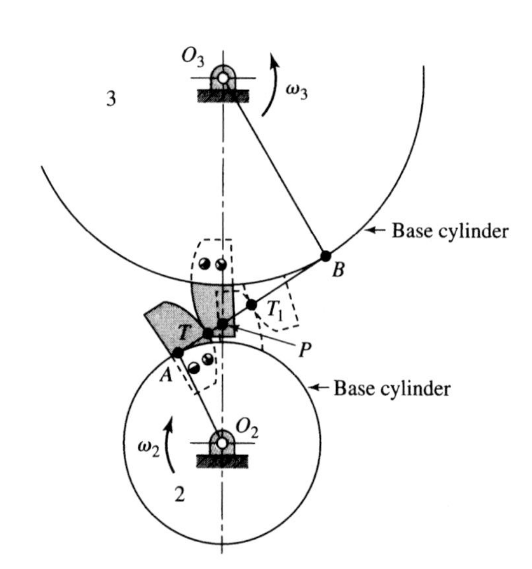

Figure 7.7 Involute action.

around the base cylinder of gear 3. If the two base cylinders are now rotated in opposite directions so as to keep the cord tight, a tracing point T traces out the involutes CD on gear 2 and EF on gear 3. The two involutes thus generated simultaneously by the single tracing point T are conjugate profiles.

Next imagine that the involutes of Fig. 7.6 are scribed on plates and that the plates are cut along the scribed curves and then bolted to the respective cylinders in the same positions. The result is illustrated in Fig. 7.7. The cord can now be removed and, if gear 2 is moved clockwise, gear 3 is caused to move counterclockwise by the camlike action of the two curved plates. The path of contact is the line AB formerly occupied by the cord. Because the line AB is the generating line for each involute, it is normal to both profiles at all points of contact. Also, it always occupies the same position because it is always tangent to both base cylinders. Therefore, point P is the pitch point. Point P does not move; therefore, the involute curves are conjugate curves and satisfy the fundamental law of gearing.

7.4 INTERCHANGEABLE GEARS; AGMA STANDARDS

A *tooth system* is the name given to a *standard*[1] that specifies the relationships among addendum, dedendum, clearance, tooth thickness, and fillet radius to attain interchangeability of gears of different tooth numbers but of the same pressure angle and the same diametral pitch or module. We should be aware of the advantages and disadvantages of such a tooth system so that we can choose the best gears for a given design and have a basis for comparison if we depart from a standard tooth profile.

For a pair of spur gears to properly mesh, they must share the same pressure angle and the same tooth size as specified by the choice of the diametral pitch or module. The numbers of teeth and the pitch diameters of the two gears in mesh need not match, but are chosen to give the desired speed ratio, as demonstrated in Eq. (7.5).

The sizes of the teeth used are chosen by selecting the diametral pitch P or module m. Standard cutters are generally available for the sizes listed in Table 7.1. Once the diametral pitch or module is chosen, the remaining dimensions of the tooth are set by the standards in Table 7.2. Tables 7.1 and 7.2 contain the standards for the spur gears most in use today, and they include the values for both SI and U.S. customary units.

Let us illustrate the design choices by an example.

TABLE 7.1 Standard gear tooth sizes

	Standard diametral pitches P, US customary, teeth/in.
Coarse	1, $1\frac{1}{4}$, $1\frac{1}{2}$, $1\frac{3}{4}$, 2, $2\frac{1}{2}$, 3, 4, 5, 6, 8, 10, 12, 14, 16, 18
Fine	20, 24, 32, 40, 48, 64, 72, 80, 96, 120, 150, 200
	Standard modules m, SI, mm/tooth
Preferred	1, 1.25, 1.5, 2, 2.5, 3, 4, 5, 6, 8, 10, 12, 16, 20, 25, 32, 40, 50
Next choice	1.125, 1.375, 1.75, 2.25, 2.75, 3.5, 4.5, 5.5, 7, 9, 11, 14, 18, 22, 28, 36, 45

TABLE 7.2 Standard tooth systems for spur gears

System	Pressure angle, ϕ (deg)	Addendum, a	Dedendum, d
Full depth	20°	$1/P$ or $1m$	$1.25/P$ or $1.25m$
Full depth	$22\frac{1}{2}$°	$1/P$ or $1m$	$1.25/P$ or $1.25m$
Full depth	25°	$1/P$ or $1m$	$1.25/P$ or $1.25m$
Stub teeth	20°	$0.8/P$ or $0.8m$	$1/P$ or $1m$

EXAMPLE 7.1

Two parallel shafts, separated by a distance (commonly referred to as the center distance) of 3.5 in, are to be connected by a gear set so that the output shaft rotates at 40% of the speed of the input shaft. Design a gearset to fit this situation.

SOLUTION

The center distance can be written as $R_2 + R_3 = 3.5$ in, and substituting the given information into Eq. (7.5) we have $\omega_3/\omega_2 = R_2/R_3 = 0.40$. Then substituting the first equation into the second equation, and rearranging, we find that $R_2 = 1.0$ in and $R_3 = 2.5$ in. Next, we must choose the size of the teeth by picking a value for the diametral pitch or module. From Eq. (7.1) we find the numbers of teeth on the two gears to be $N_2 = 2PR_2$ and $N_3 = 2PR_3$. This choice of P or m for tooth size is often iterative. First, we might choose a value of $P = 6$ teeth/in; this gives the numbers of teeth as $N_2 = 12$ teeth and $N_3 = 30$ teeth; if we choose $P = 10$ teeth/in, then we get $N_2 = 20$ teeth and $N_3 = 50$ teeth. At this time, either choice appears acceptable, and we choose $P = 10$ teeth/in. However, this choice of P (or m) must later be checked for possible undercutting, as we will study in Section 7.7, and for contact ratio, which we will study in Section 7.8, and for strength and wear of the teeth.[2]

7.5 FUNDAMENTALS OF GEAR–TOOTH ACTION

To illustrate the fundamentals we now proceed, step by step, through the actual graphical layout of a pair of spur gears. The dimensions used here are those of Example 7.1 assuming standard 20° full-depth involute tooth form as specified in Table 7.2. The various steps, in the correct order, are illustrated in Figs. 7.8 and 7.9 and are as follows.

STEP 1 Calculate the two pitch circle radii, R_2 and R_3, as in Example 7.1 and draw the two pitch circles tangent to each other, identifying O_2 and O_3 as the two shaft centers (Fig. 7.8).
STEP 2 Draw the common tangent to the pitch circles perpendicular to the line of centers and through the pitch point P (Fig. 7.8). Draw the *line of action* at an angle equal to the *pressure angle* $\phi = 20°$ from the common tangent. This line of action corresponds to the generating line discussed in Section 7.3; it is always normal to the involute curves and always passes through the pitch point. It is called the line of action or the pressure line because, assuming no friction, the resultant tooth force acts along this line.
STEP 3 Through the centers of the two gears, draw the two perpendiculars to the line of action O_2A and O_3B (Fig. 7.8). Draw the two *base circles* with radii of $r_2 = O_2A$ and $r_3 = O_3B$; these correspond to the base cylinders of Section 7.3.
STEP 4 From Table 7.2, continuing with $P = 10$ teeth/in., the addendum for both of the gears is found to be

$$a = \frac{1}{P} = \frac{1}{10 \text{ teeth/in}} = 0.10 \text{ in.}$$

Adding this to each of the pitch circle radii, draw the two addendum circles that define the top lands of the teeth on each gear. Carefully identify and label point C where the addendum circle of gear 3 intersects the line of action (Fig. 7.9). Similarly, identify and label point D where the addendum circle of gear 2 intersects the line of action. Visualizing the rotation

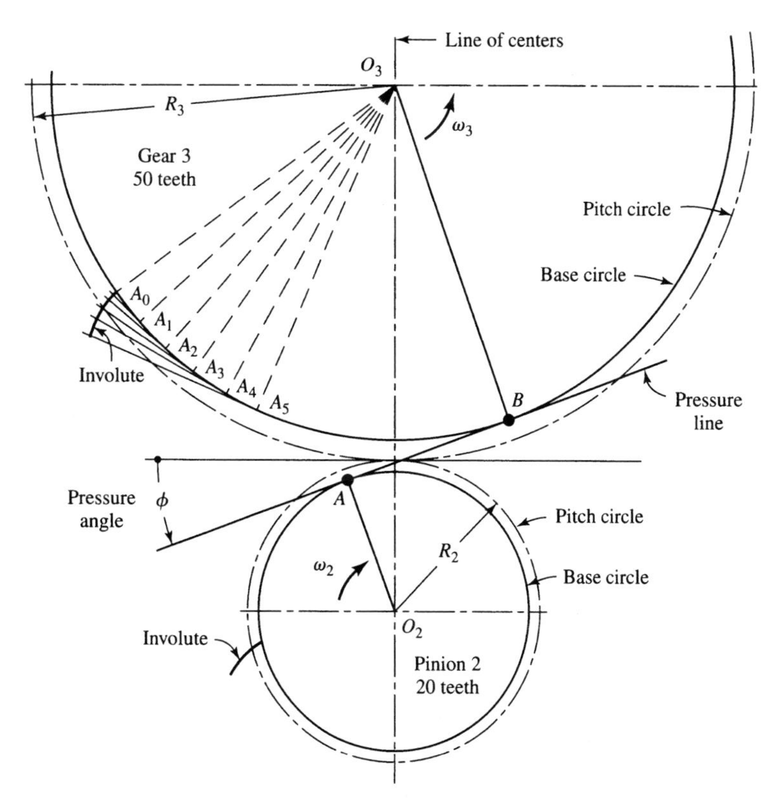

Figure 7.8 Partial gear pair layout.

of the two gears in the directions given, we see that contact is not possible before point C because the teeth of gear 3 are not of sufficient height; thus, C is the first point of contact between the teeth. Similarly, the teeth of gear 2 are too short to allow further contact after reaching point D; thus, contact between one or more pairs of mating teeth continues between C and D and then ceases.

Steps 1 through 4 are critical for verifying the choice of any gear pair. We will continue with the diagram illustrated in Fig. 7.9 when we check for interference, undercutting, and contact ratio in later sections. However, to complete our visualization of gear tooth action, let us first proceed to the construction of the complete involute tooth shapes as illustrated in Fig. 7.8.

STEP 5 From Table 7.2, the dedendum for each gear is found to be

$$d = \frac{1.25}{P} = \frac{1.25}{10 \text{ teeth/in}} = 0.125 \text{ in.}$$

Subtracting this from each of the pitch circle radii, draw the two dedendum circles that define the bottom lands of the teeth on each gear (Fig. 7.9). Note that the dedendum circles often lie quite close to the base circles; however, they have distinctly different meanings. In this example, the dedendum circle of gear 3 is larger than its base circle and the dedendum circle of the pinion 2 is smaller than its base circle. However, this is not always the case.

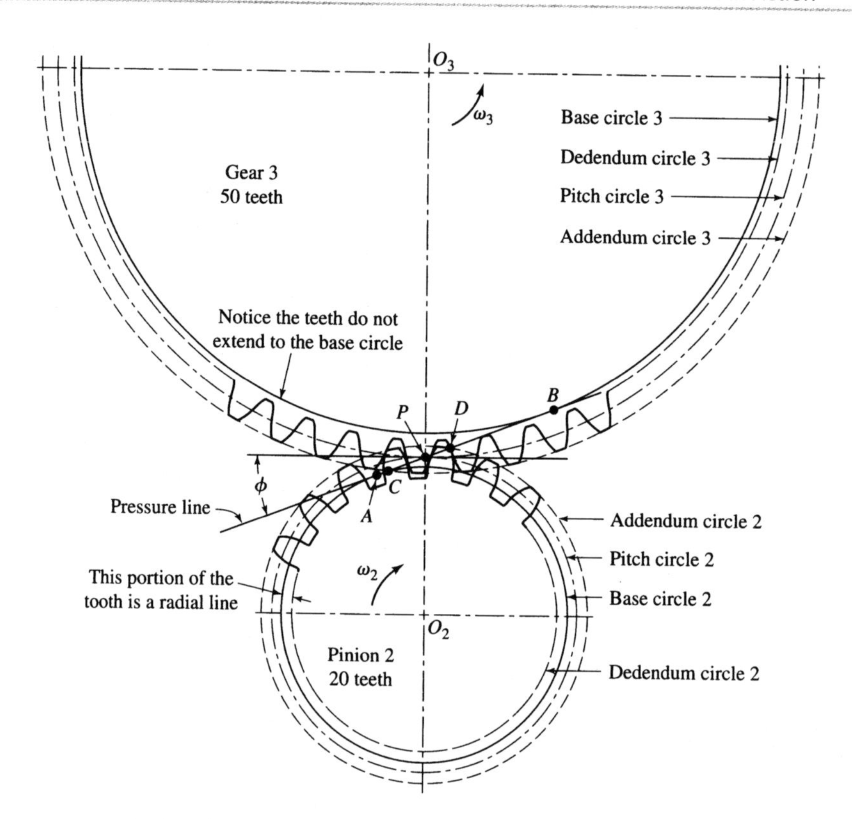

Figure 7.9 Layout of a pair of spur gears.

STEP 6 Generate an involute curve on each base circle as illustrated for gear 3 in Fig. 7.8. This is done by first dividing a portion of the base circle into a series of equal small parts, A_0, A_1, A_2, and so on. Next the radial lines O_3A_0, O_3A_1, O_3A_2, and so on are constructed, and tangents to the base circle are drawn perpendicular to each of these. The involute begins at A_0. The second point is obtained by striking an arc, with center A_1 and radius A_0A_1, up to the tangent line through A_1. The next point is found by striking a similar arc with center at A_2 and so on. This construction is continued until the involute curve is generated far enough to meet the addendum circle of gear 3. If the dedendum circle lies inside of the base circle, as is true for pinion 2 of this example, then, except for the fillet, the curve is extended inward to the dedendum circle by a radial line; this portion of the curve is not involute.
STEP 7 Using cardboard or preferably a sheet of clear plastic, cut a template for the involute curve and mark on it the center point of the corresponding gear. Note that two templates are needed because the involute curves are different for gears 2 and 3.
STEP 8 Calculate the circular pitch using Eq. (7.4).

$$p = \frac{\pi}{P} = \frac{\pi}{10 \text{ teeth/in}} = 0.314\,16 \text{ in/tooth.}$$

This distance from one tooth to the next is now marked along the pitch circle and the template is used to draw the involute portion of each tooth (Fig. 7.9). The width of a tooth and that of a tooth space are each equal to half of the circular pitch or (0.314 16 in/tooth)/2 = 0.157 08 in/tooth. These distances are marked along the pitch circle, and the same template is turned over and used to draw the opposite sides of the teeth. The portion of the tooth space between the clearance and the dedendum may be used for a fillet radius. The top and bottom lands are now drawn as circular arcs along the addendum and dedendum circles to complete the tooth profiles. The same process is performed on the other gear using the other template.

Remember that steps 5 through 8 are not necessary for the proper design of a gear set. They are only included here to help us visualize the relation between real tooth shapes and the theoretical properties of the involute curve.

Involute Rack. We may imagine a *rack* as a spur gear having an infinitely large pitch diameter. Therefore, in theory, a rack has an infinite number of teeth and its base circle is located an infinite distance from the pitch point. For involute teeth, the curves on the sides of the teeth of a rack become straight lines making an angle with the line of centers equal to the pressure angle. The addendum and dedendum distances are the same as those given in Table 7.2. Figure 7.10 illustrates an involute rack in mesh with the pinion of the previous example.

Base Pitch. Corresponding sides of involute teeth are parallel curves. The *base pitch* is the constant and fundamental distance between these curves, that is, the distance from one tooth to the next, measured along the common normal to the tooth profiles, which is the line of action (Fig. 7.10). The base pitch p_b and the circular pitch p are related as follows:

$$p_b = p \cos\phi. \tag{7.6}$$

The base pitch is a much more fundamental measurement, as we will see below.

Internal Gear. Figure 7.11 depicts the pinion of the preceding example in mesh with an *internal*, or *annular*, gear. With internal contact, both centers are on the same side of the pitch point. Thus, the positions of the addendum and dedendum circles of an internal gear are reversed with respect to the pitch circle; the addendum circle of the internal gear lies *inside* the pitch circle, whereas the dedendum circle lies *outside* the pitch circle. The base

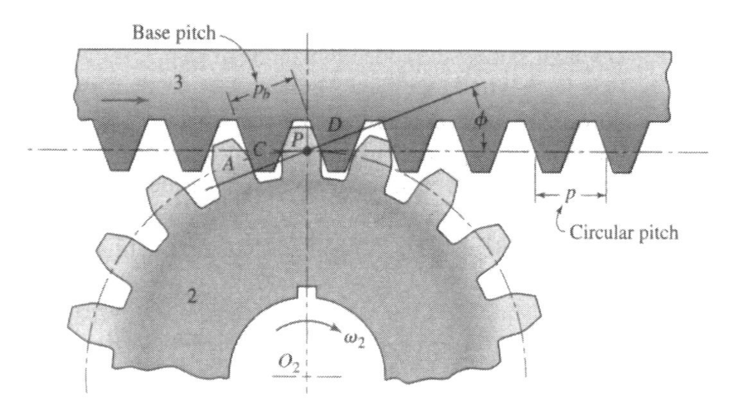

Figure 7.10 Involute pinion and rack.

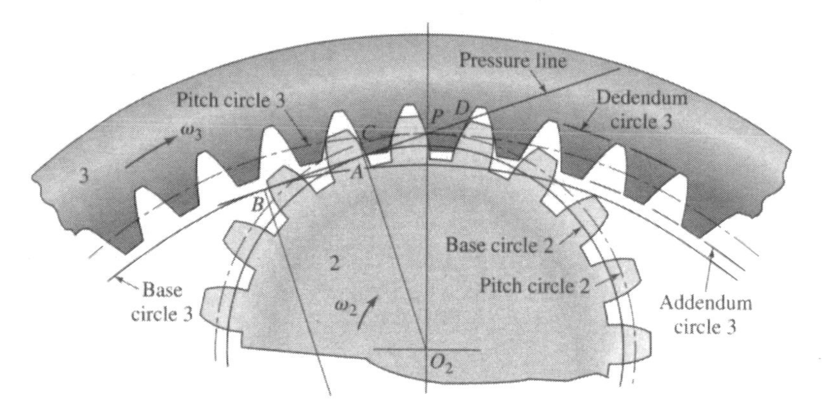

Figure 7.11 Involute pinion and internal gear.

circle lies inside the pitch circle as with an external gear, but is now near the addendum circle. Otherwise, Fig. 7.11 is constructed in the same manner as Fig. 7.9.

7.6 THE MANUFACTURE OF GEAR TEETH

There are many ways of manufacturing the teeth of gears; for example, they can be made by sand casting, shell molding, investment casting, permanent-mold casting, die casting, or centrifugal casting. They can be formed by the powder-metallurgy process, or a single bar of aluminum can be formed by extrusion and then sliced into gears. Gears that carry large loads in comparison with their sizes are usually made of steel and are cut with either *form cutters* or *generating cutters*. In form cutting, the cutter is of the exact shape of the tooth space. With generating cutters, a tool having a shape different from the tooth space is moved through several cuts relative to the gear blank to obtain the proper shape for the teeth.

Probably the oldest method of cutting gear teeth is *milling*. A form milling cutter corresponding to the shape of the tooth space, such as that illustrated in Fig. 7.12*a*, is used to machine one tooth space at a time, as illustrated in Fig. 7.12*b*, after which the gear is indexed through one circular pitch to the next position. Theoretically, with this method, a different cutter is required for each gear to be cut because, for example, the shape of the tooth space in a 25-tooth gear is different from the shape of the tooth space in, say, a 24-tooth gear. Actually, the change in tooth space shape is not very large, and eight form cutters can be used to cut any gear in the range from 12 teeth to a rack with reasonable accuracy. Of course, a separate set of form cutters is required for each pitch.

Shaping is a highly favored method of *generating* gear teeth. The cutting tool may be either a rack cutter or a pinion cutter. The operation is explained by reference to Fig. 7.13. For shaping, the reciprocating cutter is first fed into the gear blank until the pitch circles are tangent. Then, after each cutting stroke, the gear blank and the cutter roll slightly on their pitch circles. When the blank and cutter have rolled by a total distance equal to the circular pitch, one tooth has been generated and the cutting continues with the next tooth until all teeth have been cut. Shaping of an internal gear with a pinion cutter is illustrated in Fig. 7.14.

Hobbing is another method of generating gear teeth, which is quite similar to shaping them with a rack cutter. However, hobbing is done with a special tool called a *hob*, a

Figure 7.12 Manufacture of gear teeth by a form cutter. (a) A single-tooth involute hob. (b) Machining of a single tooth space. (Courtesy of Gleason Works, Rochester, NY.)

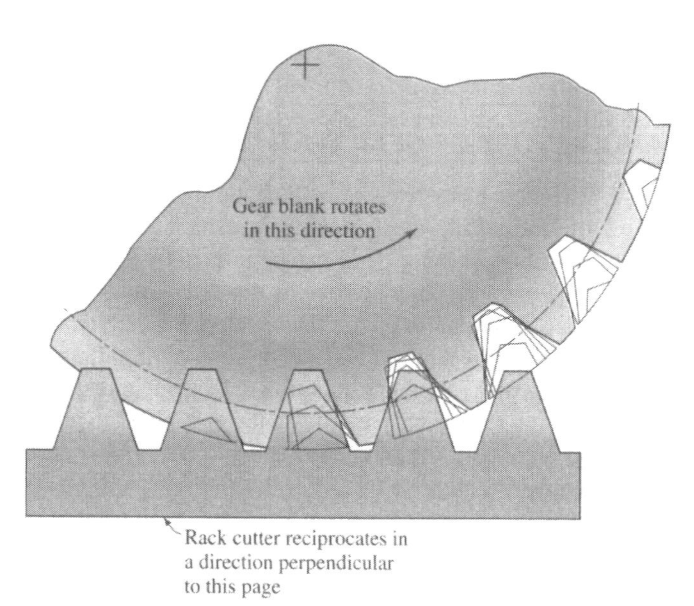

Figure 7.13 Shaping of involute teeth with a rack cutter.

cylindrical cutter with one or more helical threads quite like a screw-thread tap; the threads have straight sides like a rack. A number of different gear hobs are displayed in Fig. 7.15. A view of the hobbing of a gear is illustrated in Fig. 7.16. The hob and the gear blank are both rotated continuously at the proper angular velocity ratio, and the hob is fed slowly across the face of the blank to cut the full thickness of the teeth.

Following the cutting process, grinding, lapping, shaving, and burnishing are often used as final finishing processes when tooth profiles of very good accuracy and surface finish are desired.

Figure 7.14 Shaping of an internal gear with a pinion cutter. (Courtesy of Gleason Works, Rochester, NY.)

Figure 7.15 A variety of involute gear hobs. (Courtesy of Gleason Works, Rochester, NY.)

Figure 7.16 The hobbing of a gear. (Courtesy of Gleason Works, Rochester, NY.)

7.7 INTERFERENCE AND UNDERCUTTING

Figure 7.17 illustrates the pitch circles of the same gears used for discussion in Section 7.5. Let us assume that the pinion is the driver and that it is rotating clockwise.

We saw in Section 7.5 that for involute teeth, contact always takes place along the line of action AB. Contact begins at point C where the addendum circle of the driven gear crosses the line of action. Thus, initial contact is on the tip of the driven gear tooth and on the flank of the pinion tooth.

As the pinion tooth drives the gear tooth, contact approaches the pitch point P. Near the pitch point, contact slides *up* the flank of the pinion tooth and *down* the face of the gear tooth. At the pitch point, contact is at the pitch circles; note that P is the instant center and therefore the motion must be rolling with no slip at that point. Note also that this is the only location where the motion can be true rolling.

As the teeth recede from the pitch point, the point of contact continues to travel in the same direction as before along the line of action. Contact continues to slide *up* the face of the pinion tooth and *down* the flank of the gear tooth. The last point of contact occurs at the tip of the pinion and the flank of the gear tooth, at the intersection D of the line of action and the addendum circle of the pinion.

The *approach* phase of the motion is the period between the initial contact at point C and the pitch point P. The *angles of approach* are the angles through which the two gears rotate as the point of contact progresses from C to P. However, reflecting on the unwrapping

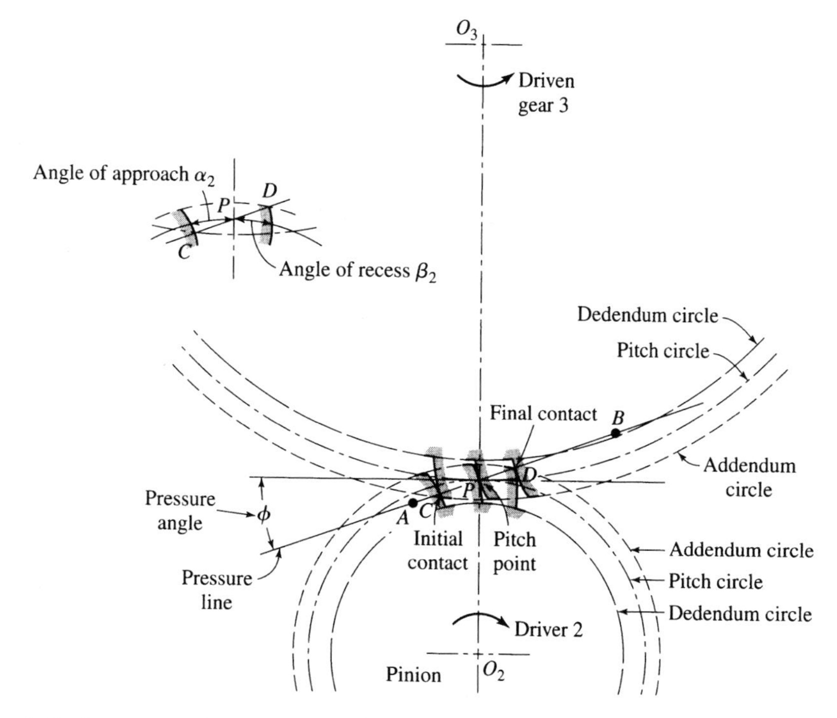

Figure 7.17 Approach and recess phases of gear tooth action.

cord analogy of Fig. 7.6, we see that the distance *CP* is equal to a length of cord unwrapped from the base circle of the pinion during the approach phase of the motion. Similarly, an equal amount of cord has wrapped onto the driven gear during that same phase. Thus, the angles of approach for the pinion and the gear, in radians, are

$$\alpha_2 = \frac{CP}{r_2} \quad \text{and} \quad \alpha_3 = \frac{CP}{r_3}. \tag{7.7}$$

The *recess* phase of the motion is the period during which contact progresses from the pitch point P to final contact at point D. The *angles of recess* are the angles through which the two gears rotate as the point of contact progresses from P to D. Again, from the unwrapping cord analogy, we find these angles, in radians, to be

$$\beta_2 = \frac{PD}{r_2} \quad \text{and} \quad \beta_3 = \frac{PD}{r_3}. \tag{7.8}$$

If the teeth come into contact such that they are not conjugate, this is called *interference*. Consider Fig. 7.18; illustrated here are two 16-tooth $14\frac{1}{2}^\circ$ pressure angle gears* with

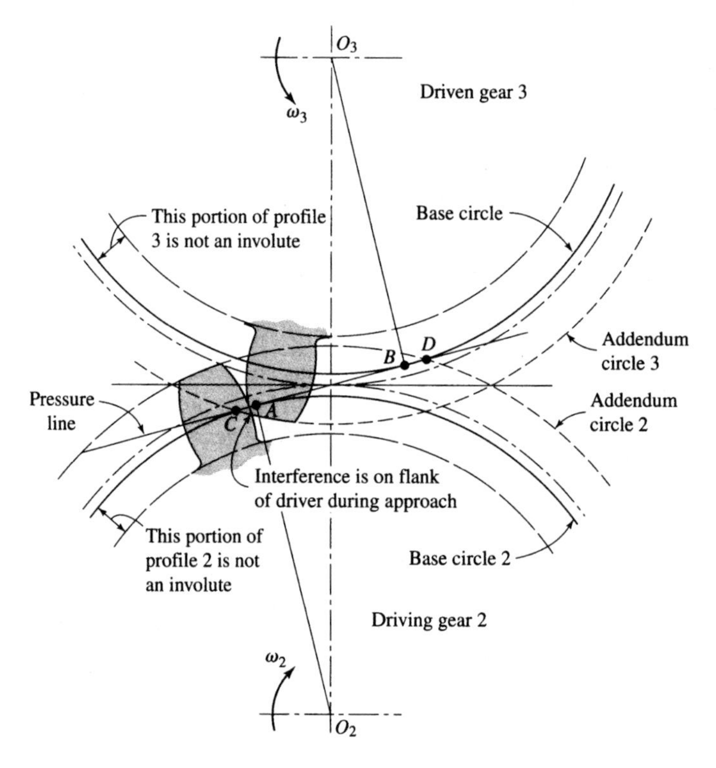

Figure 7.18 Interference in gear tooth action.

* Such gears were part of an older standard and are now obsolete. They are chosen here only to illustrate an example of interference.

full-depth involute teeth. The driver, gear 2, turns clockwise. As with previous figures, the points labeled A and B indicate the points of tangency of the line of action with the two base circles, whereas the points labeled C and D indicate the initial and final points of contact. Note that the points C and D are now *outside* of points A and B. This indicates interference.

The interference is explained as follows. Contact begins when the tip of the driven gear 3 contacts the flank of the driving tooth. In this case the flank of the driving tooth first tries to make contact with the driven tooth at point C, and this occurs *before* the involute portion of the driving tooth comes within range. In other words, contact occurs before the two teeth become tangent. The actual effect is that the nontangent tip of the driven gear interferes with and digs into the flank of the driver.

In this example a similar effect occurs again as the teeth leave contact. Contact should end at or before point B. Because for this example contact does not end until point D, the effect is for the nontangent tip of the driving tooth to interfere with and dig into the flank of the driven tooth.

When gear teeth are produced by a generating process, interference is automatically eliminated because the cutting tool removes the interfering portion of the flank. This effect is called *undercutting*. If undercutting is at all pronounced, the undercut tooth can be considerably weakened. Thus, the effect of eliminating interference by a generation process is merely to substitute another problem for the original.

The importance of the problem of teeth that have been weakened by undercutting cannot be emphasized too strongly. Of course, interference can be eliminated by using more teeth on the gears. However, if the gears are to transmit a given amount of power, more teeth can be used only by increasing the pitch diameter. This makes the gears larger, which is seldom desirable. It also increases the pitch-line velocity, which makes the gears noisier and somewhat reduces the power transmission, although not in direct proportion. In general, however, the use of more teeth to eliminate interference or undercutting is seldom an acceptable solution.

Another method of reducing interference and the resulting undercutting is to employ a larger pressure angle. The larger pressure angle creates smaller base circles, so that a greater portion of the tooth profile has an involute shape. In effect, this means that fewer teeth can be used; as a result, gears with larger pressure angle are often smaller.

Of course, the use of standard gears is far less expensive than manufacturing specially made nonstandard gears. However, as indicated in Table 7.2, gears with larger pressure angles can be found without deviating from the standards.

One more way to eliminate interference is to use gears with shorter teeth. If the addendum distance is reduced, then points C and D move inward. One way to do this is to purchase standard gears and then grind the tops of the teeth to a new addendum distance. This, of course, makes the gears nonstandard and causes concern about repair or replacement, but it can be effective in eliminating interference. Again, careful study of Table 7.2 indicates that this is possible by use of standard 20° *stub tooth* gears.

7.8 CONTACT RATIO

The zone of action of meshing gear teeth is illustrated in Fig. 7.19, where tooth contact begins and ends at the intersections of the two addendum circles with the line of action. As always, initial contact occurs at C and final contact at D. Tooth profiles drawn through

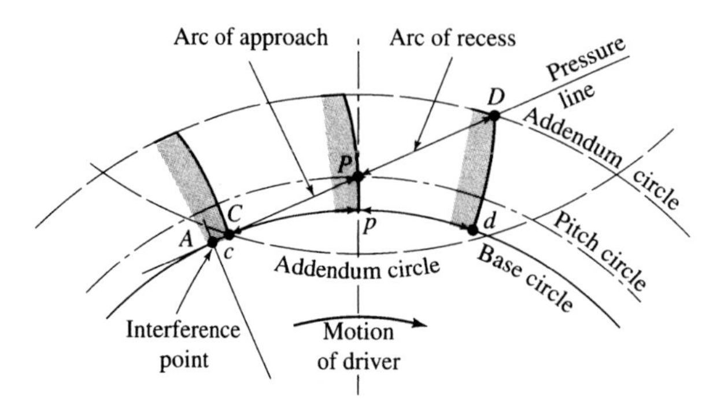

Figure 7.19

these points intersect the base circle at points c and d. Thinking back to our analogy of the unwrapping cord of Fig. 7.6, the linear distance CD, measured along the line of action, is equal to the arc length cd, measured along the base circle.

Consider a situation in which the arc length cd, or distance CD, is exactly equal to the base pitch p_b of Eq. (7.6). This means that one tooth and its space spans the entire arc cd. In other words, when a tooth is just beginning contact at C, the tooth ahead of it is just ending its contact at D. Therefore, during the tooth action from C to D there is exactly one pair of teeth in contact.

Next, consider a situation for which the arc length cd, or distance CD, is greater than the base pitch, but not much greater, say $cd = 1.1\, p_b$. This means that when one pair of teeth is just entering contact at C, the previous pair, already in contact, has not yet reached D. Thus, for a short time, there are two pairs of teeth in contact, one in the vicinity of C and the other nearing D. As meshing proceeds, the previous pair reaches D and ceases contact, leaving only a single pair of teeth in contact again, until the situation repeats itself with the next pair of teeth.

Because of the nature of this tooth action, with one, two, or even more pairs of teeth in contact simultaneously, it is convenient to define the term contact ratio m_c as

$$m_c = \frac{CD}{p_b}. \tag{7.9}$$

This is a value for which the next lower integer indicates the average number of pairs of teeth in contact. Thus, a contact ratio of $m_c = 1.35$, for example, implies that there is always at least one tooth in contact and there are two teeth in contact 35% of the time.

The minimum acceptable value of the contact ratio for smooth operation of meshing gears is $1.2 \le m_c \le 1.4$ and the recommended range of the contact ratio for most spur gearsets is $m_c > 1.4$.

The distance CD is quite convenient to measure if we are working graphically by making a drawing like Fig. 7.20 or Fig. 7.9. However, the distances CP and PD can also be determined analytically. From triangles O_3BC and O_3BP we can write

$$CP = \sqrt{(R_3 + a)^2 - (R_3 \cos\phi)^2} - R_3 \sin\phi. \tag{7.10}$$

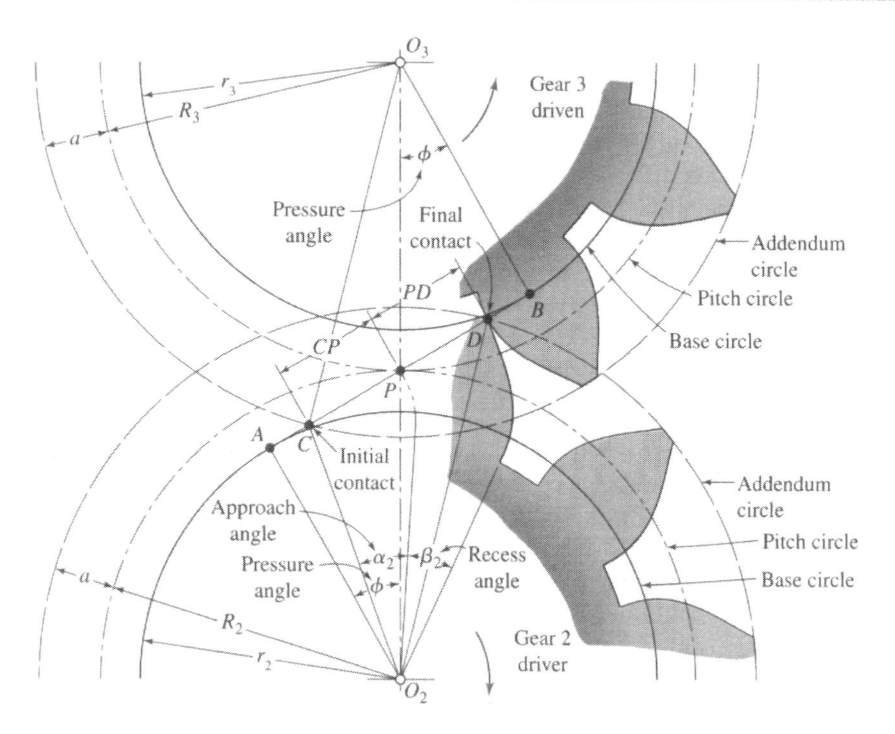

Figure 7.20

Similarly, from triangles O_2AD and O_2AP we have

$$PD = \sqrt{(R_2 + a)^2 - (R_2 \cos \phi)^2} - R_2 \sin \phi. \tag{7.11}$$

The contact ratio is then obtained by substituting the sum of Eqs. (7.10) and (7.11) into Eq. (7.9).

We should note, however, that Eqs. (7.10) and (7.11) are only valid for the conditions where

$$CP \leq R_2 \sin \phi \quad \text{and} \quad PD \leq R_3 \sin \phi \tag{7.12}$$

because proper contact cannot begin before point A or end after point B. If either of these inequalities is not satisfied, then the gear teeth have interference and undercutting results.

7.9 VARYING THE CENTER DISTANCE

Figure 7.21a illustrates a pair of meshing gears having 20° full-depth involute teeth. Because both sides of the teeth are in contact, the center distance $R_{O_3O_2}$ cannot be reduced without jamming or deforming the teeth. However, Fig. 7.21b illustrates the same pair of gears, but

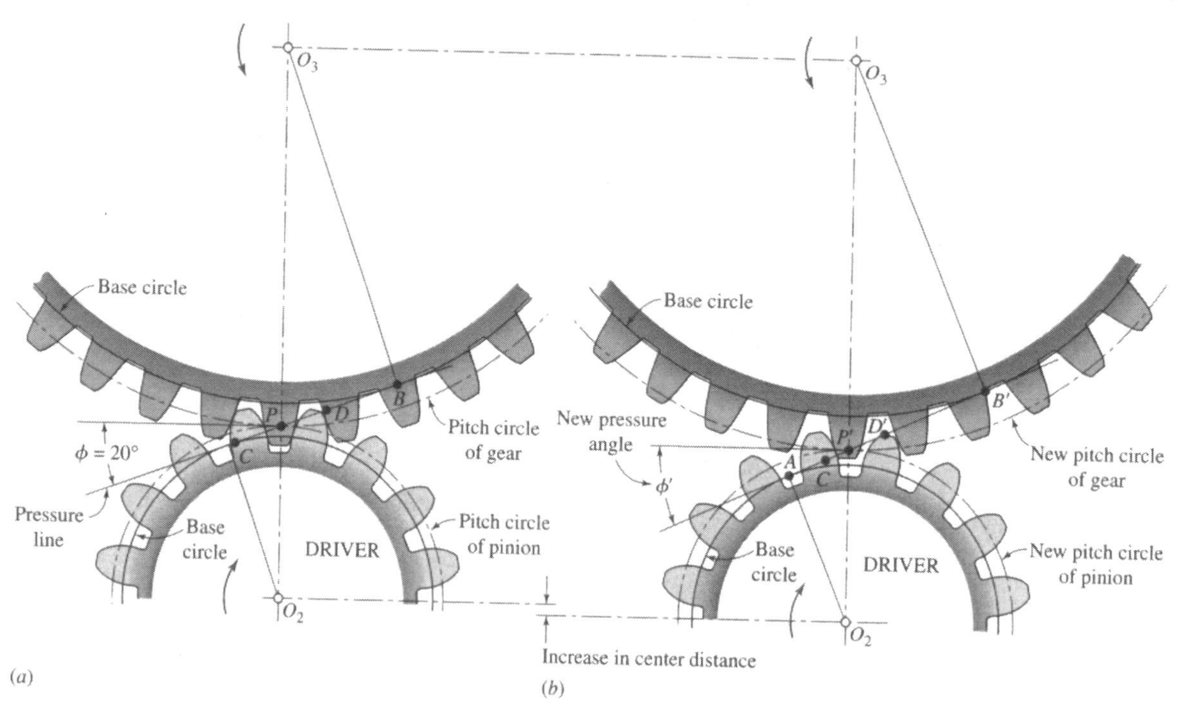

Figure 7.21 Effect of increased center distance on the action of involute gearing; mounting the gears at (*a*) normal center distance and (*b*) increased center distance.

mounted with a slightly increased distance $R_{O_3O_2}$ between the shaft centers as might happen through the accumulation of tolerances of surrounding parts. Clearance, or *backlash*, now exists between the teeth, as illustrated.

When the center distance is increased, the base circles of the two gears do not change; they are fundamental to the shapes of the gears, once manufactured. However, review of Fig. 7.6 indicates that the same involute tooth shapes still touch as conjugate curves and the fundamental law of gearing is still satisfied. However, the larger center distance results in an increase of the pressure angle and larger pitch circles passing through a new adjusted pitch point.

In Fig. 7.21*b* we can see that the triangles $O_2A'P'$ and $O_3B'P'$ are still similar to each other, although they are both modified by the change in pressure angle. Also, the distances O_2A' and O_3B' are the base circle radii and have not changed. Therefore, the ratio of the new pitch radii, O_2P' and O_3P', and the new velocity ratio remain the same as in the original design.

Another effect of increasing the center distance, observable in Fig. 7.21, is the shortening of the path of contact. The original path of contact CD in Fig. 7.21*a* is shortened to $C'D'$ in Fig. 7.21*b*. The contact ratio, Eq. (7.9), is also reduced when the path of contact $C'D'$ is shortened. Because a contact ratio of less than unity would imply periods during which no teeth would be in contact at all, the center distance must never be increased larger than that corresponding to a contact ratio of unity.

7.10 INVOLUTOMETRY

The study of the geometry of the involute curve is called *involutometry*. In Fig. 7.22 a base circle with center at O is used to generate the involute BC. AT is the generating line, ρ is the instantaneous radius of curvature of the involute, and r is the radius to point T on the curve. If we designate the radius of the base circle r_b, the line AT has the same length as the arc distance AB and so

$$\rho = r_b(\alpha + \varphi), \tag{a}$$

where α is the angle between the origin of the involute OB and the radius AT, and φ is the angle between the radius of the base circle OA and the radius OT. Because OAT is a right triangle,

$$\rho = r_b \tan\varphi. \tag{7.13}$$

Solving Eqs. (a) and (7.13) simultaneously to eliminate ρ and r_b gives

$$\alpha = \tan\varphi - \varphi,$$

which can be written

$$\text{inv}\,\varphi = \tan\varphi - \varphi. \tag{7.14}$$

and defines the *involute function*. The angle φ in this equation is the variable involute angle, given in radians. Once φ is known, inv φ can readily be determined from Eq. (7.14). The inverse problem, when inv φ is given and φ is to be found, is more difficult. One approach is to expand Eq. (7.14) in an infinite series and to employ the first several terms to obtain a numerical approximation. Another approach is to use a root-finding technique.* Here, we

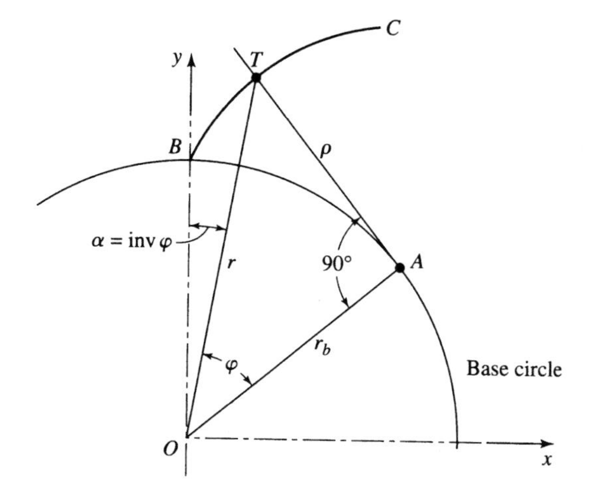

Figure 7.22

* See, for example, C. R. Mischke, *Mathematical Model Building*. Ames, IA: Iowa State University Press, 1980, chap. 4.

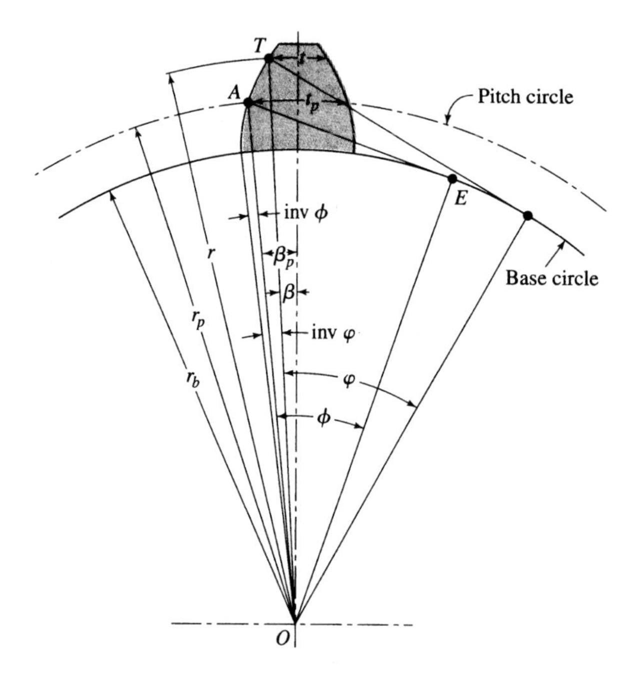

Figure 7.23

refer to Table 6 of Appendix A, in which the value of the involute function is tabulated and the angle φ can be determined directly, in degrees.

Referring again to Fig. 7.22, we see that

$$r = \frac{r_b}{\cos \varphi}. \tag{7.15}$$

To illustrate the use of the relations obtained above, let us determine the tooth dimensions of Fig. 7.23. Here, only the portion of the tooth extending above the base circle has been drawn, and the thickness of the tooth t_p at the pitch circle (point A), equal to half of the circular pitch, is given. The problem is to determine the tooth thickness at some other point, say point T. The various quantities illustrated in Fig. 7.23 are identified as follows:

r_b = radius of the base circle;
r_p = radius of the pitch circle;
r = radius at which the tooth thickness is to be determined;
t_p = tooth thickness at the pitch circle;
t = tooth thickness to be determined;
ϕ = pressure angle corresponding to the pitch circle radius r_p;
φ = involute angle corresponding to point T;
β_p = angular half-tooth thickness at the pitch circle; and
β = angular half-tooth thickness at point T.

The half-tooth thicknesses at points A and T are

$$\frac{t_p}{2} = \beta_p r_p \quad \text{and} \quad \frac{t}{2} = \beta r \tag{b}$$

so that

$$\beta_p = \frac{t_p}{2r_p} \quad \text{and} \quad \beta = \frac{t}{2r}. \tag{c}$$

From these we can write

$$\text{inv}\,\varphi - \text{inv}\,\phi = \beta_p - \beta = \frac{t_p}{2r_p} - \frac{t}{2r}. \tag{d}$$

The tooth thickness at point T is obtained by solving Eq. (d) for t:

$$t = 2r\left(\frac{t_p}{2r_p} + \text{inv}\,\phi - \text{inv}\,\varphi\right). \tag{7.16}$$

EXAMPLE 7.2

A gear has 22 teeth cut full-depth with pressure angle $\phi = 20°$, and a diametral pitch $P = 2$ teeth/in. Find the thickness of the teeth at the base circle and at the addendum circle.

SOLUTION

By the equations of Section 7.1 and Table 7.2 we find the radius of the pitch circle $r_P = 5.500$ in., the circular pitch $p = 1.571$ in/tooth, the addendum $a = 0.500$ in, and the dedendum $d = 0.625$ in.

From the right-angled triangle OEA in Fig. 7.23, the radius of the base circle can be written as

$$r_b = r_P \cos\phi = (5.500 \text{ in}) \cos 20° = 5.168 \text{ in}.$$

The thickness of the tooth at the pitch circle is

$$t_p = \frac{p}{2} = \frac{1.571 \text{ in/tooth}}{2} = 0.785\,5 \text{ in}.$$

Converting the tooth pressure angle into radians gives $\phi = 20° = 0.349$ rad. Then the involute function from Eq. (7.14) is

$$\text{inv}\,\phi = \tan 0.349 - 0.349 = 0.014\,9 \text{ rad}.$$

The involute angle at the base circle, from Eq. (7.15), is $\varphi_b = 0$. Therefore, the involute function is

$$\text{inv}\,\varphi_b = 0.$$

Substituting these results into Eq. (7.16), the tooth thickness at the base circle is

$$t_b = 2r_b\left[\frac{t_p}{2r_p} + \text{inv}\,\phi - \text{inv}\varphi_b\right] = 2(5.168 \text{ in})\left[\frac{0.785\,5 \text{ in}}{2(5.500 \text{ in})} + 0.014\,9 - 0\right] = 0.892 \text{ in}.$$

Ans.

The radius of the addendum circle is $r_a = r_P + a = 5.500 + 0.500 = 6.000$ in. Therefore, the involute angle corresponding to this radius, from Eq. (7.15), is

$$\varphi = \cos^{-1}\left(\frac{r_b}{r}\right) = \cos^{-1}\left(\frac{5.168 \text{ in}}{6.000 \text{ in}}\right) = 30.53^\circ = 0.533 \text{ rad.}$$

Thus, the involute function is

$$\text{inv } \varphi = \tan 0.533 - 0.533 = 0.056\,9 \text{ rad.}$$

Substituting these results into Eq. (7.16), the tooth thickness at the addendum circle is

$$t_a = 2r_a\left[\frac{t_p}{2r_p} + \text{inv}\phi - \text{inv}\varphi\right] = 2(6.000 \text{ in})\left[\frac{0.785\,5 \text{ in}}{2(5.500 \text{ in})} + 0.014\,9 - 0.056\,9\right]$$
$$= 0.353 \text{ in.}$$ *Ans.*

Note that the tooth thickness at the base circle is more than double the tooth thickness at the addendum circle.

7.11 NONSTANDARD GEAR TEETH

In this section we will examine the effects obtained by deviating from the specified standards and modifying such things as pressure angle, tooth depth, addendum, or center distance. Some of these modifications do not eliminate interchangeability; all of them are discussed with the intent of obtaining improved performance. Still, making such modifications probably means increased cost because modified gears will not be available and will need to be specially machined for the particular application. Of course, this will also be necessary at the time of any future repair or design modification.

The designer is often under great pressure to produce a design using gears that is small and yet will transmit a large amount of power. Consider, for example, a gearset that must have a 4:1 velocity ratio. If the smallest pinion that will carry the load has a pitch diameter of 2 in, the mating gear will have a pitch diameter of 8 in, making the overall space required for the two gears more than 10 in. On the other hand, if the pitch diameter of the pinion can be reduced by only $\frac{1}{4}$ in, the pitch diameter of the gear is reduced by a full 1 in and the overall size of the gearset is reduced by $1\frac{1}{4}$ in. This reduction assumes considerable importance when it is realized that the associated machine elements, such as shafts, bearings, and enclosure, are also reduced in size.

If a tooth of a certain pitch is required to carry the load, the only method of decreasing the pinion diameter is to use fewer teeth. However, we have already seen that problems involving interference, undercutting, and contact ratio are encountered when the tooth numbers are made too small. Thus, three principal reasons for employing nonstandard gears are to: (*i*) eliminate undercutting, (*ii*) prevent interference, and (*iii*) maintain a reasonable contact ratio. It should be noted too that if a pair of gears are manufactured of the same material, the pinion is the weaker and is subject to greater wear because each of its teeth is in contact a greater portion of the time. Therefore, any undercutting weakens the tooth that is already weaker. Thus, another objective of nonstandard gears is to gain a better balance of strength between the pinion and the gear.

As an involute curve is generated from its base circle, its radius of curvature becomes larger and larger. Near the base circle the radius of curvature is quite small, being theoretically zero at the base circle. Contact near this region of sharp curvature should be avoided if possible because of the difficulty in obtaining good cutting accuracy in areas of small curvature and because contact stresses are likely to be very high. Nonstandard gears present the opportunity of designing to avoid these sensitive areas.

Clearance Modification. A larger fillet radius at the root of the tooth increases the fatigue strength of the tooth and provides extra depth for shaving the tooth profile. Because interchangeability is not lost, the dedendum is sometimes increased to $1.300/P$ or $1.400/P$ to obtain space for a larger fillet radius.

Center-Distance Modification. When gears of low tooth numbers are paired with each other, or with larger gears, reduction in interference and improvement in the contact ratio can be obtained by increasing the center distance to greater than standard. Although such a system changes the tooth proportions and the pressure angle of the gears, the resulting tooth shapes can be generated with rack cutters (or hobs) of standard pressure angles or with standard pinion shapers. Before introducing this system, however, it will be of value to develop certain additional relations about the geometry of gears.

The first new relation is for finding the thickness of a tooth that is cut by a rack cutter (or hob) when the pitch line of the rack cutter is displaced or offset a distance e from the pitch circle of the gear being cut. What we are doing here is moving the rack cutter away from the center of the gear being cut. Stated another way, suppose the rack cutter does not cut as deeply into the gear blank and the teeth are not cut to full depth. This produces teeth that are thicker than the standard, and this thickness will now be determined. Figure 7.24*a* illustrates the problem, and Fig. 7.24*b* illustrates the solution. The increase of tooth thickness at the pitch circle is $2e \tan \phi$, so that

$$t = 2e \tan \phi + \frac{p}{2}, \tag{7.17}$$

where ϕ is the pressure angle of the rack cutter and t is the thickness of the modified gear tooth measured on its pitch circle.

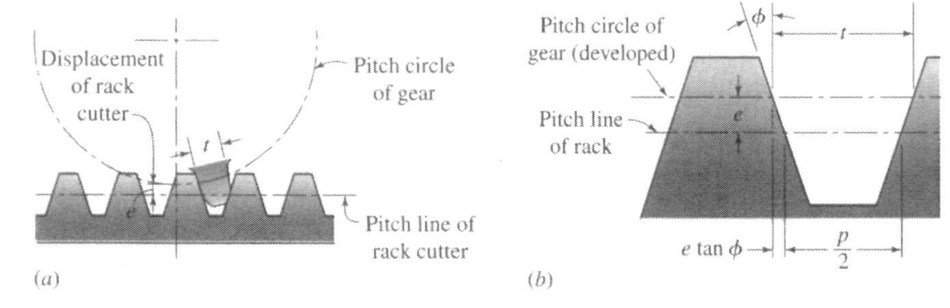

Figure 7.24

Now suppose that two gears of different tooth numbers have both been cut with the cutter offset from their pitch circles as in the previous paragraph. Because the teeth of both have been cut with offset cutters, they will mate at a modified pressure angle and with modified pitch circles and consequently modified center distances. The word modified is used here in the sense of being nonstandard. Our problem is to determine the radii of these modified pitch circles and the value of the modified pressure angle.

In the following notation, the word *standard* refers to values that would have been obtained had the usual, or standard, systems been employed to obtain the dimensions:

ϕ = pressure angle of generating rack cutter;
ϕ' = modified pressure angle at which gears will mate;
R_2 = standard pitch radius of pinion;
R_2' = modified pitch radius of pinion when meshing with given gear;
R_3 = standard pitch radius of gear;
R_3' = modified pitch radius of gear when meshing with given pinion;
t_2 = thickness of pinion tooth at standard pitch radius R_2;
t_2' = thickness of pinion tooth at modified pitch radius R_2';
t_3 = thickness of gear tooth at standard pitch radius R_3; and
t_3' = thickness of gear tooth at modified pitch radius R_3'.

From Eq. (7.16), the thickness of a gear tooth at the standard pitch radius and at the modified pitch radius can be written, respectively, as

$$t_2' = 2R_2'\left(\frac{t_2}{2R_2} + \text{inv}\,\phi - \text{inv}\,\phi'\right) \tag{a}$$

and

$$t_3' = 2R_3'\left(\frac{t_3}{2R_3} + \text{inv}\,\phi - \text{inv}\,\phi'\right). \tag{b}$$

Note that the sum of these two thicknesses must be the new circular pitch. Therefore, using Eq. (7.3), we can write

$$t_2' + t_3' = p' = \frac{2\pi R_2'}{N_2}. \tag{c}$$

Since the pitch diameters of a pair of mating gears are proportional to their tooth numbers then

$$R_3 = \frac{N_3}{N_2}R_2 \quad \text{and} \quad R_3' = \frac{N_3}{N_2}R_2'. \tag{d}$$

Substituting Eqs. (*a*), (*b*), and (*d*) into Eq. (*c*) and rearranging gives

$$\text{inv}\,\phi' = \frac{N_2\left(t_2' + t_3'\right) - 2\pi R_2}{2R_2\left(N_2 + N_3\right)} + \text{inv}\,\phi. \tag{7.18}$$

This equation gives the modified pressure angle ϕ' at which a pair of gears will operate when the tooth thicknesses on their standard pitch circles are modified to t_2' and t_3'.

Although the base circle of a gear is fundamental to its shape and fixed once the gear is generated, gears have no pitch circles until a pair of them is brought into contact. Bringing a pair of gears into contact creates a pair of pitch circles that are tangent to each other at the modified pitch point. Throughout this discussion, the idea of a pair of so-called standard pitch circles has been used to define a certain point on the involute curves. These standard pitch circles, as we have seen, are the ones that would come into existence when the gears are paired *if the gears are not modified from the standard dimensions*. On the other hand, the base circles are fixed circles that are not changed by tooth modifications. The base circle remains the same whether the tooth dimensions are changed or not, so we can determine the base circle radius using either the standard pitch circle or the new pitch circle. Thus, from Eq. (7.15), we can write

$$R_2 \cos\varphi = R_2' \cos\varphi'$$

Therefore, the modified pitch radius of the pinion can be written as

$$R_2' = \frac{R_2 \cos\phi}{\cos\phi'}. \tag{7.19}$$

Similarly, the modified pitch radius of the gear can be written as

$$R_3' = \frac{R_3 \cos\phi}{\cos\phi'}. \tag{7.20}$$

Equations (7.19) and (7.20) give the values of the actual pitch radii when the two gears with modified teeth are brought into mesh without backlash. The new center distance is, of course, the sum of these two radii.

All necessary relations have now been developed to create nonstandard gears with changes in the center distance. The use of these relations is now illustrated by an example.

Figure 7.25 is a drawing of a 20° pressure angle, 1 tooth/in diametral pitch, 12-tooth pinion generated with a rack cutter to full depth with a standard clearance of $0.250/P$. In

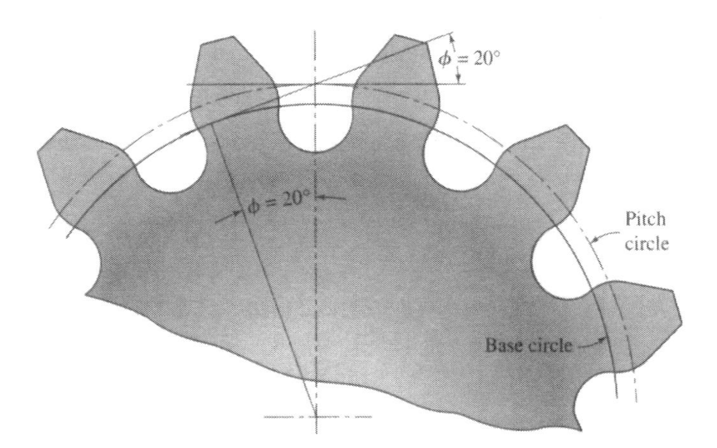

Figure 7.25 Standard 20° pressure angle, 1-tooth/in diametral pitch, 12-tooth full-depth involute gear showing undercut.

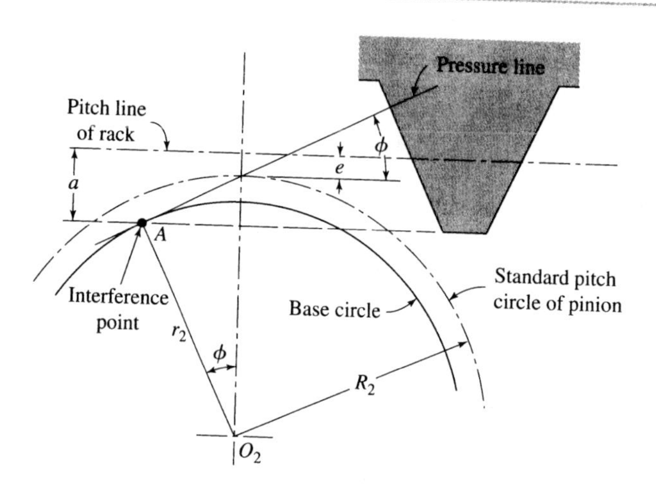

Figure 7.26 Offset of a rack cutter to cause its addendum line to pass through the interference point.

the 20° full-depth system, interference is severe when the number of teeth is less than 14. The resulting undercutting is evident in the drawing.

In an attempt to eliminate undercutting, improve the tooth action, and increase the contact ratio, suppose that this pinion were not cut to full depth; suppose instead that the rack cutter were only allowed to cut to a depth for which its addendum passes through the interference point A of the pinion being cut—that is, the point of tangency of the 20° line of action and the base circle—as illustrated in Fig. 7.26. From Eq. (7.15) we know that

$$r_2 = R_2 \cos\phi. \tag{e}$$

Then, from Fig. 7.26, the depth of the cut would be offset from the standard by

$$e = a + r_2 \cos\phi - R_2. \tag{f}$$

Substituting Eq. (e) into Eq. (f), the offset can be written as

$$e = a + R_2 \cos^2\phi - R_2 = a - R_2 \sin^2\phi. \tag{7.21}$$

If the offset is any less than this, then the rack will cut below the interference point A and will result in undercutting.

EXAMPLE 7.3

A 12-tooth pinion with pressure angle $\varphi = 20°$ and diametral pitch $P = 1$ tooth/in is to be mated with a standard 40-tooth gear. If the pinion were cut to full depth, then Eq. (7.9). demonstrates that the contact ratio would be 1.41, but there would be undercutting as indicated in Fig. 7.25. Instead, let the 12-tooth pinion be cut from a larger blank using center-distance modifications. Determine the cutter offset, the modified pressure angle, the modified pitch radii of the pinion and gear, the modified center distance, the modified outside radii of the pinion and gear, and the contact ratio. Has the contact ratio increased significantly?

SOLUTION

Designating the pinion as subscript 2 and the gear as 3, then with $P = 1$ tooth/in and $\phi = 20°$, the following values are determined:

$$p = 3.142 \text{ in/tooth}, \ R_2 = 6 \text{ in}, \ R_3 = 20 \text{ in}, \ N_2 = 12 \text{ teeth},$$
$$N_3 = 40 \text{ teeth}, \quad \text{and} \quad t_3 = 1.571 \text{ in}.$$

For a standard rack cutter, from Table 7.2, the addendum is $a = 1/P = 1.0$ in.

From Eq. (7.21) the rack cutter will be offset by

$$e = 1.0 - 6.0 \sin^2 20° = 0.298 \text{ in.} \qquad \textit{Ans.}$$

Then the thickness of the pinion tooth at the 6 in pitch circle, using Eq. (7.17), is

$$t_2' = 2e \tan\phi + \frac{p}{2} = 2\,(0.298 \text{ in}) \tan 20° + \frac{3.142 \text{ in}}{2} = 1.788 \text{ in.}$$

The pressure angle at which this (and only this) gearset will operate is determined from Eq. (7.18), that is

$$\text{inv}\,\phi' = \frac{N_2\,(t'_2 + t'_3) - 2\pi R_2}{2R_2\,(N_2 + N_3)} + \text{inv}\phi$$
$$= \frac{12(1.788 \text{ in} + 1.571 \text{ in}) - 2\pi(6.0 \text{ in})}{2(6.0 \text{ in})\,(12 + 40)} + \text{inv } 20° = 0.019\,08 \text{ rad.}$$

From Appendix A, Table 6, we find that the new pressure angle is $\varphi' = 21.65°$. *Ans.*

Using Eqs. (7.19) and (7.20), the modified pitch radii are determined to be

$$R_2' = \frac{R_2 \cos\phi}{\cos\phi'} = \frac{(6.0 \text{ in}) \cos 20°}{\cos 21.65°} = 6.066 \text{ in.} \qquad \textit{Ans.}$$

$$R_3' = \frac{R_3 \cos\phi}{\cos\phi'} = \frac{(20.0 \text{ in}) \cos 20°}{\cos 21.65°} = 20.220 \text{ in.} \qquad \textit{Ans.}$$

So the modified center distance is

$$R_2' + R_3' = 6.066 + 20.220 = 26.286 \text{ in.} \qquad \textit{Ans.}$$

Note that the center distance has not increased as much as the offset of the rack cutter.

Standard clearance of $0.25/P$ results from the standard dedendums equal to $1.25/P$ as indicated in Table 7.2. So the root radii of the two gears are

$$\text{Root radius of pinion} = 6.298 - 1.250 = 5.048 \text{ in.}$$
$$\text{Root radius of gear} = 20.000 - 1.250 = 18.750 \text{ in.}$$
$$\text{Sum of root radii} = 23.798 \text{ in.}$$

The difference between this sum and the center distance is the working depth plus twice the clearance. Because the clearance is 0.25 in for each gear, the working depth is

$$\text{Working depth} = 26.286 - 23.798 - 2(0.250) = 1.988 \text{ in.}$$

The outside radius of each gear is the sum of the root radius, the clearance, and the working depth, that is

$$\text{Outside radius of pinion} = 5.048 + 0.250 + 1.988 = 7.286 \text{ in.} \qquad \textit{Ans.}$$

$$\text{Outside radius of gear} = 18.750 + 0.250 + 1.988 = 20.988 \text{ in.} \qquad \textit{Ans.}$$

The result is illustrated in Fig. 7.27, and the pinion is seen to have a stronger looking form than the one of Fig. 7.25. Undercutting has been completely eliminated.

The contact ratio can be obtained from Eqs. (7.9) through (7.11). The following quantities are needed:

$$\text{Outside radius of pinion} = R_2' + a = 7.286 \text{ in.}$$

$$\text{Outside radius of gear} = R_3' + a = 20.988 \text{ in.}$$

$$r_2 = R_2 \cos\phi = (6.000 \text{ in}) \cos 20^\circ = 5.638 \text{ in.}$$

$$r_3 = R_3 \cos\phi = (20.000 \text{ in}) \cos 20^\circ = 18.794 \text{ in.}$$

$$p_b = p \cos\phi = (3.141\,6 \text{ in / tooth}) \cos 20^\circ = 2.952 \text{ in/tooth.}$$

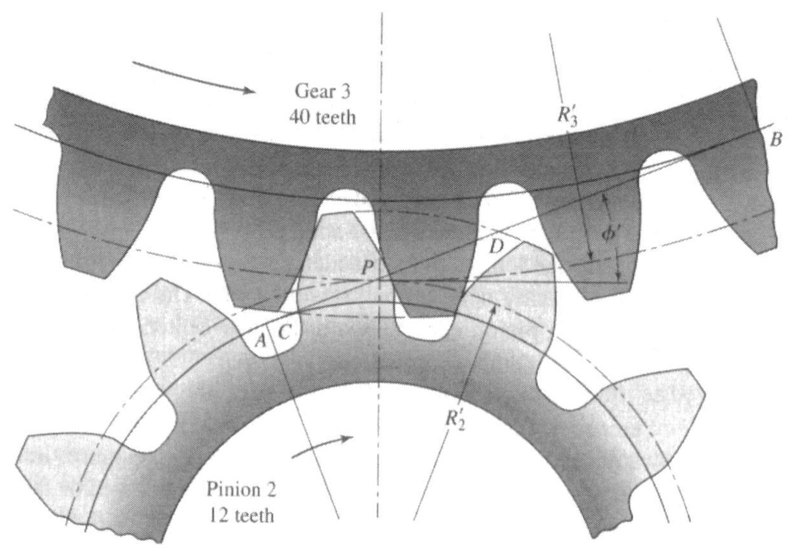

Figure 7.27

Therefore, for Eqs. (7.10) and (7.11) we have

$$CP = \sqrt{(R_3' + a)^2 - r_3^2} - R_3' \sin\phi$$
$$= \sqrt{(20.988\text{ in})^2 - (18.794\text{ in})^2} - (20.220\text{ in})\sin 21.65^\circ = 1.883\text{ in}$$
$$PD = \sqrt{(R_2' + a)^2 - r_2^2} - R_2' \sin\phi$$
$$= \sqrt{(7.286\text{ in})^2 - (5.638\text{ in})^2} - (6.066\text{ in})\sin 21.65^\circ$$
$$= 2.377\text{ in}$$

Finally, from Eq. (7.9), the contact ratio is

$$m_c = \frac{CP + PD}{p_b} = \frac{1.883\text{ in} + 2.377\text{ in}}{2.952\text{ in/tooth}} = 1.443\text{ teeth avg.} \qquad \textit{Ans.}$$

Therefore, the contact ratio has increased only slightly (approximately a 2% increase). The modification, however, is justified because of the elimination of undercutting, which results in a substantial improvement in the strength of the teeth.

Long-and-Short-Addendum Systems It often happens in the design of machinery that the center distance between a pair of gears is fixed by some other design consideration or feature of the machine. In such a case, modifications to obtain improved performance cannot be made by varying the center distance.

In the previous section we saw that improved action and tooth shape can be obtained by backing the rack cutter away from the gear blank during forming of the teeth. The effect of this withdrawal is to create the active tooth profile farther away from the base circle. Examination of Fig. 7.27 indicates that more dedendum could be used on the gear (not the pinion) before the interference point is reached. If the rack cutter is advanced into the gear blank by a distance equal to the withdrawal from the pinion blank, more of the gear dedendum will be used and at the same time the center distance will not be changed. This is called the *long-and-short-addendum system*.

In the long-and-short-addendum system there are no changes in the pitch circles and consequently none in the pressure angle. The effect is to move the contact region away from the pinion center toward the gear center, thus shortening the approach action and lengthening the recess action.

The characteristics of the long-and-short-addendum system can be explained by reference to Fig. 7.28. Figure 7.28*a* illustrates a conventional (standard) set of gears having a dedendum equal to the addendum plus the clearance. Interference exists, and the tip of the gear tooth will have to be relieved as illustrated or the pinion will be undercut. This is indicated because the addendum circle crosses the line of action at C, outside of the tangency or interference point A; hence, the distance AC is a measure of the degree of interference.

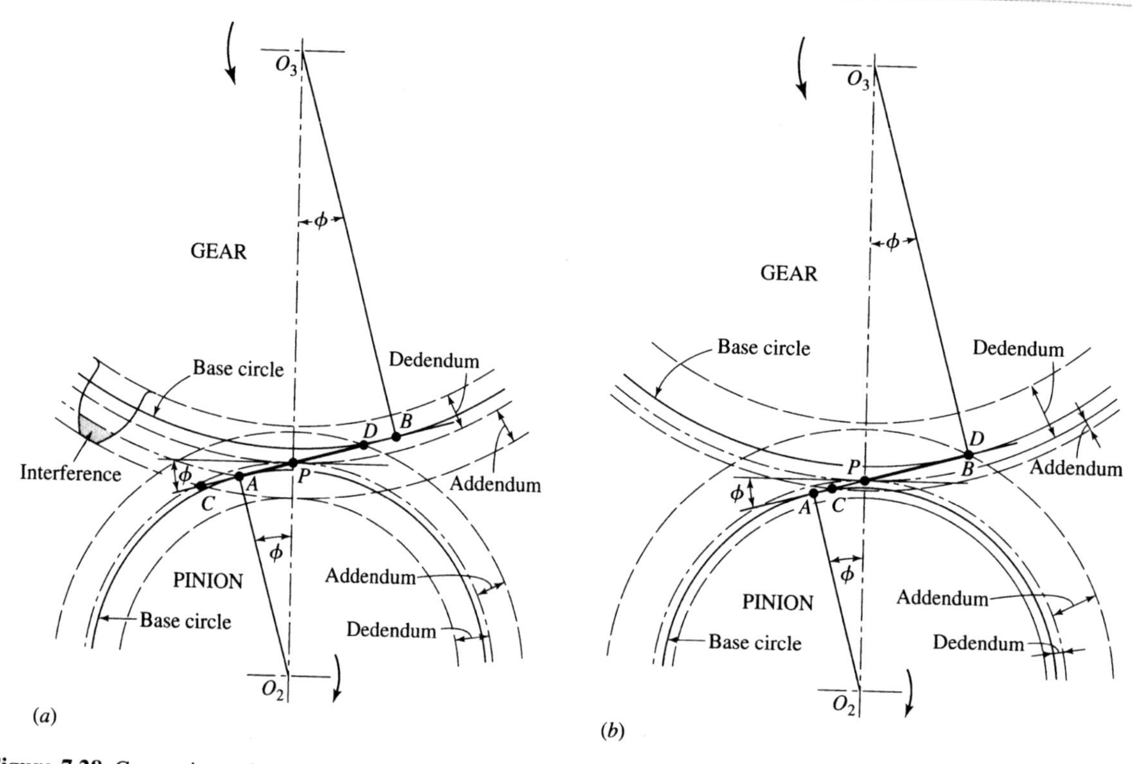

Figure 7.28 Comparison of standard gears and gears cut by the long-and-short-addendum system: (*a*) gear and pinion with standard addendum and dedendum; (*b*) gear and pinion with long-and-short addendum.

To eliminate the undercutting or interference, the pinion addendum may be enlarged, as in Fig. 7.28*b*, until the addendum circle of the pinion passes through the interference point (point *B*) of the gear. In this manner, we shall be using all of the gear–tooth profile. The same whole depth may be retained; hence, the dedendum of the pinion may be reduced by the same amount that the addendum is increased. This means that we must also lengthen the gear dedendum and shorten the dedendum of the mating pinion. With these changes the path of contact is the line *CD* of Fig. 7.28*b*. It is longer than the path *AD* of Fig. 7.28*a*, and so the contact ratio is higher. Note too that the base circles, the pitch circles, the pressure angle, and the center distance have not changed. Both gears can be cut with standard cutters by advancing the cutter into the gear blank, for this modification by a distance equal to the amount of withdrawal from the pinion blank. Finally, note that the blanks from which the pinion and gear are cut must now be of different diameters than the standard blanks.

The tooth dimensions for the long-and-short-addendum system can be determined using the equations developed in the previous sections.

A less obvious advantage of the long-and-short-addendum system is that more recess action than approach action is obtained. The approach action of gear teeth is analogous to pushing a piece of chalk across a blackboard; the chalk screeches. But when the chalk is pulled across a blackboard, analogous to the recess action, it glides smoothly. Thus, recess action is always preferable because of the smoothness and the lower frictional forces.

7.12 REFERENCES

[1] Standards are defined by the American Gear Manufacturers Association (AGMA) and the American National Standards Institute (ANSI). The AGMA standards may be quoted or extracted in their entirety, provided that an appropriate credit line is included—for example, "Extracted from AGMA Information Sheet—Strength of Spur, Helical, Herringbone, and Bevel Gear Teeth (AGMA 225.01) with permission of the publisher, the American Gear Manufacturers Association, 1500 King Street, Suite 201, Alexandria, VA 22314." These standards have been used extensively in Chapter 7 and in Chapter 8.

[2] The strength and wear of gears are covered in texts such as *Shigley's Mechanical Engineering Design,* 8th ed., R.G. Budynas and J.K. Nisbett, New York: McGraw–Hill, 2008.

PROBLEMS

7.1 Find the diametral pitch of a pair of gears having 32 and 84 teeth, respectively, whose center distance is 3.625 in.

7.2 Find the number of teeth and the circular pitch of a 6-in pitch diameter gear whose diametral pitch is 9 teeth/in.

7.3 Determine the module of a pair of gears having 18 and 40 teeth, respectively, whose center distance is 58 mm.

7.4 Find the number of teeth and the circular pitch of a gear whose pitch diameter is 200 mm if the module is 8 mm/tooth.

7.5 Find the diametral pitch and the pitch diameter of a 40-tooth gear whose circular pitch is 3.50 in/tooth.

7.6 The pitch diameters of a pair of mating gears are 3.50 and 8.25 in, respectively. If the diametral pitch is 16 teeth/in, how many teeth are there on each gear?

7.7 Find the module and the pitch diameter of a gear whose circular pitch is 40 mm/tooth if the gear has 36 teeth.

7.8 The pitch diameters of a pair of gears are 60 and 100 mm, respectively. If their module is 2.5 mm/tooth, how many teeth are there on each gear?

7.9 What is the diameter of a 33-tooth gear if its circular pitch is 0.875 in/tooth?

7.10 A shaft carries a 30-tooth, 3-teeth/in diametral pitch gear that drives another gear at a speed of 480 rev/min. How fast does the 30-tooth gear rotate if the shaft center distance is 9 in?

7.11 Two gears having an angular velocity ratio of 3:1 are mounted on shafts whose centers are 136 mm apart. If the module of the gears is 4 mm/tooth, how many teeth are there on each gear?

7.12 A gear having a module of 4 mm/tooth and 21 teeth drives another gear at a speed of 240 rev/min. How fast is the 21-tooth gear rotating if the shaft center distance is 156 mm?

7.13 A 4-tooth/in diametral pitch, 24-tooth pinion is to drive a 36-tooth gear. The gears are cut on the 20° full-depth involute system. Find and tabulate the addendum, dedendum, clearance, circular pitch, base pitch, tooth thickness, pitch circle radii, base circle radii, length of paths of approach and recess, and contact ratio.

7.14 A 5-tooth/in diametral pitch, 15-tooth pinion is to mate with a 30-tooth internal gear. The gears are 20° full-depth involute. Make a drawing of the gears showing several teeth on each gear. Can these gears be assembled in a radial direction? If not, what remedy should be used?

7.15 A $2\frac{1}{2}$-teeth/in diametral pitch 17-tooth pinion and a 50-tooth gear are paired. The gears are cut on the 20° full-depth involute system. Find the angles of approach and recess of each gear and the contact ratio.

7.16 A gearset with a module of 5 mm/tooth has involute teeth with $22\frac{1}{2}°$ pressure angle and 19 and

31 teeth, respectively. They have 1.0m for the addendum and 1.25m for the dedendum.* Tabulate the addendum, dedendum, clearance, circular pitch, base pitch, tooth thickness, base circle radius, and contact ratio.

7.17 A gear with a module of 8 mm/tooth and 22 teeth is in mesh with a rack; the pressure angle is 25°. The addendum and dedendum are 1.0m and 1.25m, respectively.* Find the lengths of the paths of approach and recess and determine the contact ratio.

7.18 Repeat Problem 7.15 using the 25° full-depth system.

7.19 Draw a 2-tooth/in diametral pitch, 26-tooth, 20° full-depth involute gear in mesh with a rack.

(a) Find the lengths of the paths of approach and recess and the contact ratio.

(b) Draw a second rack in mesh with the same gear but offset 1/8 in further away from the gear center. Determine the new contact ratio. Has the pressure angle changed?

7.20 through 7.24 Shaper gear cutters have the advantage that they can be used for either external or internal gears and also that only a small amount of runout is necessary at the end of the stroke. The generating action of a pinion shaper cutter can easily be simulated by employing a sheet of clear plastic. Figure P7.20 illustrates one tooth of a 16-tooth pinion cutter with 20° pressure angle as it can be cut from a plastic sheet. To construct the cutter, lay out the tooth on a sheet of drawing paper. Be sure to include the clearance at the top of the tooth. Draw radial lines through the pitch circle spaced at distances equal to one fourth of the tooth thickness, as illustrated in Fig. P7.20. Next, fasten the sheet of plastic to the drawing and scribe the cutout, the pitch circle, and the radial lines onto the sheet. Then remove the sheet and trim the tooth outline with a razor blade. Then use a small piece of fine sandpaper to remove any burrs.

To generate a gear with the cutter, only the pitch circle and the addendum circle need be drawn. Divide the pitch circle into spaces equal to those used on the template and construct radial lines through them. The tooth outlines are then obtained by rolling the template pitch circle upon that of the gear and drawing the cutter tooth lightly for each position. The resulting generated tooth upon the gear will be evident. The following problems all employ a standard 1-tooth/in diametral pitch 20° full-depth template constructed as described above. In each case you should generate a few teeth and estimate the amount of undercutting.

Table P7.20 to P7.24

Problem no.	P7.20	P7.21	P7.22	P7.23	P7.24
No. of teeth	10	12	14	20	36

7.25 A 10-mm/tooth module gear has 17 teeth, a 20° pressure angle, an addendum of 1.0m, and a dedendum of 1.25m.* Find the thickness of the teeth at the base circle and at the addendum circle. What is the pressure angle corresponding to the addendum circle?

7.26 A 15-tooth pinion has 1½-tooth/in diametral pitch 20° full-depth involute teeth. Calculate the thickness of the teeth at the base circle. What are the tooth thickness and the pressure angle at the addendum circle?

7.27 A tooth is 0.785 in thick at a pitch circle radius of 8 in. and a pressure angle of 25°. What is the thickness at the base circle?

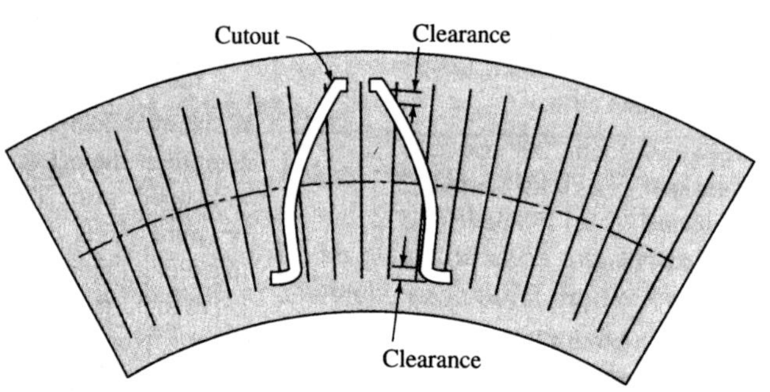

Figure P7.20

* In SI, tooth sizes are given in modules, m, and $a = 1.0\,\text{m}$ means 1 module, not 1 meter.

7.28 A tooth is 1.571 in thick at the pitch radius of 16 in and a pressure angle of 20°. At what radius does the tooth become pointed?

7.29 A 25° full-depth involute, 12-tooth/in diametral pitch pinion has 18 teeth. Calculate the tooth thickness at the base circle. What are the tooth thickness and pressure angle at the addendum circle?

7.30 A nonstandard 10-tooth 8-tooth/in diametral pitch involute pinion is to be cut with a $22\frac{1}{2}°$ pressure angle. What maximum addendum can be used before the teeth become pointed?

7.31 The accuracy of cutting gear teeth can be measured by fitting hardened and ground pins in diametrically opposite tooth spaces and measuring the distance over the pins. For a 10-tooth/in diametral pitch 20° full-depth involute system 96 tooth gear:

(a) Calculate the pin diameter that will contact the teeth at the pitch lines if there is to be no backlash.

(b) What should be the distance measured over the pins if the gears are cut accurately?

7.32 A set of interchangeable gears with 4-tooth/in diametral pitch is cut on the 20° full-depth involute system. The gears have tooth numbers of 24, 32, 48, and 96. For each gear, calculate the radius of curvature of the tooth profile at the pitch circle and at the addendum circle.

7.33 Calculate the contact ratio of a 17-tooth pinion that drives a 73-tooth gear. The gears are 96-tooth/in diametral pitch and cut on the 20° full-depth involute system.

7.34 A 25° pressure angle 11-tooth pinion is to drive a 23-tooth gear. The gears have a diametral pitch of 8 teeth/in and have involute stub teeth. What is the contact ratio?

7.35 A 22-tooth pinion mates with a 42-tooth gear. The gears have full-depth involute teeth, have a diametral pitch of 16 teeth/in, and are cut with a $17\frac{1}{2}°$ pressure angle.* Find the contact ratio.

7.36 The center distance of two 24-tooth, 20° pressure angle, full-depth involute spur gears with diametral pitch of 2 teeth/in is increased by 0.125 in over the standard distance. At what pressure angle do the gears operate?

7.37 The center distance of two 18-tooth, 25° pressure angle, full-depth involute spur gears with diametral pitch of 3 teeth/in is increased by 0.0625 in over the standard distance. At what pressure angle do the gears operate?

7.38 A pair of mating gears have 24 teeth/in. diametral pitch and are generated on the 20° full-depth involute system. If the tooth numbers are 15 and 50, what maximum addendums may they have if interference is not to occur?

7.39 A set of gears is cut with a $4\frac{1}{2}$-in/tooth circular pitch and a $17\frac{1}{2}°$ pressure angle.* The pinion has 20 full-depth teeth. If the gear has 240 teeth, what maximum addendum may it have to avoid interference?

7.40 Using the method described for Problems 7.20 through 7.24, cut a 1-tooth/in diametral pitch 20° pressure angle full-depth involute rack tooth from a sheet of clear plastic. Use a nonstandard clearance of 0.35/P to obtain a stronger fillet. This template can be used to simulate the generating action of a hob. Now, using the variable-center-distance system, generate an 11-tooth pinion to mesh with a 25-tooth gear without interference. Record the values found for center distance, pitch radii, pressure angle, gear blank diameters, cutter offset, and contact ratio. Note that more than one satisfactory solution exists.

7.41 Using the template cut in Problem 7.40, generate an 11-tooth pinion to mesh with a 44-tooth gear with the long-and-short-addendum system. Determine and record suitable values for gear and pinion addendum and dedendum and for the cutter offset and contact ratio. Compare the contact ratio with that of standard gears.

7.42 A pair of involute spur gears with 9 and 36 teeth are to be cut with a 20° full-depth cutter with diametral pitch of 3 teeth/in.

(a) Determine the amount that the addendum of the gear must be decreased to avoid interference.

(b) If the addendum of the pinion is increased by the same amount, determine the contact ratio.

7.43 A standard 20° pressure angle full-depth involute 1-tooth/in diametral pitch 20-tooth pinion drives a 48-tooth gear. The speed of the pinion is 500 rev/min. Using the position of the point of contact along the line of action as the abscissa, plot a curve indicating the sliding velocity at all points of contact. Note that the sliding velocity changes sign when the point of contact passes through the pitch point.

---

* Such gears came from an older standard and are now obsolete.

Uicker et al., 4th Edition – Chapter 8

Textbook: Theory of Machines and Mechanisms

Chapter Title: Helical Gears, Bevel Gears, Worms and Worm Gears

Therefore, the pitch radius of the pinion is

$$R_2 = 0.737\,6R_3.$$

This, along with the given shaft center distance, $R_2 + R_3 = 8.63$ in., gives the pitch radius of the pinion $R_2 = 3.663$ in. and the pitch radius of the gear $R_3 = 4.967$ in. Choosing a normal diametral pitch of $P_n = 6$ teeth/in., the numbers of teeth on the pinion and the gear, respectively, are

$$N_2 = 2P_nR_2 \cos\psi_2 = 2(6\ \text{teeth/in})(3.663\ \text{in})\cos 35° = 36\ \text{teeth} \qquad \textit{Ans.}$$

and

$$N_3 = 2P_nR_3 \cos\psi_3 = 2(6\ \text{teeth/in})(4.967\ \text{in})\cos 25° = 54\ \text{teeth}. \qquad \textit{Ans.}$$

8.8 STRAIGHT-TOOTH BEVEL GEARS

When rotational motion is transmitted between shafts whose axes intersect, some form of bevel gears is usually used. Bevel gears have pitch surfaces that are cones, with their cone axes matching the two shaft rotation axes, as illustrated in Fig. 8.8. The gears are mounted so that the apexes of the two pitch cones are coincident with the point of intersection of the shaft axes. These pitch cones roll together without slipping.

Although bevel gears are often made for an angle of 90° between the shafts, they can be designed for almost any angle. When the shaft intersection angle is other than 90°, the gears are called *angular bevel gears*. For the special case where the shaft intersection angle is 90° and both gears are of equal size, such bevel gears are called *miter gears*. A pair of miter gears is illustrated in Fig. 8.9.

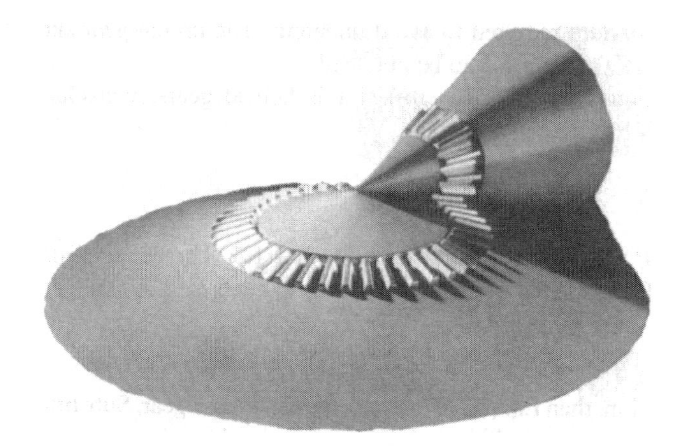

Figure 8.8 The pitch surfaces of bevel gears are cones that have only rolling contact. (Courtesy of Gleason Works, Rochester, NY.)

Figure 8.9 A pair of miter gears in mesh. (Courtesy of Gleason Works, Rochester, NY.)

For straight-tooth bevel gears, the true shape of a tooth is obtained by taking a spherical section through the tooth, where the center of the sphere is at the common apex, as illustrated in Fig. 8.8. As the radius of the sphere increases, the same number of teeth is projected onto a larger surface; therefore, the size of the teeth increases as larger spherical sections are taken. We have seen that the action and contact conditions for spur gear teeth may be viewed in a plane taken at right angles to the axes of the spur gears. For bevel gear teeth, the action and contact conditions should properly be viewed on a spherical surface (instead of a plane). We can even think of spur gears as a special case of bevel gears in which the spherical radius is infinite, thus producing a plane surface on which the tooth action is viewed. Figure 8.10 is typical of many straight-tooth bevel gear sets.

It is standard practice to specify the pitch diameter of a bevel gear at the large end of the teeth. In Fig. 8.11, the pitch cones of a pair of bevel gears are drawn and the pitch radii are given as R_2 and R_3, respectively, for the pinion and the gear. The cone angles γ_2 and γ_3 are defined as the pitch angles, and their sum is equal to the shaft intersection angle Σ, that is

$$\Sigma = \gamma_2 + \gamma_3.$$

The first-order kinematic coefficient, the angular velocity ratio between the shafts, is obtained in the same manner as for spur gears and is

$$\left|\theta'_{3/2}\right| = \left|\frac{\omega_3}{\omega_2}\right| = \frac{R_2}{R_3} = \frac{N_2}{N_3}. \tag{8.12}$$

In the kinematic design of bevel gears, the tooth numbers of the two gears and the shaft angle are usually given, and the corresponding pitch angles are to be determined. Although

Figure 8.10 A pair of straight-tooth bevel gears. (Courtesy of Gleason Works, Rochester, NY.)

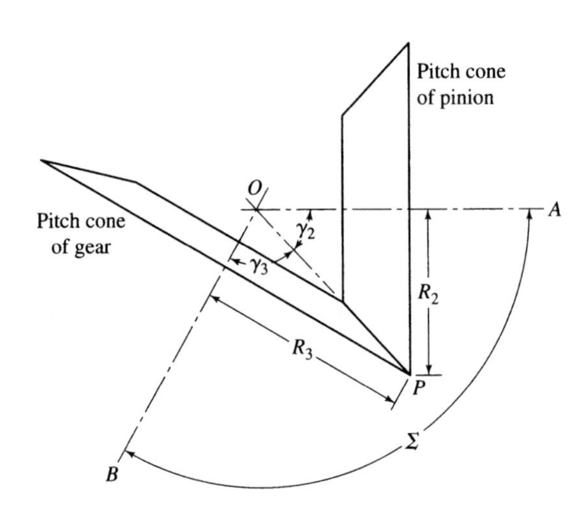

Figure 8.11 Pitch cones of bevel gears.

they can be determined graphically, the analytical approach gives more exact values. From Fig. 8.11 the distance OP may be written as

$$OP = \frac{R_2}{\sin \gamma_2} = \frac{R_3}{\sin \gamma_3}, \tag{a}$$

so that

$$\sin \gamma_2 = \frac{R_2}{R_3} \sin \gamma_3 = \frac{R_2}{R_3} \sin(\Sigma - \gamma_2) \tag{b}$$

or

$$\sin \gamma_2 = \frac{R_2}{R_3} (\sin \Sigma \cos \gamma_2 - \cos \Sigma \sin \gamma_2). \tag{c}$$

Dividing both sides of this equation by $\cos \gamma_2$ and rearranging gives

$$\tan \gamma_2 = \frac{R_2}{R_3} (\sin \Sigma - \cos \Sigma \tan \gamma_2).$$

Then, rearranging this equation gives

$$\tan \gamma_2 = \frac{\sin \Sigma}{(R_3/R_2) + \cos \Sigma} = \frac{\sin \Sigma}{(N_3/N_2) + \cos \Sigma}. \tag{8.13}$$

Similarly,

$$\tan \gamma_3 = \frac{\sin \Sigma}{(N_2/N_3) + \cos \Sigma}. \tag{8.14}$$

For a shaft angle of $\Sigma = 90°$, the above expressions reduce to

$$\tan \gamma_2 = \frac{N_2}{N_3} \tag{8.15}$$

and

$$\tan \gamma_3 = \frac{N_3}{N_2}. \tag{8.16}$$

The projection of bevel gear teeth onto the surface of a sphere would indeed be a difficult and time-consuming task. Fortunately, an approximation that reduces the problem to that of ordinary spur gears is common. This approximation is called *Tredgold's approximation* and, as long as the gear has eight or more teeth, it is accurate enough for practical purposes. It is in almost universal use, and the terminology of bevel gear teeth has evolved around it.

In Tredgold's method, a *back cone* is formed of elements that are perpendicular to the elements of the pitch cone at the large end of the teeth, as illustrated in Fig. 8.12. The length of a back-cone element is called the back-cone radius. Now an equivalent spur gear is constructed whose pitch radius R_e is equal to the back-cone radius. Thus, from a pair of bevel gears, using Tredgold's approximation, we can obtain a pair of equivalent spur

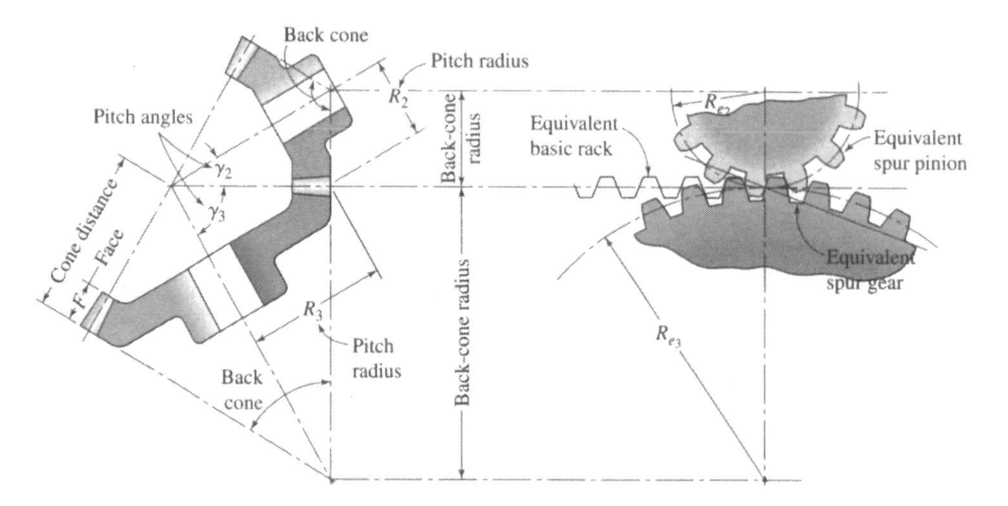

Figure 8.12 Tredgold's approximation.

gears that are then used to define the tooth profiles. They can also be used to determine the tooth action and the contact conditions, just as for ordinary spur gears, and the results will correspond closely to those for the bevel gears.

From the geometry of Fig. 8.12, the equivalent pitch radii are

$$R_{e2} = \frac{R_2}{\cos \gamma_2}, \quad R_{e3} = \frac{R_3}{\cos \gamma_3}. \tag{8.17}$$

The number of teeth on each of the equivalent spur gears is

$$N_e = \frac{2\pi R_e}{p}, \tag{8.18}$$

where p is the circular pitch of the bevel gear measured at the large end of the teeth. Usually the equivalent spur gears will *not* have integral numbers of teeth.

8.9 TOOTH PROPORTIONS FOR BEVEL GEARS

Practically all straight-tooth bevel gears manufactured today use a 20° pressure angle. It is not necessary to use an interchangeable tooth form because bevel gears cannot be interchanged. For this reason, the long-and-short-addendum system, described in Section 7.11, is used. The proportions are tabulated in Table 8.2.

Bevel gears are usually mounted on the outboard side of the bearings because the shaft axes intersect, which means that the effect of shaft deflection is to pull the small end of the teeth away from mesh, causing the larger end to take more of the load. Thus, the load across the tooth is variable; for this reason, it is desirable to design a fairly short tooth. As indicated in Table 8.2, the face width is usually limited to about one third of the cone distance. We note also that a short face width simplifies the tooling problems in cutting bevel gear teeth.

action between these gears is a combination of rolling and sliding along a straight line and has much in common with that of worm gears (see Section 8.13).

8.13 WORMS AND WORM GEARS

A *worm* is a machine member having a screw-like thread, and worm teeth are frequently spoken of as threads. A worm meshes with a conjugate gear-like member called a *worm wheel* or a *worm gear*. Figure 8.20 illustrates a worm and worm gear in an application. These gears are used with nonintersecting shafts that are usually at a shaft angle of 90°, but there is no reason why shaft angles other than 90° cannot be used if a design demands it.

Worms in common use have from one to four teeth and are said to be *single-threaded*, *double-threaded*, and so on. As we will see, there is no definite relation between the number of teeth and the pitch diameter of a worm. The number of teeth on a worm gear is usually much higher and, therefore, the angular velocity of the worm gear is usually much lower than that of the worm. In fact, often, one primary application for a worm and worm gear is to obtain a very large angular velocity reduction, that is, a very low first-order kinematic coefficient. In keeping with this low velocity ratio, the worm gear is usually the driven member of the pair and the worm is usually the driving member.

A worm gear, unlike a spur or helical gear, has a face that is made concave so that it partially wraps around, or envelops, the worm, as illustrated in Fig. 8.21. Worms are sometimes designed with a cylindrical pitch surface or they may have an hourglass shape, such that the worm also wraps around or partially encloses the worm gear. If an enveloping worm gear is mated with a cylindrical worm, the set is said to be *single enveloping*. When the worm is hourglass shaped, the worm and worm gearset is said to be *double enveloping* because each member partially wraps around the other; such a worm is sometimes called a *Hindley worm*. The nomenclature of a single-enveloping worm and worm gearset is illustrated in Fig. 8.21.

A worm and worm gear combination is similar to a pair of mating crossed-helical gears except that the worm gear partially envelops the worm. For this reason, they have line contact instead of the point contact found in crossed-helical gears and are thus able to transmit more power. When a double-enveloping worm and worm gearset is used, even more power can be transmitted, at least in theory, because contact is distributed over an area on both tooth surfaces.

Figure 8.20 A single-enveloping worm and worm gear set. (Courtesy of Gleason Works, Rochester, NY.)

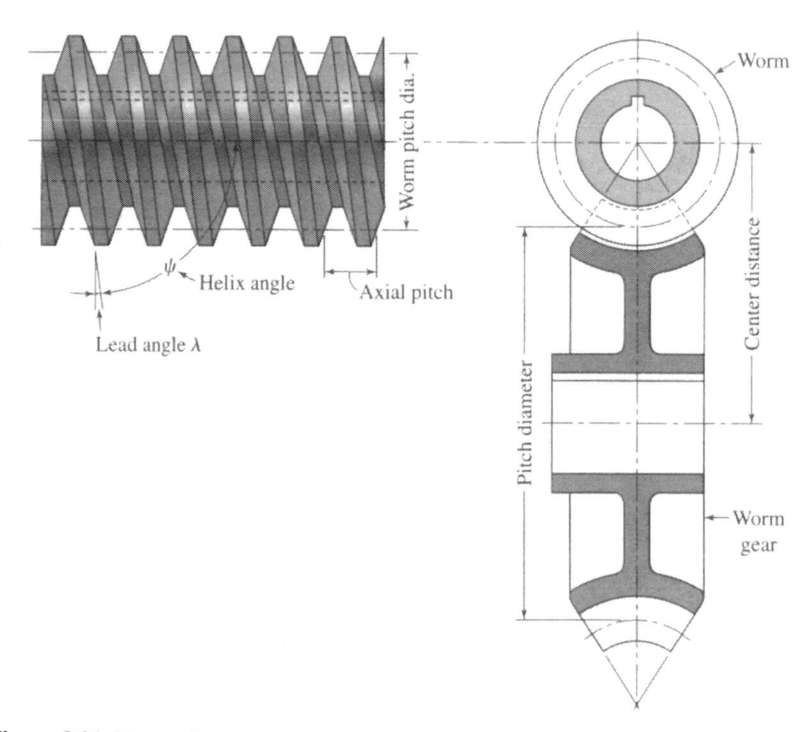

Figure 8.21 Nomenclature of a single-enveloping worm and worm gearset.

In a single-enveloping worm and worm gearset it makes no difference whether the worm rotates on its own axis and drives the gear by a screwing action or whether the worm is translating along its axis and drives the worm gear through rack action. The resulting motion and contact are the same. For this reason, a single-enveloping worm need not be accurately mounted along its axis. However, the worm gear should be accurately mounted along its rotation axis; otherwise, its pitch surface is not properly aligned with the worm axis. In a double-enveloping worm and worm gearset, both members are throated and therefore both must be accurately mounted in all directions to obtain correct contact.

A mating worm and worm gear with a 90° shaft angle have the same hand of helix, but the helix angles are usually very different. The helix angle on the worm is usually quite large (at least for one or two teeth) and quite small on the worm gear. On the worm, the *lead angle* is the complement of the helix angle, as illustrated in Fig. 8.21. Because of this, it is customary to specify the lead angle for the worm and specify the helix angle for the worm gear. This is convenient because these two angles are equal for a 90° shaft angle.

In specifying the pitch of a worm and worm gearset, it is usual to specify the axial pitch of the worm and the circular pitch of the worm gear. These are equal if the shaft angle is 90°. It is common to employ even fractions, such as $1/4$, $3/8$, $1/2$, $3/4$, 1, and $1\,1/4$ in/tooth, for the circular pitch of the worm gear; there is no reason, however, why the AGMA standard diametral pitches used for spur gears (Table 7.1) should not also be used for worm gears.

The pitch radius of a worm gear is determined in the same manner as that of a spur gear, that is,

$$R_3 = \frac{N_3 p}{2\pi}, \tag{8.19}$$

where all values are defined in the same manner as for spur gears, but refer to the parameters of the worm gear.

The pitch radius of the worm may have any value, but it should be the same as that of the hob used to cut the worm gear teeth. The relation between the pitch radius of the worm and the center distance, as recommended by AGMA, is

$$R_2 = \frac{(R_2 + R_3)^{0.875}}{4.4}, \tag{8.20}$$

where the quantity (R_2+R_3) is the center distance in inches. This equation gives proportions that result in good power capacity. The AGMA standard also states that the denominator of Eq. (8.20) may vary from 3.4 to 6.0 without appreciably affecting the power capacity. Equation (8.20) is not required, however; other proportions will also serve well and, in fact, power capacity may not always be the primary consideration. However, there are a lot of variables in worm gear design, and the equation is helpful in obtaining trial dimensions.

The *lead* of a worm has the same meaning as for a screw thread and is the axial distance through which a point on the helix will move when the worm is turned through one revolution. Thus, in equation form, the lead of the worm is given by

$$l = p_x N_2, \tag{8.21}$$

where p_x is the axial pitch and N_2 is the number of teeth (threads) on the worm. The lead and the *lead angle* are related as follows,

$$\lambda = \tan^{-1}\left(\frac{l}{2\pi R_2}\right), \tag{8.22}$$

where λ is the lead angle, as illustrated in Fig. 8.21.

The teeth on a worm are usually cut in a milling machine or on a lathe. Worm gear teeth are most often produced by hobbing. Except for clearance at the top of the hob teeth, the worm should be an exact duplicate of the hob to obtain conjugate action. This also means that, where possible, the worm should be designed using the dimensions of existing hobs.

The pressure angles used on worms and worm gearsets vary widely and should depend approximately on the value of the lead angle. Good tooth action is obtained if the pressure angle is made large enough to eliminate undercutting of the worm gear tooth on the side at which the contact ends. Recommended values are given in Table 8.3.

A satisfactory tooth depth that has about the right relation to the lead angle is obtained by making the depth a proportion of the normal circular pitch. Using an addendum of $1/P = p_n/\pi$, as for full-depth spur gears, we obtain the following proportions for worms

TABLE 8.3 Recommended pressure angles for worm and worm gear sets

Lead angle λ	Pressure angle ϕ
0°–16°	14½°
16°–25°	20°
25°–35°	25°
35°–45°	30°

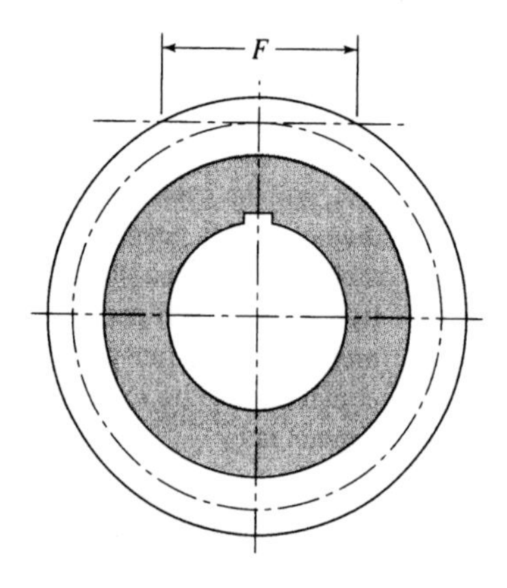

Figure 8.22 Face width of a worm gear.

and worm gears.

$$\text{Addendum} = 1.000/\text{P} = 0.318\,3p_\text{n}$$
$$\text{Whole depth} = 2.000/\text{P} = 0.636\,6p_\text{n}$$
$$\text{Clearance} = 0.157/\text{P} = 0.050\,7p_\text{n}$$

The face width of the worm gear should be obtained as illustrated in Fig. 8.22. This makes the face of the worm gear equal to the length of a tangent to the worm pitch circle between its points of intersection with the addendum circle.

8.14 NOTES

1. The equation of an ellipse with its center at the origin of an xy coordinate system with a and b as its semimajor and semiminor axes, respectively, is

$$\frac{x^2}{a^2} + \frac{y^2}{b^2} = 1. \qquad (a)$$

Also, the formula for radius of curvature is

$$\rho = \frac{\left[1 + (dy/dx)^2\right]^{3/2}}{d^2y/dx^2}. \qquad (b)$$

Using these two equations, it is not difficult to find the radius of curvature corresponding to $x = 0, y = b$. The result is

$$\rho = a^2/b. \tag{c}$$

Then, referring to Fig. 8.3, we substitute $a = R/\cos\psi$ and $b = R$ into Eq. (c) and obtain Eq. (8.5).

PROBLEMS

8.1 A pair of parallel-axis helical gears has 14½° normal pressure angle, diametral pitch of 6 teeth/in, and 45° helix angle. The pinion has 15 teeth, and the gear has 24 teeth. Calculate the transverse and normal circular pitch, the normal diametral pitch, the pitch radii, and the equivalent tooth numbers.

8.2 A set of parallel-axis helical gears are cut with a 20° normal pressure angle and a 30° helix angle. They have diametral pitch of 16 teeth/in and have 16 and 40 teeth, respectively. Find the transverse pressure angle, the normal circular pitch, the axial pitch, and the pitch radii of the equivalent spur gears.

8.3 A parallel-axis helical gearset is made with a 20° transverse pressure angle and a 35° helix angle. The gears have diametral pitch of 10 teeth/in and have 15 and 25 teeth, respectively. If the face width is 0.75 in, calculate the base helix angle and the axial contact ratio.

8.4 A set of helical gears is to be cut for parallel shafts whose center distance is to be about 3.5 in to give a velocity ratio of approximately 1.8. The gears are to be cut with a standard 20° pressure angle hob whose diametral pitch is 8 teeth/in. Using a helix angle of 30°, determine the transverse values of the diametral and circular pitches and the tooth numbers, pitch radii, and center distance.

8.5 A 16-tooth helical pinion is to run at 1 800 rev/min and drive a helical gear on a parallel shaft at 400 rev/min. The centers of the shafts are to be spaced 11.0 in apart. Using a helix angle of 23° and a pressure angle of 20°, determine the values for the tooth numbers, pitch radii, normal circular and diametral pitch, and face width.

8.6 The catalog description of a pair of helical gears is as follows: 14½° normal pressure angle, 45° helix angle, diametral pitch of 8 teeth/in, 1.0-in face width, and normal diametral pitch of 11.31 teeth/in. The pinion has 12 teeth and a 1.500-in pitch diameter, and the gear has 32 teeth and a 4.000-in pitch diameter. Both gears have full-depth teeth, and they may be purchased either right or left handed. If a right-hand pinion and left-hand gear are placed in mesh, find the transverse contact ratio, the normal contact ratio, the axial contact ratio, and the total contact ratio.

8.7 In a medium-size truck transmission a 22 tooth clutch-stem gear meshes continuously with a 41-tooth countershaft gear. The data indicate normal diametral pitch of 7.6 teeth/in, 18½° normal pressure angle, 23½° helix angle, and a 1.12-in face width. The clutch-stem gear is cut with a left-hand helix, and the countershaft gear is cut with a right-hand helix. Determine the normal and total contact ratio if the teeth are cut full depth with respect to the normal diametral pitch.

8.8 A helical pinion is right handed, has 12 teeth, has a 60° helix angle, and is to drive another gear at a velocity ratio of 3.0. The shafts are at a 90° angle, and the normal diametral pitch of the gears is 8 teeth/in. Find the helix angle and the number of teeth on the mating gear. What is the shaft center distance?

8.9 A right-hand helical pinion is to drive a gear at a shaft angle of 90°. The pinion has 6 teeth and a 75° helix angle and is to drive the gear at a velocity ratio of 6.5. The normal diametral pitch of the gear is 12 teeth/in. Calculate the helix angle and the number of teeth on the mating gear. Also determine the pitch radius of each gear.

8.10 Gear 2 in Fig. P8.10 is to rotate clockwise and drive gear 3 counterclockwise at a velocity ratio of 2. Use a normal diametral pitch of 5 teeth/in, a shaft center distance of about 10 in, and the same helix angle for

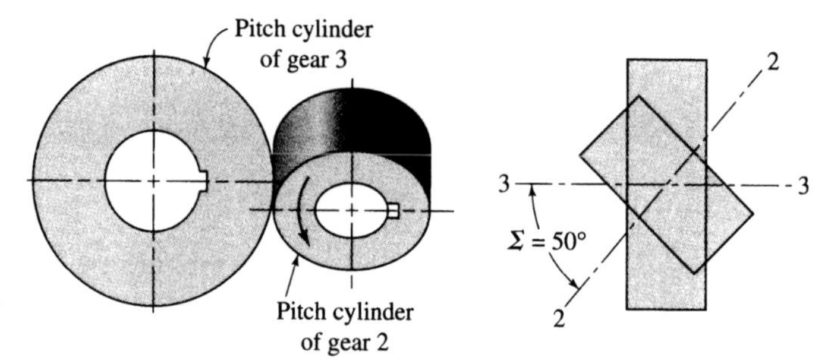

Figure P8.10

both gears. Find the tooth numbers, the helix angles, and the exact shaft center distance.

8.11 A pair of straight-tooth bevel gears is to be manufactured for a shaft angle of 90°. If the driver is to have 18 teeth and the velocity ratio is to be 3:1, what are the pitch angles?

8.12 A pair of straight-tooth bevel gears has a velocity ratio of 1.5 and a shaft angle of 75°. What are the pitch angles?

8.13 A pair of straight-tooth bevel gears is to be mounted at a shaft angle of 120°. The pinion and gear are to have 15 and 33 teeth, respectively. What are the pitch angles?

8.14 A pair of straight-tooth bevel gears with diametral pitch of 2 teeth/in have 19 and 28 teeth, respectively. The shaft angle is 90°. Determine the pitch diameters, pitch angles, addendum, dedendum, face width, and pitch diameters of the equivalent spur gears.

8.15 A pair of straight-tooth bevel gears with diametral pitch of 8 teeth/in have 17 and 28 teeth, respectively, and a shaft angle of 105°. For each gear, calculate the pitch radius, pitch angle, addendum, dedendum, face width, and equivalent tooth numbers. Make a sketch of the two gears in mesh. Use standard tooth proportions as for a 90° shaft angle.

8.16 A worm having 4 teeth and a lead of 1.0 in drives a worm gear at a velocity ratio of 7.5. Determine the pitch diameters of the worm and worm gear for a center distance of 1.75 in.

8.17 Specify a suitable worm and worm gear combination for a velocity ratio of 60 and a center distance of 6.50 in. Use an axial pitch of 0.500 in/tooth.

8.18 A triple-threaded worm drives a worm gear having 40 teeth. The axial pitch is 1.25 in, and the pitch diameter of the worm is 1.75 in. Calculate the lead and lead angle of the worm. Find the helix angle and pitch diameter of the worm gear.

8.19 A triple-threaded worm with a lead angle of 20° and an axial pitch of 0.400 in/tooth drives a worm gear with a velocity reduction of 15 to 1. Determine the following for the worm gear: (*a*) the number of teeth, (*b*) the pitch radius, and (*c*) the helix angle. (*d*) Determine the pitch radius of the worm. (*e*) Compute the center distance.

Uicker et al., 4th Edition – Chapter 9

Textbook: Theory of Machines and Mechanisms
Chapter Title: Mechanism Trains

9 Mechanism Trains

Mechanisms arranged in combinations so that the driven member of one mechanism is the driver for another mechanism are called *mechanism trains*. With certain exceptions, to be explored here, the analysis of such trains can proceed in serial fashion using the methods developed in the previous chapters.

9.1 PARALLEL-AXIS GEAR TRAINS

In Chapter 3, we learned that the first-order *kinematic coefficient* is the term used to describe the ratio of the angular velocity of the driven member to that of the driving member. Thus, for example, in a four-bar linkage with link 2 as the driving or input member and link 4 as the driven or output member, we have

$$\theta'_{42} = \frac{\omega_4}{\omega_2} = \frac{d\theta_4/dt}{d\theta_2/dt} = \frac{d\theta_4}{d\theta_2} \qquad (a)$$

where it is noted that, as in Chapter 5, we adopt the second subscript to explicitly indicate the number of the driving or input member. This second subscript is important in Chapter 9 because many mechanism trains have more than one degree of freedom.

In this section, where we deal with serially connected gear trains, we prefer to write Eq. (*a*) as

$$\theta'_{LF} = \frac{\omega_L}{\omega_F} = \frac{d\theta_L/dt}{d\theta_F/dt} = \frac{d\theta_L}{d\theta_F}, \qquad (9.1)$$

where ω_L is the angular velocity of the *last* gear and ω_F is the angular velocity of the *first* gear in the train because, usually, the last gear is the output and is the driven gear and the first is the input and driving gear.

The term θ'_{LF} in Eq. (9.1) is the first-order kinematic coefficient, called the *speed ratio* by some or the *train value* by others. Equation (9.1) is often written in the more convenient form:

$$\omega_L = \theta'_{LF}\omega_F. \tag{9.2}$$

Next, we consider a pinion 2 driving a gear 3. The speed of the driven gear is

$$\omega_3 = \pm\frac{R_2}{R_3}\omega_2 = \pm\frac{N_2}{N_3}\omega_2, \tag{b}$$

where, for each gear, R is the radius of the pitch circle, N is the number of teeth, and ω is either the angular velocity or the angular displacement completed during a chosen time interval.

For parallel-shaft gearing, the directions can be kept track of by following the vector sense, that is, by specifying that angular velocity is positive when counterclockwise as seen from a chosen side. For parallel-shaft gearing we shall use the following sign convention: If the last gear of a parallel-shaft gear train rotates in the same sense as the first gear, then θ'_{LF} is positive; if the last gear rotates in the opposite sense to the first gear, then θ'_{LF} is negative. This sign convention approach is not as easy, however, when the gear shafts are not parallel, as in bevel, crossed-helical, or worm gearing. In such cases, it is often simpler to track the directions by visually inspecting a sketch of the train.

The gear train illustrated in Fig. 9.1 is made up of five gears in series. Applying Eq. (*b*) three times, we find the speed of gear 6 to be

$$\omega_6 = -\frac{N_5}{N_6}\frac{N_4}{N_5}\frac{N_2}{N_3}\omega_2.$$

Here we note that gear 5 is an idler; that is, its tooth numbers cancel in Eq. (*c*) and hence the only purpose served by gear 5 is to change the direction of rotation of gear 6. We further note that gears 5, 4, and 2 are drivers, whereas gears 6, 5, and 3 are driven members. Thus, Eq. (9.1) can also be written

$$\theta'_{LF} = \pm\frac{\text{product of driving tooth numbers}}{\text{product of driven tooth numbers}}. \tag{9.3}$$

Note also that because they are proportional, pitch radii can be used in Eq. (9.3) just as well as tooth numbers.

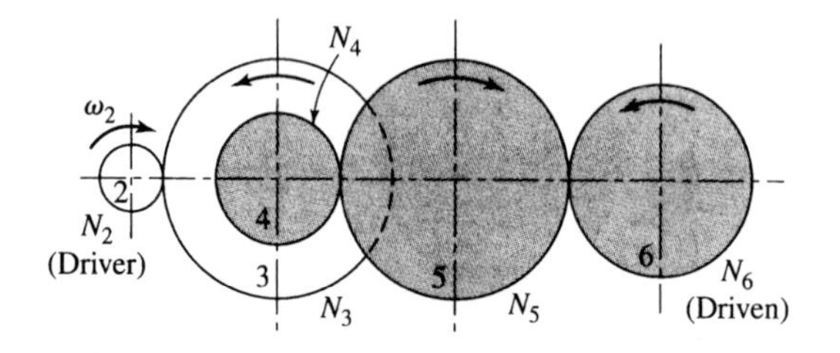

Figure 9.1

9.2 EXAMPLES OF GEAR TRAINS

In speaking of gear trains it is convenient to describe a train having only one gear on each axis as a *simple* gear train. A *compound* gear train then is one that has two or more gears on one or more axes, such as the train illustrated in Fig. 9.1. Another example of a compound gear train is illustrated in Fig. 9.2. Figure 9.2 illustrates a transmission for a small- or medium-size truck, which has four speeds forward and one in reverse.

The compound gear train illustrated in Fig. 9.3 is composed of bevel, helical, and spur gears. The helical gears are crossed, and so their direction of rotation depends upon their hand.

A *reverted* gear train is one in which the first and last gears have collinear axes of rotation, such as the one illustrated in Fig. 9.4. This produces a compact arrangement and is

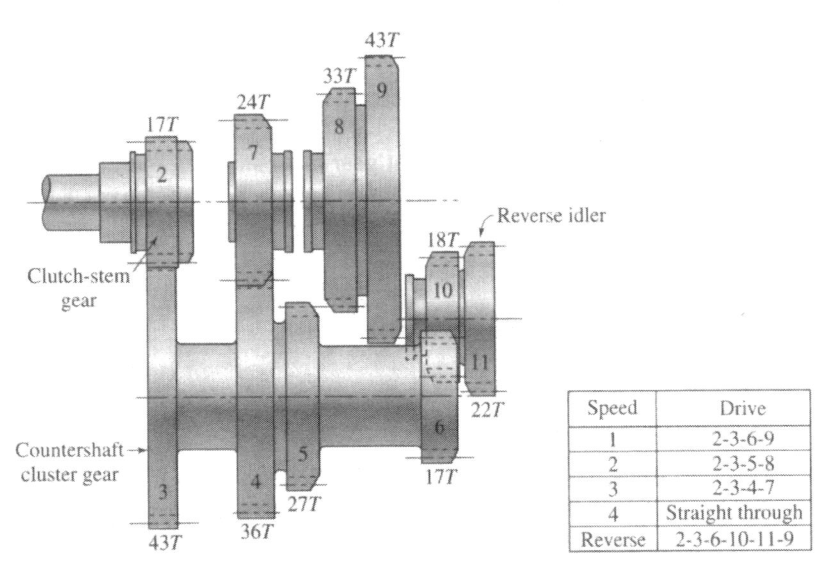

Speed	Drive
1	2-3-6-9
2	2-3-5-8
3	2-3-4-7
4	Straight through
Reverse	2-3-6-10-11-9

Figure 9.2 A truck transmission with gears having diametral pitch of 7 teeth/in. and pressure angle of 22.5°.

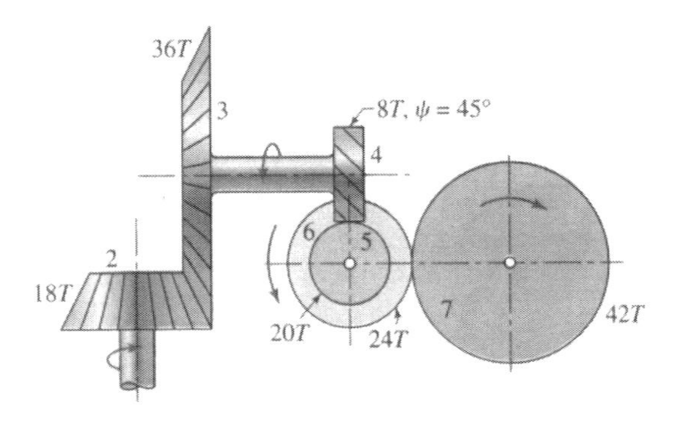

Figure 9.3 A gear train composed of bevel, crossed-helical, and spur gears.

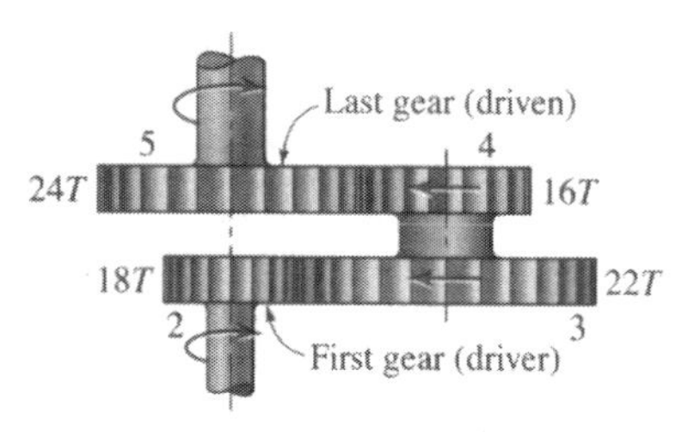

Figure 9.4 A reverted gear train.

used in such applications as speed reducers, clocks (to connect the hour hand to the minute hand), and machine tools. As an exercise, it is suggested that you seek out a suitable set of diametral pitches for each pair of gears illustrated in Fig. 9.4 so that the first and last gears have the same axis of rotation with all gears properly engaged.

9.3 DETERMINING TOOTH NUMBERS

When notable power is transmitted through a speed reduction unit, the speed ratio of the last pair of meshing gears is usually chosen larger than that of the first gear pair because the torque is greater at the low-speed end. In a given amount of space, more teeth can be used on gears of lesser pitch; hence, a greater speed reduction can be obtained at the high-speed end.

Without examining the problem of tooth strength, suppose we wish to use two pairs of gears in a train to obtain an overall kinematic coefficient of $\theta'_{LF} = 1/12$. Let us also impose the restriction that the tooth numbers must not be less than 15 and that the reduction in the first pair of gears should be about twice that of the second pair. This means that

$$\theta'_{52} = \frac{N_4}{N_5}\frac{N_2}{N_3} = \frac{1}{12}, \tag{a}$$

where N_2/N_3 is the kinematic coefficient of the first gear pair and N_4/N_5 is that of the second pair. Because the kinematic coefficient of the first pair should be half that of the second, Eq. (a) can be written as

$$\left(\frac{N_4}{N_5}\right)\left(\frac{N_4}{2N_5}\right) = \frac{1}{12} \tag{b}$$

or

$$\frac{N_4}{N_5} = \sqrt{\frac{1}{6}} = 0.408\,248 \tag{c}$$

to six decimal places. The following tooth numbers are seen to be close:

$$\frac{15}{37}\ \frac{16}{39}\ \frac{18}{44}\ \frac{20}{49}\ \frac{22}{54}\ \frac{24}{59}.$$

Of these, $N_4/N_5 = 20/49$ is the closest approximation, but note that

$$\theta'_{52} = \left(\frac{N_4}{N_5}\right)\left(\frac{N_2}{N_3}\right) = \left(\frac{20}{49}\right)\left(\frac{20}{98}\right) = \frac{400}{4802} = \frac{1}{12.005},$$

which is very close to $1/12$. On the other hand, the choice of $N_4/N_5 = 18/44$ gives exactly

$$\theta'_{52} = \left(\frac{N_4}{N_5}\right)\left(\frac{N_2}{N_3}\right) = \left(\frac{18}{44}\right)\left(\frac{22}{108}\right) = \frac{396}{4752} = \frac{1}{12}.$$

In this case, the reduction in the first gear pair is not exactly twice the reduction in the second gear pair. However, this consideration is usually of only minor importance.

The problem of specifying tooth numbers and the number of pairs of gears to give a kinematic coefficient with a specified degree of accuracy has interested many people. Consider, for instance, the problem of specifying a set of gears to have a kinematic coefficient of $\theta'_{LF} = \pi/10$ accurate to eight decimal places, while we know that π is an irrational number and cannot be expressed as a ratio of integers.

9.4 EPICYCLIC GEAR TRAINS

Figure 9.5 illustrates an elementary epicyclic gear train together with its schematic diagram as suggested by Lévai.* The train consists of a *central gear* 2 and an *epicyclic gear* 4, which produces epicyclic motion for its points by rolling around the periphery of the central gear. A *crank arm* 3 contains the bearings for the epicyclic gear to maintain the gears in mesh.

Epicyclic trains are also called *planetary* or *sun-and-planet* gear trains. In this nomenclature, gear 2 of Fig. 9.5 is called the *sun gear*, gear 4 is called the *planet gear*, and crank 3 is called the *planet carrier*. Figure 9.6 illustrates the train of Fig. 9.5 with two redundant planet gears added. This produces better force balance; also, adding more planet gears

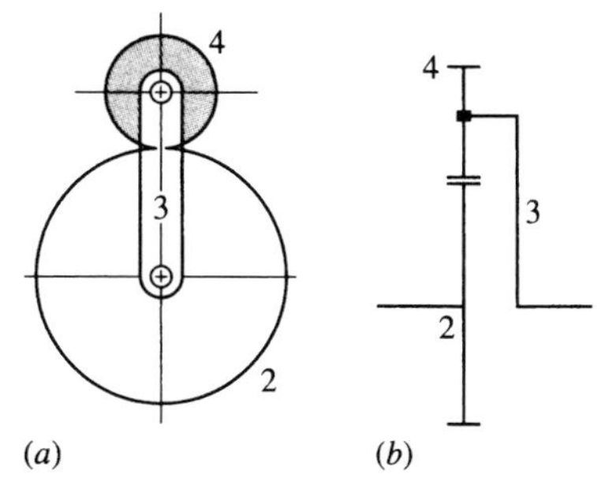

Figure 9.5 (*a*) The elementary epicyclic gear train; (*b*) its schematic diagram.

---

* Literature devoted to epicyclic gear trains is rather scarce; however, see G. R. Pennock and J. J. Alwerdt, "Duality Between the Kinematics of Gear Trains and the Statics of Beam Systems," *Mechanism and Machine Theory*, 42, no. 11 (2007):1527–46. For a comprehensive study in the English language, see Z. L. Lévai, *Theory of Epicyclic Gears and Epicyclic Change-Speed Gears*, Budapest: Technical University of Building, Civil, and Transport Engineering, 1966. This book lists 104 references.

Lieu and Sorby, 2nd Edition – Chapter 15

Textbook: Visualization, Modelling, and Graphics for Engineering Design
Chapter Title. Working in a Team Environment

Chapter 15

Working in a Team Environment

Chapter Introduction

Objectives

After completing this chapter, you should be able to

- Understand the benefits of working in a team
- Organize team projects and team member responsibilities
- Communicate and work in a team
- Assess the strengths and weaknesses of your team
- Apply strategies for improving team performance
- Solve problems with team members
- Work in a team effectively

15.01 Introduction

Assume your instructor has assigned a "design" project for the semester. Moreover, your instructor has assigned everyone in the class to work in teams. You probably have worked on team projects in the past. Perhaps your experiences were positive; however, you may have been frustrated at times because not everyone on the team put forth the same amount of effort in completing the project. You may be a bit skeptical about the viability and value of working on a team.

In real life, most engineering projects are accomplished by teams. Think of the space missions. The astronauts would not be able to travel into space without the effort of thousands of engineers, technicians, ground crew, and other support staff. As another example, think of the last movie you watched. The movie "team" included the director, producer, cast, camera operator, costumer, sound effects crew, and many other people who pulled together to develop the successful final product. Automobiles, computers, iPods, cell phones, and countless other everyday products were designed, produced, packaged, marketed, and distributed through a team effort. So if you really want to have a successful career as an engineer or a technologist, you have no choice but to learn to work in a team environment. In fact, you may even find that you enjoy the team atmosphere as you begin to appreciate the way individuals pull together and complement one another to create products that are innovative and timely.

While working on team projects in high school or college classes, you may have experienced less than satisfactory results. In those often dysfunctional teams, task assignments may not have been distributed evenly or team members may not have delivered on promises to complete their tasks. While being on a dysfunctional team is an experience you would like to avoid, being a part of a good team is a rewarding experience. This chapter will outline reasons for working on a team, ways to get started on the road to a successful team, strategies for making your team effective and efficient, and tools for dealing with problems that will inevitably confront you and your team. As with most activities, being prepared for the challenges and opportunities, and having a plan for how to deal with them will improve your chances of success and lead to an enjoyable and productive team experience. Further, learning how to develop successful team environments will prepare you for a future technological career. As you read this chapter, pay particular attention to the sections on organization, time management, and communication, because they are the keys to a successful project—as an individual or as a team.

15.02 Why Work in a Group?

Many of the projects suggested in these chapters are difficult, if not impossible, to do by yourself. Amost all of them can be enhanced by the interpersonal interaction that comes from being part of a good team and the environment that comes with being a member of a team. Obviously, more team members means more bodies at work (and the likelihood of a better final product); but a team is more than a group of people who divide up a task into manageable parts. The diversity of different life and professional experiences of team members leads to a larger group of ideas and a variety of approaches in solving problems. Team discussions can generate ideas, expand options, and improve the final product. Even questions from naysayers are helpful—both in clarifying ideas and identifying fatal flaws—before a great deal of time and effort is spent on an idea that ultimately leads to a dead end.

In addition, employers value employees (whether for summer jobs, internships, co-op positions, or full-time jobs after graduation) who can work as a member of a team and who are team players. An engineer rarely works alone. Most projects require a range of training and skills beyond what the most skilled engineer can be expected to do. Even if a project is small enough for a single person to design it from start to finish, engineers rely on information collected or generated by others and need others to fabricate or construct what they design; often engineers rely on others to make sure the project has been built correctly and to identify problems.

Working on a team does not involve personal relationships; a team is based on professional relationships that require you to respect and value the skill sets that other members of the team bring to the project. A team is a group of colleagues who work together to complete an assigned task.

15.03 What Does It Mean to Be a Team Player?

Being a good team member takes work. Most people are used to working on their own—making decisions, prioritizing tasks, and being accountable for their own work. Working with others requires a different approach than working alone. To be a successful part of a team, you need to consider several issues. You should be prepared *not* to be in charge of everything. For some people, this requires a great deal of effort; for other people, it is less taxing. At times, you will be the supervisor; other times, you will be supervised. You need to be flexible and understand that a team consisting only of leaders (or only of followers) is not likely to perform well.

Also be prepared to have some interesting (and some frustrating) encounters with your new work mates. Be prepared to exchange points of view and to learn from those around you. Everyone on a team is responsible for success and is accountable for failure.

Most importantly, prepare to learn how to be a team member. Share your strengths with the team and be willing to contribute. Remember, the combined efforts of all team members should yield a better outcome than the efforts of one individual. Learn new team skills and be adaptable.

Many teams have problems when everyone tries to be in charge or when no one tries to be in charge. The result can be the same: uneven distribution of work, incomplete work, missed deadlines, subpar performance, and frustration. Even though a team is a united effort, each individual is accountable for the overall performance of the team.

Individuals generally react differently in groups than they do on their own. If you miss deadlines or produce inferior work as an individual, you can expect to be held accountable if your work habits are the same when you are part of a team. Conversely, if you produce high-quality work on your own and do the same as part of a team, you will be rewarded accordingly. Remember that team members are accountable first for their individual performance and second for the group's performance. Keep everyone informed of your progress.

15.04 Differences between Teaming in the Classroom and Teaming in the Real World

Selecting personnel and identifying skills are the most important tasks of assembling a team to work on a real-world project. While it may be advantageous to pick people who have worked together previously and who have established a good working relationship, you need to make sure that all of the skills required for the completion of the project are represented by at least one person on the team. For example, if the team is designing a building, the team must have a member who understands, among other things, the:

- Design of a foundation.
- Design of the structure.

- Design of elevator and/or escalator systems.
- Design of air-conditioning systems.

Additional skills are likely to be on the list if the building is to be made of reinforced concrete, or if it is to be constructed in Alaska, California, or Louisiana.

In the real world and in the classroom, the goal is to complete a successful project on time and within budget. However, the skills and training of potential team members in the classroom are all virtually the same (unlike the real world). Furthermore, the *primary* goal in the classroom is for each member of the team to learn about each task required in the project. Whereas a mechanical engineer is not expected to teach other members of a building-design team how the air-conditioning system works or why the particular components were selected, each member of a classroom team is expected to explain her part of the team project. Unless the team members complete all of the tasks together, each member must teach the rest of the team what she did on her part of the project.

15.05 Team Roles

For your team to operate like a "well-oiled machine," you need to understand that the members must fill specific **team roles** (The roles that team members fill to ensure maximum effectiveness for a team.) , if effective collaborative work is to result. Typically, well-functioning teams have a leader, a timekeeper, and a note taker at a minimum. If there are additional team members, assigning someone to the role of devil's advocate is also a good idea. The **team leader** (The person who calls the meetings, sets the agenda, and maintains the focus of team meetings.) does just that—she leads. This does not mean the team leader dictates or makes all of the decisions for the group. The team leader sets the meeting time, sets the agenda for the meeting, and generally keeps the meeting moving. The team leader also makes sure the team stays on target and remains focused on the task at hand.

The **note taker** (The person who records the actions discussed and taken at team meetings and then prepares the formal written notes for the meeting.) keeps a written record of the team's progress. He or she records what tasks have been assigned to whom and records the expected completion dates of the tasks. The note taker is responsible for sending the minutes of the meeting to all team members. The minutes are a written record of what transpired during the meeting and serve as a reminder of who is responsible for completing what task(s).

The **timekeeper** (The person who keeps track of the meeting agenda, keeping the team on track to complete all necessary items within the allotted time frame.) makes sure the schedule is maintained and that meetings do not run over the allotted time. If meetings routinely last longer than planned, team members may skip them or resent coming to them —either of which leads to less productive team encounters.

Finally, the role of the **devil's advocate** (The team member who challenges ideas to ensure that all options are considered by the group.) is to challenge ideas without being too overbearing or unpleasant. The devil's advocate makes sure that all options are considered and that ideas are sound. However, a devil's advocate should not challenge ideas just for the sake of the challenge; doing this can annoy teammates and detract from the overall effectiveness of the team's operation.

Depending on your personality, you might be naturally inclined toward one role over another. For example, you may naturally be a critic who performs the role of devil's advocate very well. In the classroom setting, you should try out other team roles, so you can develop additional team skills. You may need to hone your note taking skills, and filling that role on the team may help your personal development. In classroom projects, team members can rotate roles so everyone has a chance to experience each role. By performing roles that are unfamiliar to you, you learn to appreciate people who work in these roles. Developing an appreciation for and respecting the skills of the other members of your team are the first steps toward your becoming an effective team member.

15.06 Characteristics of an Effective Team

Most successful teams either knowingly or unknowingly operate by certain ground rules that contribute to overall team productivity. Some of these ground rules are described in subsequent sections of this chapter, giving you the opportunity to learn the rules and adapt them to your particular setting and project.

15.06.01 Decisions Made by Consensus

For a diverse team, it will be nearly impossible to get 100 percent agreement on all of the decisions. Trying to achieve this unreachable goal will lead to frustration and poor productivity for the team. **consensus** (A process of decision making where an option is chosen that everyone supports.) means finding an option that all team members will support. It does not mean that all team members would select that option as their first choice, although some of them probably would. When making decisions by consensus, everyone on the team is invited to voice an opinion. Some people may be naturally shy and unwilling to speak up. The team leader should note when a team member has not voiced an opinion and invite the individual to speak up. Silence should not be interpreted as agreement—often it is not. Another important aspect in making team decisions is to consider the data carefully. Decisions made based on feelings, where data are ignored, are usually not optimal.

5.06.02 Everyone Participates

No member of a team should be allowed to sit back and watch others do the work. As mentioned previously, it is important for every member of the team to voice an opinion during meetings. It is just as important for every member of the team to participate in the work of the team. Tasks should be assigned to members based on talents/skills, and no one should be allowed to choose not to do something. The leader is responsible for making sure that every member participates equally in the work of the team. This does not mean that every task needs to be divided into equal parts, but it does mean the *overall* work should be divided evenly.

5.06.03 Professional Meetings

Team meetings should be productive and engaging. If they are ineffective and a waste of time, team members will likely skip them or not participate fully. This, in turn, will lead to poor-quality work from the team. Team meetings work best when a procedure has been established for the conduct of the meeting and an agenda has been created in advance. The agenda should be prepared by the team leader and e-mailed to participants in advance of team meetings. As a rule of thumb, the **agenda** (The list of topics for discussion/action at a team meeting.) should include

(1) a review of progress to date,

(2) a review and possible revision of the project schedule, and

(3) new task assignments for team members as needed.

This list is not exhaustive—your agenda will be dictated somewhat by the project you are working on. The timekeeper is responsible for making sure the team follows the agenda so that all items on the agenda are completed within the time allotted for the meeting. Punctuality at team meetings is a necessary ingredient for success. A person who shows up late for a meeting is not being fair to or respectful of her teammates. Attendees should do their best to be on time.

15.07 Project Organization—Defining Tasks and Deliverables

Your team was likely formed to work on a project, maybe even a design project. In a subsequent chapter, you will learn about the design process and its various stages. For now, understand that design is an iterative process that proceeds from stage to stage until completion. At some point in the process, you may have to return to an earlier stage to redo something the team thought was completed. Redoing earlier work is a normal part of the design process, especially in cases where you are trying something new and do not know if your idea or solution will actually work. During each stage, you meet as a team to review your progress and to determine what needs to be done next. In the early stages of the project, you probably made a list of all of the tasks that needed to be done. You should review this list at each meeting because the list will likely change as the project evolves.

Once you are sure you have a complete list of what needs to be done before the next meeting, it is time to make a list of tasks and assign responsibility to the person who will be completing each task. When reviewing the items on the list of tasks to accomplish, you need to determine which tasks depend on the outcome of other tasks. For example, if you are going to machine a part, you need to create a drawing of the part. To create the drawing, you may need to create a solid model of it in your CAD system. Thus, it would be unreasonable to expect someone to do the machining *before* the modeling work has been done—the task of machining depends on the outcome of the modeling task.

Another consideration when assigning tasks is to determine which tasks can be done by an individual and which require a group effort. If someone on your team has a difficult schedule to work around, you may want to assign individual tasks to that person most of the time (but not always) to accommodate their schedule. In addition to looking at the individual/group effort required of each task, you also should try to estimate the amount of time each task will require for completion. If one group member ends up completing a task that requires ten hours while two other members complete tasks that require one hour each, resentment is likely to build, thereby hindering group progress. However, as stated earlier in this chapter, division of labor for the project should be balanced overall. So if the person assigned the ten-hour task has been a slacker on previous assignments, perhaps that person should complete a significant task the next time one is assigned.

When assigning tasks, you should try to match the talents and abilities of each team member to the requirements of the task. However, you want to rotate duties so one person is not burdened with all of the writing or all of the modeling or all of the calculating (similar to the way the team roles are rotated so everyone has the opportunity to experience each role). As you are thinking about the assignment of tasks, ask yourself the following questions:

- Which team member is *best qualified* to do the task?
- Who is *able* to do the task in terms of either time or skill?
- Who is *willing* to do the task?

You need to make choices between assigning a task to a person who can accomplish it and assigning the task to the best person. The best outcome may not result from the best person being assigned to a task. If the person is overloaded as a result of the task assignment, she may not do a very good job or may not be able to complete the task. Balancing task assignments is key to producing the best possible project. A project may not be the best one a team can produce; rather, it is the best project a team can produce within the limits of available resources.

15.08 Time Management—Project Scheduling

Once you have organized your team and started work on your project, you should begin developing a plan for completing the project. Think of the plan as being dynamic, not static, since you are likely to be making changes to the plan as the project unfolds. Think of the initial plan as a flowchart of activities or a calendar of events that should include items such as who was assigned to work on each task and where each task fits in the overall project. When examining the various tasks that make up the final project, think about the interrelationships between tasks. What task precedes/follows each task? How does information flow from one task to another? In the previous example, modeling was the first task to be accomplished, which was followed by the creation of a drawing, culminating in a fabricated part. The task that precedes and follows each task is well-defined, but the method of communicating information between tasks may not be as straightforward. Ideally, this information flows seamlessly through the CAD software; however, you may need to run file translator routines to move information between tasks.

Perhaps the most important activity in project scheduling is to determine how each task fits within the overall project and when each task should be completed for your project to end successfully. Usually, when you are organizing your project, you can begin at the beginning or you can begin at the end. If you start at the beginning, the organization of the project can be done in a cyclic manner. As a team, ask the following questions:

1. What needs to be done first?
2. What do we need to know before we can do WhatNeedsToBeDoneFirst (WNTBDF)?
3. Now that we have a new WNTBDF, repeat Step 2.

Sometimes it is more efficient to begin by considering the *final* deliverable for the project (product, design, prototype, sketch, etc.). You then work backward through the process, identifying what needs to happen before a specific task can be accomplished. Another way of thinking about this is to consider all of the other tasks that must be completed before a specific task can be accomplished. When organizing your project from the beginning or the end, you also need to consider who will be completing each task—you cannot establish a timeline without considering the realities of everyone's schedule.

Sometimes the result of this activity is to discover that the timeline for your well-planned project does not match the deadlines established by your client (or instructor).

In this case, you must revisit the task list and eliminate tasks or compress the time to complete each task. In others words, you must determine how good of a project you can deliver in the time allowed. Even in the real world, you do not always have enough time to produce the best design (or the client does not have enough money to build the best product). The goal is to produce the best product within the constraints given. These constraints are usually time, money, materials, and talent.

By now, you may have realized that the schedule for your project and the organization of tasks in your project (presented in the previous section) are linked. The following sections include information on two tools you may find useful as you organize and plan your project: the Gantt chart and the critical path method.

15.08.01 Gantt Charts

When working in teams, it is essential to establish a well-thought-out plan for completing the project. If you are working on a project as an individual, the planning stage is not nearly as critical, since no one else is depending on you to complete a task to an acceptable level of quality within a certain time frame. One useful tool to help you organize your project, assigning a timeline for the completion of various tasks, is a Gantt chart. A **Gantt chart** (A tool for scheduling a project timeline.) is a table that lists the tasks in the leftmost column that must be completed, and it identifies the dates across the top row by which each task must be completed. Shading indicates the times for working on each of the project tasks. Without a detailed plan that lays out due dates and establishes a timeline, most projects get bogged down in trivial details and important tasks are delayed. This will often put the success of the project in jeopardy of being completed on time. Figure 15.01 shows a simple Gantt chart for a student project in reverse engineering, a topic that will be discussed in more detail in a subsequent chapter.

Figure 15.01

Gantt chart for reverse engineering project.

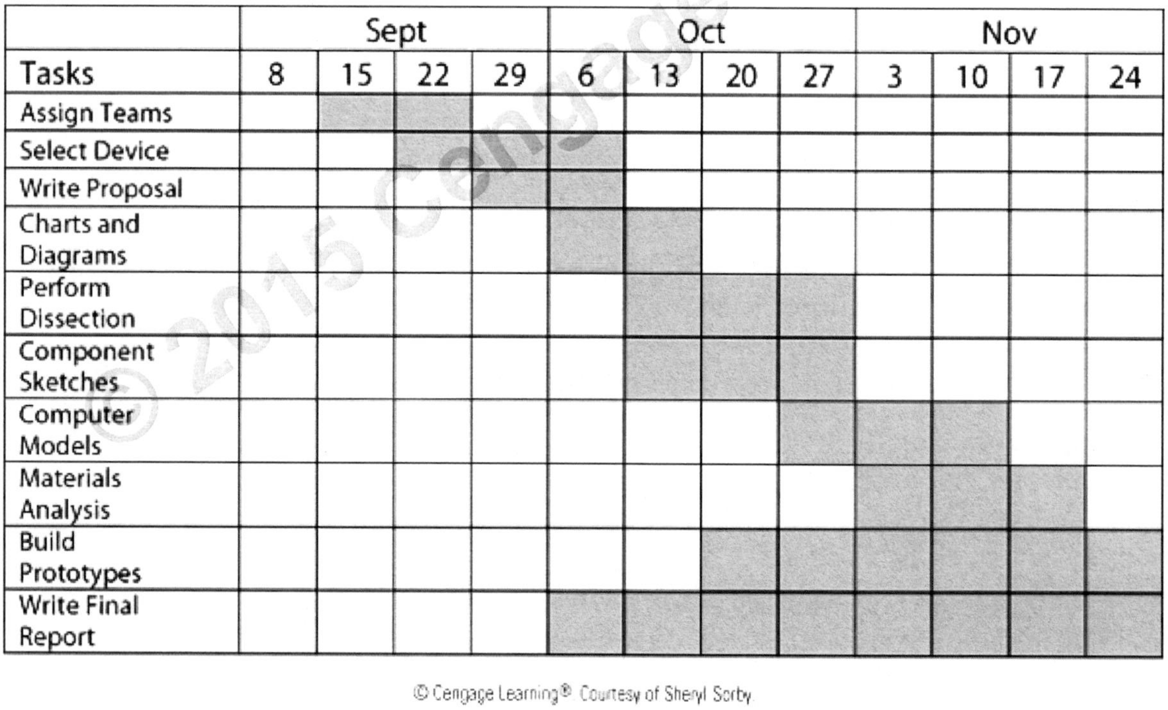

	Sept				Oct				Nov			
Tasks	8	15	22	29	6	13	20	27	3	10	17	24
Assign Teams												
Select Device												
Write Proposal												
Charts and Diagrams												
Perform Dissection												
Component Sketches												
Computer Models												
Materials Analysis												
Build Prototypes												
Write Final Report												

Note that each of the major task headings can be broken down further and a Gantt chart created for each major task. For example, Figure 15.02 shows a new Gantt chart created just for the last task listed in the previous Gantt chart.

Figure 15.02

Gantt chart for report writing task.

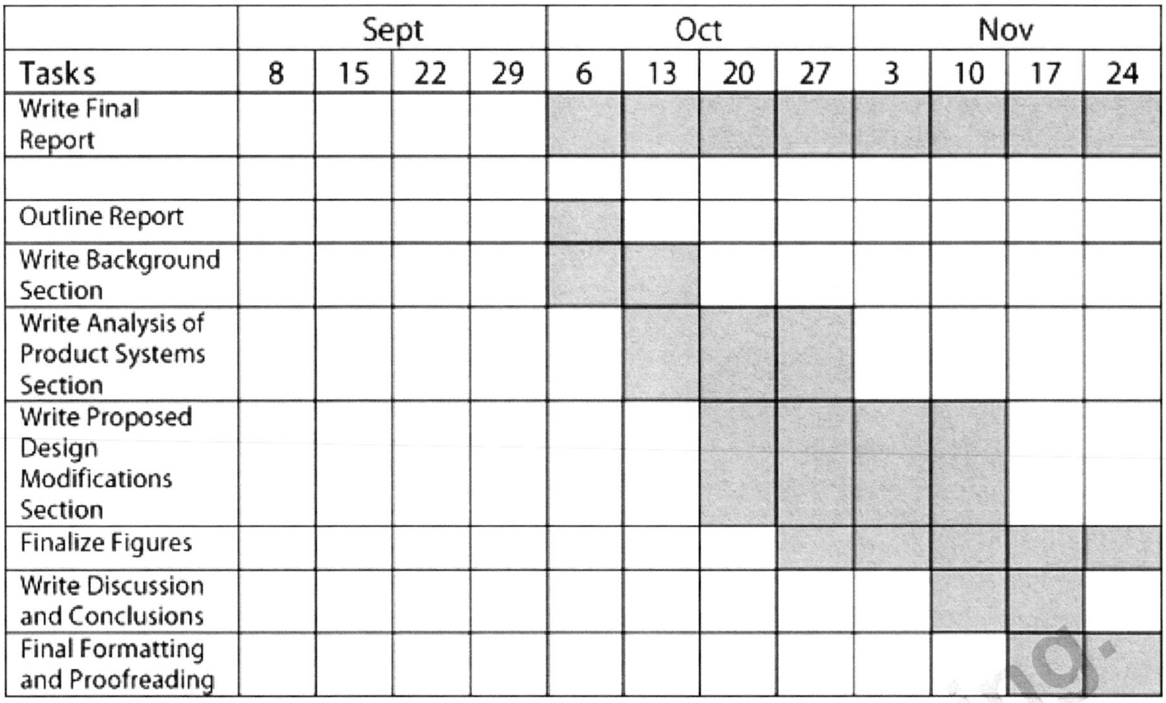

	Sept				Oct				Nov			
Tasks	8	15	22	29	6	13	20	27	3	10	17	24
Write Final Report												
Outline Report												
Write Background Section												
Write Analysis of Product Systems Section												
Write Proposed Design Modifications Section												
Finalize Figures												
Write Discussion and Conclusions												
Final Formatting and Proofreading												

© Cengage Learning®. Courtesy of Sheryl Sorby.

5.08.02 Critical Path Method

The **critical path method (CPM)** (A tool for determining the least amount of time in which a project can be completed.) is used in project scheduling to determine the least amount of time needed to complete a given project. CPM is also used to determine which activities are most critical to the on-time completion of the project (hence, the name) and which activities are not as critical to the overall project schedule. The **critical path** (The sequence of activities in a project that have the longest duration.) includes the sequence of the activities that have the longest duration. When these activities are strung together on the critical path, you can determine the shortest possible duration of the project. Activities that are not on the critical path can be allowed to "float" with regard to schedule, which will not impact completion of the overall project.

The CPM is like a flowchart for the project. It helps everyone visualize what has been accomplished, in what stage of the project the team is (what percentage is complete, whether the project is ahead or behind schedule, etc.), and what needs to be done next at each step of the project. If names are associated with each task, the CPM can also serve as a reminder of who is waiting for a finished task before that person can begin the next task. The information required to construct a CPM diagram is:

- A list of all activities required to complete the project.
- The amount of time each activity will take to complete.

- The dependencies between tasks (i.e., what task relies on the completion of another task before it can be started).

As an example of a CPM, consider a project broken down into six major activities. The task breakdown is characterized as follows:

Activity	Duration	Depends on
1	2 days	—
2	4 days	1
3	6 days	2
4	5 days	1
5	2 days	2, 4
6	8 days	3, 5

The critical path diagram for this project can be constructed as shown in Figure 15.03. Note that arrows are used to show forward progress through the project, and dependencies between tasks are shown graphically. For this example, there are three possible "paths" through the project from start to finish, as shown in Figure 15.04. The first path includes Activities 1, 2, 3, and 6. This path has a total duration of 20 days. The second path includes Activities 1, 2, 5, and 6 with a duration of 16 days. The final path includes Activities 1, 4, 5, and 6 with a duration of 17 days. The critical path is the first one (with Activities 1, 2, 3, and 6) because this path has the longest duration. Based on the critical path, the *shortest* possible completion schedule for the project is 20 days; anything less than this is impossible.

Figure 15.03

Critical path diagram.

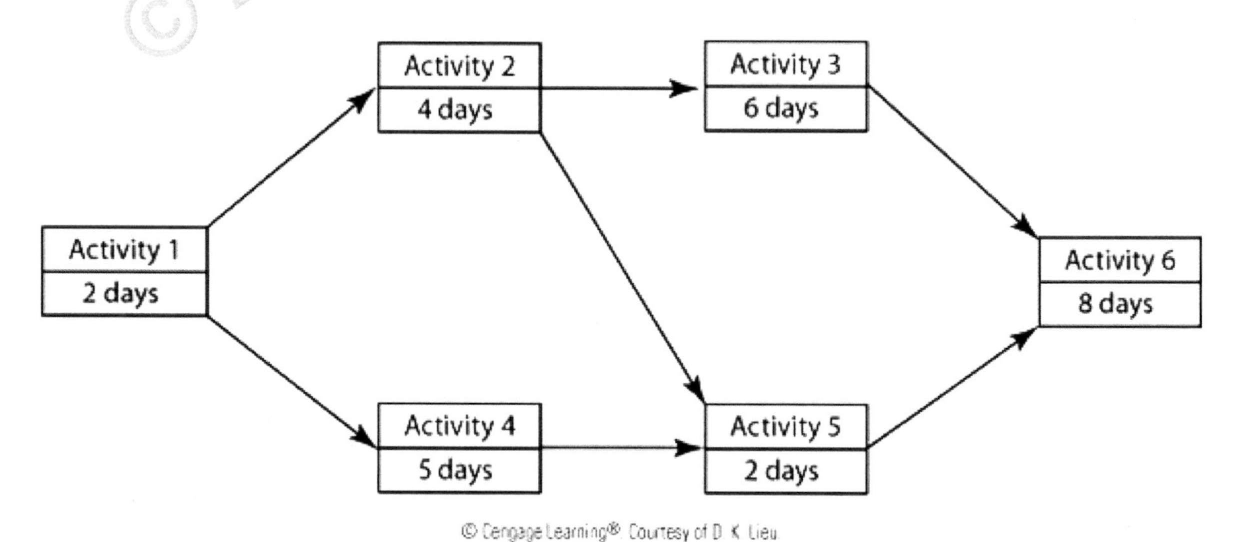

Figure 15.04

Three possible paths through project completion.

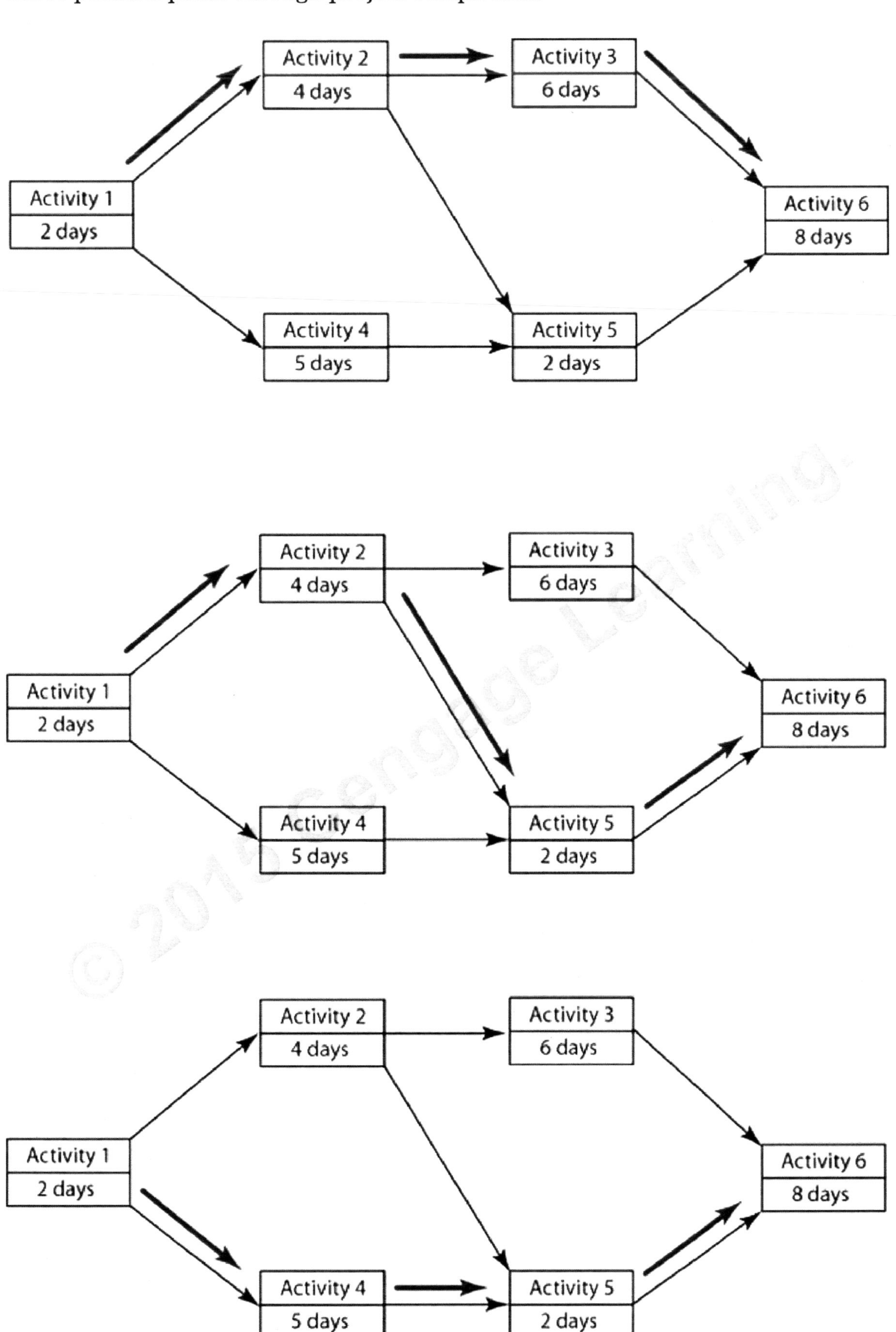

Each activity on the critical path is now a critical activity; that is, if any of the activities are not completed on time, the overall time needed to complete the project will be lengthened. This means that Activities 4 and 5 (the only activities not on the critical path) have some float time—if they are not completed on time, the overall project will still be on schedule. If you are a project manager, this information shows you where to concentrate your efforts. If it looks as though Activity 4 is beginning to interfere with Activity 3, you can suspend work on Activity 4 for a short while to make sure Activity 3 proceeds unhindered. Or if Activity 2 starts to flounder, you can shift resources away from Activity 4 to make sure the critical activity (Activity 2) is completed on schedule. Because completion of any project is dynamic rather than static, as with Gantt charts, you should review your critical path diagram periodically to ensure that it still accurately reflects the realities of your project.

15.09 Communication

Following is a discussion about communication between the team and the outside world and communication among members of the team. Each member of the team must communicate openly and honestly during task assignments (Do I agree with the plan? Do I have other commitments that will interfere with the timeline? Can I commit enough time to the task? Do I have the skills required to complete the task alone?). A team member's silence is usually interpreted as agreement with the plan, or at least acceptance of the plan. Most problems between team members result from a lack of open and honest communication.

15.09.01 Agreeing How to Communicate

E-mailing, chatting, and text messaging are useful modes of communication, and you have probably used all of them; however, regular face-to-face meetings are essential for a team's successful communication. Electronic modes of communication can be used, but nothing takes the place of face-to-face meetings. Because no one has time for an unnecessary or poorly conducted meeting, when team members do meet, they want the meetings to be productive. Decide early in the project how often the team needs to convene to conduct its business. Team members have other obligations, and the team project will get only a portion of their time. Having a regularly scheduled meeting time enables all team members to plan their activities around project time.

Not all information is shared in meetings; and the options of e-mail, notes, memos, and voice mail are appropriate under the right circumstances. All team members need to agree on how to communicate.

In your initial team meeting, find out how the team wants to handle communication. How much time do members have for meetings? Does everyone have e-mail? How often do they check it? Is voice mail reliable? The value of written records of team meetings and decisions cannot be overemphasized. By investing whatever time is needed to agree on how to communicate, team members will save time during the course of the project.

Documentation is essential after meetings. The note taker's responsibility is to summarize the conduct of the meeting and communicate his or her notes to the rest of the group. Sometimes the team keeps a bound notebook of team-meeting notes, with the notebook passed from one note taker to the next (when team roles are being rotated). Usually, the note taker will convert meeting notes to an electronic document and e-mail them to all team members, summarizing the following:

- Did we say what needed to be said?
- Did we begin and end on time?
- What was decided?
- What tasks were assigned?
- Who is responsible for each task?
- What questions still need to be answered?
- What must be done before the next meeting?

After the note taker e-mails the document out for review, you should look over the notes carefully, making comments if any points conflict with how you remember the meeting. If you do not take the time to review the notes or if you do not bother to raise objections with the notes as written, the notes will become the permanent record of what transpired at the meeting. Once again, silence is interpreted as agreement. You cannot come back later to say that you did not know you were responsible for completing a task when that assignment was included in the meeting notes that you agreed (either actively or passively) were complete and accurate.

15.09.02 Communicating Outside Meetings

As the team's level of trust increases, the need for face-to-face meetings may decline. Alternative modes of communication can be used to keep members informed of progress, changes, and the need for team meetings. The team should also decide how members unable to meet with the entire group will communicate with the team, as well as how the team will communicate meeting results to absent members. Progress reports should include information about what is or is not happening. Sometimes what is not happening is as important as what is actually taking place. A progress report should also point out tasks that need immediate attention or changes that need to be made. One of the most important parts of a progress report is recommendations for what needs to be done to get a task or an activity back on track.

A **process check** (A method for resolving differences and making adjustments in team performance.) , another way of communicating in face-to-face meetings or in electronic forms of communication, can help move the team (or individuals) toward improved performance. Usually, a process check is a reflection of:

- What the team members did well that they want to continue doing.
- What did not go as well as the team would have liked.
- What the team can do to improve the things they want to improve and not detract from the things they think are going well.

The team should periodically conduct a process check at the beginning or end of a team meeting; doing so will help resolve people's differences and reinforce good feelings. Either way, a process check facilitates a professional, functioning team.

15.09.03 Communicating with the Outside World

A key to effective teamwork is effective communication. In the previous sections, you learned how to communicate with other team members. In reality, teams also need to be able to communicate with the outside world. In a university setting, the outside world may include instructors and classmates. In the working world, the outside world may consist of bosses, coworkers, clients, and/or the general public. The modes with which your team might need to communicate include progress reports, final reports, final design documentation, and project presentations.

When preparing a progress report, focus on the status of the project and any obstacles the team has encountered. Typically, a progress report is prepared for an instructor or boss, so the points must be clear and direct. Progress reports are meant to give an overview of the progress since the last report, so shorter is probably better as long as all necessary details are included. If your team has identified additional resources that are necessary for the successful completion of the project, these resources should be identified in the progress report as well. In final reports to the client or boss, you should provide the details of your design. Final reports are typically several pages in length (depending on the project) and outline the choices you made, the analysis you performed, and any test results you obtained. Design documentation usually includes drawings, specifications, and the details of any analysis you performed. Your design documentation should show how you arrived at the answer(s) you did. Presentations should show the highlights of your project. Usually, you will be given separate instructions from your professor about her expectations for a group presentation. Make sure you adhere to any guidelines you are given. If you are going to make a group presentation, you should practice at least once as a team to make sure you stay within the time limit and cover all necessary points.

For classroom projects, you will probably need to convey to your instructor how each person on the team contributed to the overall project. Specifically, you may be asked to address the following:

- Who contributed?
- What did each person contribute?
- How much credit does each person deserve?

Be honest and fair in your appraisal of your teammates' work on the project, as well as in your appraisal of your own work. Giving someone a pass that does not deserve one is not fair to the rest of the team, nor is judging someone too harshly.

Chapter 15: Working in a Team Environment: 15.10 Tools for Dealing with Personnel Issues
Book Title: Visualization, Modeling, and Graphics for Engineering Design
Printed By: Jessica D'Eall (jessica.davies@nelson.com)

15.10 Tools for Dealing with Personnel Issues

The following sections will give you tools for avoiding problems usually attributed to team members who become difficult to work with because of a personality trait that does not adapt well to teamwork or because the team member is not properly committed to completing the assignment. Keeping everyone motivated is best done proactively rather than reactively.

15.10.01 Team Contract

A **team contract** (The rules under which a team agrees to operate (also known as a code of conduct, an agreement to cooperate, or rules of engagement).) (also known as a code of conduct, an agreement to cooperate, or rules of engagement) is a formal written document, which should be readily available during all team meetings. The document should be established only after careful, thorough, and honest discussion by all team members. The contract lists the rules the team agrees to live by. Often, team members need to revise a contract once they spend enough time working together that they (and others on the team) discover pet peeves. A good method for establishing a contract is to ask each member to bring to a team meeting a list of two or three of the biggest problems they encountered while members of previous teams. At the meeting, the team should consider each item in a round-robin fashion, until all members' lists have been exhausted. The resulting contract is a list of rules/agreements that, if followed, will incorporate the items on each person's list. This means that if everyone follows the rules in the contract, no one will violate any pet peeve or cause any of the previously experienced team-related problems.

Changes to the contract may be required as the team progresses, because well-meaning members, in spite of their best intentions, revert to inappropriate behavior. However, this does not have to mean an end to the team. Although revised contracts often include rewards and penalties to assist members in bringing about the desired behavior, you want to avoid coercing a member into accepting a contract. It is imperative that everyone on the team be treated with respect and that all disagreements are viewed as legitimate. Benjamin Franklin, upon the signing of the Declaration of Independence, said, "We must ... all hang together or, most assuredly, we shall all hang separately." This quote applies to teams as well.

15.10.02 Publication of the Rules

Once you have determined how to operate as a team, write down the agreed-upon rules. Figure 15.05 shows a sample team contract established by a student project group.

Figure 15.05

Sample team contract.

Sample Contract

1. All members will attend meetings or notify the team by e-mail or phone in advance of anticipated absences.
2. All members will be fully engaged in team meetings and will not work on other assignments during meetings.
3. All members will complete assigned tasks by agreed-upon deadlines.
4. Major decisions will be subject to group discussion and consensus or majority vote.
5. The roles of recorder and timekeeper will rotate on a weekly rotational basis (all members will take their turn — NO EXCEPTIONS).
6. The team meetings will occur only at the regularly scheduled (weekly) time or with at least a two-day notice.

Once you have created the contract, make sure every team member receives a copy and use the ground rules as a tool for effective communication and teamwork. Establishing, revisiting, and revising (if necessary) a contract can start the team out on the right foot or keep the team on track as crunch time materializes. Make sure all members support the rules. If you find that the rules are not working, change them, making sure everyone agrees to the new set of rules.

5.10.03 Signature Sheet and Task Credit Matrix

You should include a signature sheet with every assignment, whether it is requested by your instructor or not. On the sheet, explain that the individuals' signatures mean that the individuals participated in the assignment, have a general understanding of the entire submission, and are deserving of the credit indicated on the task credit matrix.

The **task credit matrix** (A table that lists all team members and their efforts on project tasks.) is a table that lists each member on the team, explains each task, and indicates how the credit should be shared among the members of the team. If a team member does not deserve credit for the project, he or she should not be allowed to sign the signature sheet and a zero should be entered in the task credit matrix. It is not expected that each member will have contributed to each task. What is expected is that the team split the work fairly and that each individual deserves credit. Note that the credit does not have to be equal, although ideally (and usually) it is. Your team may decide to weight different tasks, based on the amount of effort or contribution required. You also may decide to split the tables, one for the task breakdown (to aid the instructor in knowing to whom to direct questions) and a second expressing the team's desire for distribution of points or credit. Often, instructors will allow the team this privilege.

Chapter Review

5.11 Chapter Summary

In this chapter, you learned about the importance of teams in engineering and in working on student projects. You learned about organizing a team and assigning roles for efficient, effective team meetings. The need to rotate roles among members was also discussed. You learned about the keys to successful team meetings and about organizing projects for optimal teamwork. The critical path method and Gantt charts were introduced as tools to help keep a project moving forward and for maintaining a reasonable project schedule. Finally, you learned about the importance of communication when working on a team. Two types of communication were discussed: internal communication among team members and external communication for working with other people, such as bosses and instructors.

Chapter Review

5.12 Glossary of Key Terms

agenda (The list of topics for discussion/action at a team meeting.)

consensus (A process of decision making where an option is chosen that everyone supports.)

critical path (The sequence of activities in a project that have the longest duration.)

critical path method (CPM) (A tool for determining the least amount of time in which a project can be completed.)

devil's advocate (The team member who challenges ideas to ensure that all options are considered by the group.)

Gantt chart (A tool for scheduling a project timeline.)

note taker (The person who records the actions discussed and taken at team meetings and then prepares the formal written notes for the meeting.)

process check (A method for resolving differences and making adjustments in team performance.)

task credit matrix (A table that lists all team members and their efforts on project tasks.)

team contract (The rules under which a team agrees to operate (also known as a code of conduct, an agreement to cooperate, or rules of engagement).)

team leader (The person who calls the meetings, sets the agenda, and maintains the focus of team meetings.)

team roles (The roles that team members fill to ensure maximum effectiveness for a team.)

timekeeper (The person who keeps track of the meeting agenda, keeping the team on track to complete all necessary items within the allotted time frame.)

ACKNOWLEDGEMENTS

Engineering 1C03: Engineering Design and Graphics
Doyle, T.
Engineering 1C03: Engineering Design and Graphics, Doyle, T.
© 2016 Doyle, Thomas
Reprinted with permission.

Chapter 4: Orthographic Projection
Giesecke, F. et al.
Modern Graphics Communication, Giesecke, F. et al.
© 2010 Pearson Education - NJ
Reprinted with permission.

Chapter 1: Gear-Design Trends
Dudley, D.
Handbook of Practical Gear Design, Dudley, D.
© 1984 McGraw Hill Companies
Reprinted with permission.

Chapter 17: Advanced Visualization Techniques
Lieu, D. and Sorby, S.
Visualization, Modeling, and Graphics for Engineering Design, Lieu, D. and Sorby, S.
© 2009 Cengage Learning Nelson Education
Reprinted with permission.

Chapter 7: Spur Gears
Uiker, J. et al.
Theory of Machines and Mechanisms, Uiker, J. et al.
© 2010 Oxford University Press (US)
Reprinted with permission.

Excerpts from: Chapter 8 Helical Gears, Bevel Gears, Worms, and Worm Gears
Uiker, J. et al.
Theory of Machines and Mechanisms, Uiker, J. et al.
© 2010 Oxford University Press (US)
Reprinted with permission.

Excerpts from: Chapter 9 Mechanism Trains
Uiker, J. et al.
Theory of Machines and Mechanisms, Uiker, J. et al.
© 2010 Oxford University Press (US)
Reprinted with permission.

Chapter 15: Working in a Team Environment

Lieu, D. and Sorby, S.

Visualization, Modeling, and Graphics for Engineering Design, Lieu, D. and Sorby, S.